Biomechanics of Contemporary Implants and Prosthesis: Modeling, Experiments, and Clinical Application

Biomechanics of Contemporary Implants and Prosthesis: Modeling, Experiments, and Clinical Application

Editor

Oskar Sachenkov

MDPI • Basel • Beijing • Wuhan • Barcelona • Belgrade • Manchester • Tokyo • Cluj • Tianjin

Editor
Oskar Sachenkov
Institute of Mathematics and
Mechanics
Kazan Federal University
Kazan
Russia

Editorial Office
MDPI
St. Alban-Anlage 66
4052 Basel, Switzerland

This is a reprint of articles from the Special Issue published online in the open access journal *Materials* (ISSN 1996-1944) (available at: www.mdpi.com/journal/materials/special_issues/Contemporary_Implants_Prosthesis).

For citation purposes, cite each article independently as indicated on the article page online and as indicated below:

LastName, A.A.; LastName, B.B.; LastName, C.C. Article Title. *Journal Name* **Year**, *Volume Number*, Page Range.

ISBN 978-3-0365-5962-9 (Hbk)
ISBN 978-3-0365-5961-2 (PDF)

© 2022 by the authors. Articles in this book are Open Access and distributed under the Creative Commons Attribution (CC BY) license, which allows users to download, copy and build upon published articles, as long as the author and publisher are properly credited, which ensures maximum dissemination and a wider impact of our publications.

The book as a whole is distributed by MDPI under the terms and conditions of the Creative Commons license CC BY-NC-ND.

Contents

About the Editor vii

Preface to "Biomechanics of Contemporary Implants and Prosthesis: Modeling, Experiments, and Clinical Application" ix

Daniel Schaffarzick, Karl Entacher, Dietmar Rafolt and Peter Schuller-Götzburg
Temporary Protective Shoulder Implants for Revision Surgery with Bone Glenoid Grafting
Reprinted from: *Materials* **2022**, *15*, 6457, doi:10.3390/ma15186457 1

Anna A. Kamenskikh, Lyaysan Sakhabutdinova, Nataliya Astashina, Artem Petrachev and Yuriy Nosov
Numerical Modeling of a New Type of Prosthetic Restoration for Non-Carious Cervical Lesions
Reprinted from: *Materials* **2022**, *15*, 5102, doi:10.3390/ma15155102 17

Shirish M. Ingawale and Tarun Goswami
Design and Finite Element Analysis of Patient-Specific Total Temporomandibular Joint Implants
Reprinted from: *Materials* **2022**, *15*, 4342, doi:10.3390/ma15124342 39

Liaisan Saleeva, Ramil Kashapov, Farid Shakirzyanov, Eduard Kuznetsov, Lenar Kashapov and Viktoriya Smirnova et al.
The Effect of Surface Processing on the Shear Strength of Cobalt-Chromium Dental Alloy and Ceramics
Reprinted from: *Materials* **2022**, *15*, 2987, doi:10.3390/ma15092987 73

Alex G. Kuchumov, Aleksandr Khairulin, Marina Shmurak, Artem Porodikov and Andrey Merzlyakov
The Effects of the Mechanical Properties of Vascular Grafts and an Anisotropic Hyperelastic Aortic Model on Local Hemodynamics during Modified Blalock–Taussig Shunt Operation, Assessed Using FSI Simulation
Reprinted from: *Materials* **2022**, *15*, 2719, doi:10.3390/ma15082719 93

William Solórzano-Requejo, Carlos Ojeda and Andrés Díaz Lantada
Innovative Design Methodology for Patient-Specific Short Femoral Stems
Reprinted from: *Materials* **2022**, *15*, 442, doi:10.3390/ma15020442 117

Adrian Sauer, Maeruan Kebbach, Allan Maas, William M. Mihalko and Thomas M. Grupp
The Influence of Mathematical Definitions on Patellar Kinematics Representations
Reprinted from: *Materials* **2021**, *14*, 7644, doi:10.3390/ma14247644 149

Muhammad Imam Ammarullah, Ilham Yustar Afif, Mohamad Izzur Maula, Tri Indah Winarni, Mohammad Tauviqirrahman and Imam Akbar et al.
Tresca Stress Simulation of Metal-on-Metal Total Hip Arthroplasty during Normal Walking Activity
Reprinted from: *Materials* **2021**, *14*, 7554, doi:10.3390/ma14247554 165

Anna Kamenskikh, Alex G. Kuchumov and Inessa Baradina
Modeling the Contact Interaction of a Pair of Antagonist Teeth through Individual Protective Mouthguards of Different Geometric Configuration
Reprinted from: *Materials* **2021**, *14*, 7331, doi:10.3390/ma14237331 175

Leonid Maslov, Alexey Borovkov, Irina Maslova, Dmitriy Soloviev, Mikhail Zhmaylo and Fedor Tarasenko
Finite Element Analysis of Customized Acetabular Implant and Bone after Pelvic Tumour Resection throughout the Gait Cycle
Reprinted from: *Materials* **2021**, *14*, 7066, doi:10.3390/ma14227066 187

Pavel Bolshakov, Nikita Kharin, Ramil Kashapov and Oskar Sachenkov
Structural Design Method for Constructions: Simulation, Manufacturing and Experiment
Reprinted from: *Materials* **2021**, *14*, 6064, doi:10.3390/ma14206064 205

About the Editor

Oskar Sachenkov

Head of the Department of Computer Mathematics and Informatics, N.I. Lobachevsky Institute of Mathematics and Mechanics, Kazan Federal University, Kazan, Russia. Scientific interests are mostly in the biomechanical field, especially the development of bone tissue mechanical models, patient-based biomechanical simulations, the development of CT-based models and CT-based finite elements, and endoprosthesis structural design.

Preface to "Biomechanics of Contemporary Implants and Prosthesis: Modeling, Experiments, and Clinical Application"

Modern medicine is now more oriented towards patient-based treatments. Taking into account individual biological features allows for increasing the quality of the healing process. Opportunities for modern hardware and software allow not only the complex behavior of implants and prostheses to be simulated, but also take into account any peculiarities of the patient. Moreover, the development of additive manufacturing expands the opportunities for materials. Technical limits for composite materials, biomaterials, and metamaterials are decreasing. On the other hand, there is a need for more detailed analyses of biomechanics research. A deeper understanding of the technological processes of implants, and the mechanobiological interactions of implants and organisms will potentially allow us to raise the level of medical treatment. Modern trends of the biomechanics of contemporary implants and prostheses, including experimental and mathematical modeling and clinical application, are discussed in this book.

Oskar Sachenkov
Editor

Article

Temporary Protective Shoulder Implants for Revision Surgery with Bone Glenoid Grafting

Daniel Schaffarzick [1], Karl Entacher [2], Dietmar Rafolt [3] and Peter Schuller-Götzburg [4,*]

1. ECS Schaffarzick—Engineering/Consulting/Service, Sankt-Peter-Straße 15/2, A-5061 Elsbethen, Austria
2. Department of Information Technology and Systems Management, Salzburg University of Applied Science, Urstein Süd 1, A-5412 Puch, Austria
3. Center for Medical Physics and Biomedical Engineering, Medical University of Vienna, Vienna General Hospital, Währinger Gürtel 18-20, A-1090 Wien, Austria
4. Department of Prosthetic Dentistry, University Dental Clinic Vienna, Sensengasse 2a, A-1090 Vienna, Austria
* Correspondence: peter.schuller-goetzburg@meduniwien.ac.at; Tel.: +43-676-5339832

Citation: Schaffarzick, D.; Entacher, K.; Rafolt, D.; Schuller-Götzburg, P. Temporary Protective Shoulder Implants for Revision Surgery with Bone Glenoid Grafting. *Materials* **2022**, *15*, 6457. https://doi.org/10.3390/ma15186457

Academic Editor: Oskar Sachenkov

Received: 11 July 2022
Accepted: 7 September 2022
Published: 17 September 2022

Publisher's Note: MDPI stays neutral with regard to jurisdictional claims in published maps and institutional affiliations.

Copyright: © 2022 by the authors. Licensee MDPI, Basel, Switzerland. This article is an open access article distributed under the terms and conditions of the Creative Commons Attribution (CC BY) license (https://creativecommons.org/licenses/by/4.0/).

Abstract: This article describes the development of a temporary protective glenoid prosthesis placed between the augmentation and humeral head prosthesis during the healing phase of shoulder prosthesis revision with necessary reconstruction of the bony structure of the glenoid. The glenoid protection prosthesis ensures the fixation of the augmentation material and protects the screws from contact with the metallic humeral head prosthesis. Another approach of the development is a reduction of the resorption of the augmentation by targeted mechanical stimulation of the tissue. The aim should be to achieve significantly improved conditions for the implantation of a new glenoid component at the end of the healing phase of the augmentation material in comparison to the current standard method. The development of the protective prosthesis was carried out according to specific needs and includes the collection of requirements and boundary conditions, the design and technical detailing of the implant, the verification of the development results as well as the validation of the design. For the verification, FEM simulations (Finite Element Analysis) were performed to estimate the mechanical stability in advance. Mechanical tests to confirm the stability and abrasion behavior have been carried out and confirm the suitability of the protective implant. The result of the present work is the detailed technical design of two variants of a glenoid protective prosthesis "GlenoProtect" for use in revision procedures on shoulder joints—with large-volume defects on the glenoid—treated by arthroplasty and the necessity of augmenting the glenoid, including a description of the surgical procedure for implantation.

Keywords: glenoid implant; implant development; finite element analysis; 3D modelling; abrasion test; glenoid defect

1. Introduction

The annual number of shoulder arthroplasties increases continuously. In 2017, an estimated number of about 800,000 patients were living in the United States with a shoulder replacement with a prevalence of 0.258%, increasing markedly from 1995 (0.031%) and 2005 (0.083%) [1]. In 2008, 27,000 shoulder arthroplasties (total endoprostheses only) were performed [2]. Although the number of shoulder arthroplasties performed is still far below that of knee or hip arthroplasties, the indications for shoulder arthroplasties are much more varied and range from rheumatoid arthritis, degenerative arthroses and osteonecrosis to posttraumatic osteoarthritis or osteoarthritis [3,4]. Due to the increasing demand for total endoprostheses to the shoulder, the revision of the glenoid is becoming increasingly important and demonstrates the need for basic research and development in this field.

Problems with arthroplasty or reimplantation of glenoid prostheses consist of the large-volume combined with central-peripheral glenoid defects. Due to the destroyed glenoid with bony defects, immediate insertion of a prosthesis is not possible because primary stability cannot be guaranteed by anchoring the prosthesis. So far, there have been several attempts to correct the bony defects with augmentations, however with moderate success [5–12].

A common problem was identified in connection with the screws used to fix the augmentation material. In one study, complications were reported in 78% of the observed cases of screw fracture, bent screws or metallic abrasion [13]. The main reason given for performing revision surgery is loosening of the glenoid prosthesis in 32% to 39% of cases [14–16], followed by instability in 23% to 30% and periprosthetic fracture in 11% of cases [14,17]. The success rate over shorter periods of time (<5 years) is given as up to 98% to 99%, which drops drastically to between 33% and 51.5% with a follow-up of >10 years [18–21]. Eccentric loads and the resulting high micro-movements at the boundary between bone cement and cortical and trabecular bone play a major role in loosening the glenoid component. Over a longer period, a phenomenon occurs which is referred to in pertinent literature as the "Rocking Horse Phenomenon" [22,23] and can subsequently lead to bone resorption and massive bony defects in the area of glenoid anchorage. When the glenoid is loosened, patients often suffer increasingly from load-dependent pain after an interval that is sometimes very long and symptom-free [24]. In the case of a confirmed loosening of the glenoid prosthesis, there is in principle an indication for a one-stage changes of the glenoid component, provided that stable anchoring is possible. However, the biggest problems with revisions of the glenoid with a new prosthesis (implant) are large bone defects or deficits in which both the cancellous part and the cortical layer are affected [25]. In these cases, insertion of a new glenoid component is often not possible because it cannot be sufficiently attached to the scapula and glenoid.

In addition to other options for augmentation with calcium phosphates, bio-glass, hydrogels, human bone allografts and biocomposites made by bioactive elements [26], there are some experimental studies about cell-instructive bioengineering procedures to support and restore preexisting bone repair and osseointegration [27,28]. An option that is commonly used as a standard is still a two-stage procedure by means of building-up the defective glenoid with autologous bony augmentations [5–7,10,11,29,30]. The defective glenoid component is removed, and large-volume bone deficits are built up with cortigospongious chips from the iliac crest.

In the case of extensive defects and thus large augmentations, it is necessary to fix the augmentation with screws. After sufficient osseous integration of the cortigospongious augmentation material in the glenoid (about 3 months), a new glenoid component is inserted into the bone material [31,32]. A retrospective examination of 16 revision procedures of shoulder arthroplasties with glenoid build-up showed a settlement or atrophy of the augmentation in all cases [25]. The observed shrinkage is 5 mm in 3 patients, 5 to 10 mm in six patients and more than 10 mm in two patients. All four cases of the investigated group in which the augmentation was fixed with screws showed such a large shrinkage in which the screws were exposed and touched the metallic humeral head of the humeral prosthesis. In one case, the screw even broke due to the additional load.

In cases where the screws used to fix the augmentation material comes into contact with the metallic humeral head, abrasion may occur, which may subsequently lead to complications such as metallosis [33,34].

In this paper, we propose the application of temporary protection implants for the augmentation of shoulder prosthesis revision procedures. This paper describes the development of two kinds of temporary prosthesis which support the correction of bony defects and shall significantly improve the result of glenoid augmentation. This is intended to ensure a high degree of primary stability during insertion of a glenoid prosthesis, even during revision, and consequently to achieve more effective rehabilitation.

The goals of this study are as follows: to protect screw heads against direct contact with the metal joint ball of the implant (humeral head), to prevent screws from being unscrewed, to form a sliding partner during the healing phase, better "cohesion" of the bone fragments and to provide a more even force introduction and even pressure on the augmentation material (functional load). In the following sections, the material and methods used in this study are described.

2. Materials and Methods

GlenoProtect is a glenoid component made of a suitable material that was developed to protect the augmentation in shoulder prosthesis revision procedures. Two variants have been developed to ensure the above-mentioned capabilities.

2.1. Variant 1: Multidirectional Angle-Stable Screw Connection (Rigid Fixation)

The screws have a thread in the head area and can be placed at an angle to the implant (multidirectional). To achieve angular stability, the thread cuts into the implant material and thus fixes the screw. The inclined position of the individual screws results in greater stabilization and resistance to dents (medial/proximal) and withdrawal (Figure 1).

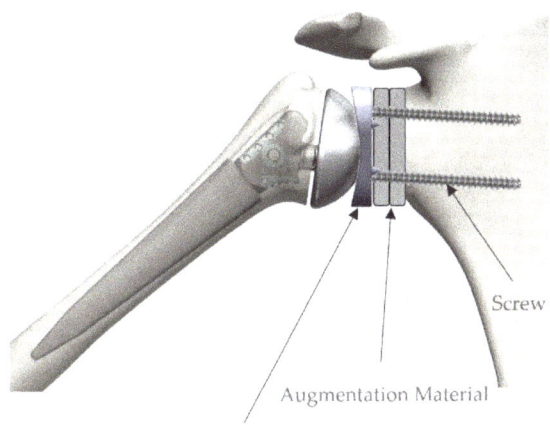

Figure 1. Rigid fixation of the augmentation material.

Resorption is the greatest unknown factor despite the protection of the augmentation. The resorption leads to a reduction of the functional pressure on the augmentation material and possibly to an increased resorption. The original design was adapted and improved in detail in several processing and optimization steps, with the aim of maximizing mechanical stability while reducing the profile cross-section (thickness or height of the implant). For this purpose, the originally designed curved oblong holes for screw connection were replaced by circular holes, as the intended cutting of the thread of the screw heads did not provide sufficient stability in this case. In addition, the radius of the implant surface was adjusted to achieve an even lower implant height. The result of the detailed design can be seen in Figure 2.

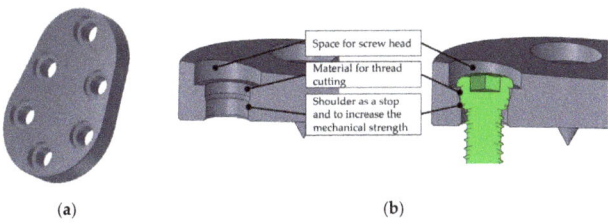

Figure 2. Result of detailed design of rigid variant: (**a**) three-dimensional view; (**b**) sectional view.

Particular attention was paid to the design of the fixing holes. The exact geometry of these holes is important to ensure that the screw heads are held securely and that the screw heads do not project beyond the implant surface. The implant or the fixing holes were designed for the use of osteosynthesis screws "Locking Screw 3.5 mm" from the Small Fragment Locking Compression Plate (LCP) System (DePuy Synthes Companies, Zuchwil, Switzerland, and Warsaw, IN, USA, https://www.jnjmedicaldevices.com/en-EMEA/companies/depuy-synthes, accessed on 8 July 2022). The screws can be introduced in multiple directions, i.e., at different angles to each other, in order to adapt to the respective anatomical situations and to increase the mechanical stability of the fixation (Figure 3).

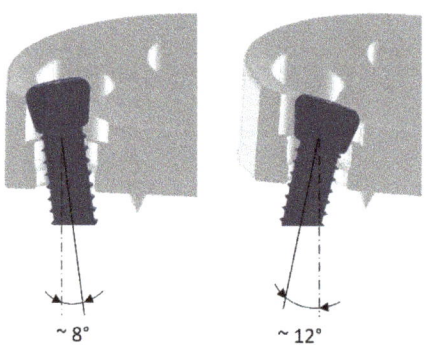

Figure 3. Multidirectional screw positioning possibility.

2.2. Variant 2: Dynamic Fixation (Angular Stable Pins)

To compensate for the risk of functional underloading of the augmentation material, a dynamic system is proposed. Instead of screws, pins are used which only have a thread in the head area that cuts into the implant material. The pins should be placed at a right angel to the implant backside (defined by the geometry), as shown in Figures 4 and 5.

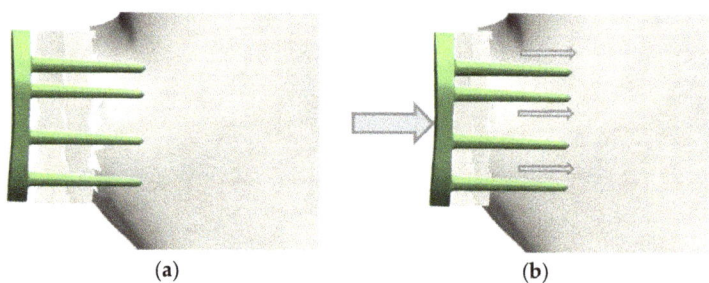

Figure 4. Dynamic fixation using angle-stable pins: (**a**) initial post-operative situation; (**b**) dynamic load results in compression and functional stimulus of the augmented bone.

Only shear and torsional forces in the medial or sagittal plane are absorbed. The entire unit can sink if the bone block atrophies and the dynamic load and thus a functional stimulus is maintained. This is to reduce bone resorption. The dynamic version can be fitted with up to five pins, and the most appropriate pins should be used depending on the specific circumstances.

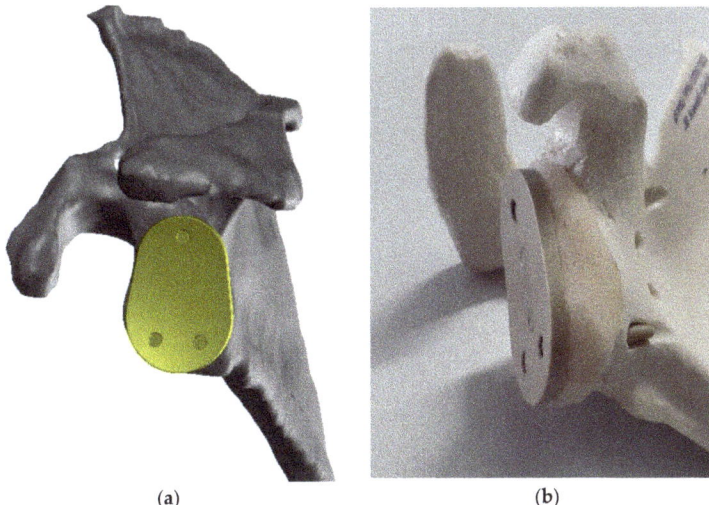

(a) (b)

Figure 5. Visualization of dynamic variant of GlenoProtect on glenoid/scapula (**a**) digital model for simulation purpose (bone defect, augmentation not shown) and (**b**) analog model for development purpose (foam material simulates the bone augmentation).

As an implant material, the use of a plastic material should be considered as the most sensible option. Metallic materials are critical due to the sliding pairing with the also metallic humeral head [35,36]. PE (polyethylene) and PEEK (polyetheretherketone) are possible plastics [37–39]. PE is the most commonly used material for primary glenoid components, where ultra-high molecular weight polyethylene (UHMWPE) is used. PE has good sliding properties, is easy to process and is comparatively inexpensive. With PEEK, the raw material is more expensive, however, it has more suitable mechanical properties than PE. Thus, the modulus of elasticity is closer to that of bone. This results in a better or natural distribution of force on the bone [40,41]. PEEK is a high-performance biomaterial suitable for long-term implantation. It is used to manufacture a wide variety of medical devices and human implants (dentistry, orthopedics and traumatology) and is used in a variety of ways in these applications [42–45]. Due to its chemical composition, PEEK is a very pure and inert material. Extensive biocompatibility tests do not provide any evidence of cytotoxicity, systemic toxicity, irritation or acroscopic reactions [31–35]. In addition, the very low levels of residues and extractable metal ions minimize the potential risk of allergic reactions commonly associated with nickel or other metal ions. In addition, PEEK can be sterilized using all common methods. PEEK is suitable for gamma, ethylene oxide and saturated steam sterilization [46].

In order to determine the mechanical strength, corresponding simulations were carried out. The selected implant variants were analyzed using FEM (Finite Element Method), also known as FEA (Finite Element Analysis) [47]. The simulation program Abaqus (Dassault Systèmes, Vélizy-Villacoublay, France, http://www.3ds.com, accessed on 8 July 2022) was used. A material approved for medical applications in the human body and repeatedly proven in implants is PEEK-Optima from Invibio Ltd. (for properties, see Table 1: Material properties PEEK, Invibio Ltd., Lancashire, UK). These material properties were used as a basis for the FEM analysis.

Table 1. Material properties PEEK.

Property	Units	PEEK-OPTIMA
Tensile Strength (Yield)	MPa (ksi)	115
Tensile Elongation (Break)	%	20
Flexural Modulus	GPa	4
Flexural Strength	MPa	170
Izod Impact (Unnotched)	kJm^{-2}	Does not break
Izod Impact (Notched)	kJm^{-2}	4.7

The load and force assumptions for the simulation were selected by Westerhoff and Bergmann [48] according to the measurements at the Julius Wolff Institute. For the FEM analysis, the data set "Lifting a weight of 10 kg" was used, since the measurement results here show the highest values. The data give a maximum value of 1500 N for the force acting normally on the implant surface. After consultation with shoulder surgeons, a maximum realistic force of 500 N is to be assumed for the intended application. The simulation was performed with 1500 N as well as with 500 N. In the FEM simulation, static forces/loads are applied and the resulting stresses, displacements (elastic) or plastic deformations are determined. For this purpose, a so-called substitute model must be created, which represents a section of the entire situation, reflecting the relevant force application. For the calculation of the stresses in the protective implant, the replacement model was in the form of a hemisphere representing the humeral head with an equally distributed load application.

For the simulation of the rigid version, a joint ball made of stainless steel, with a diameter of 48 mm, and its support centered on the implant was modelled (Figure 6).

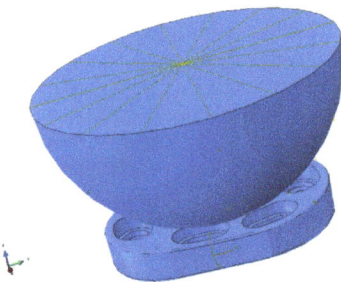

Figure 6. Modeling of the load for the FEM analysis.

Fixation of the implant was assumed with four screws. The posterior implant surface was assumed without bone support, so the implant is only held in place by the screws. Thus, the determined displacement (corresponds to "deflection") of the implant can act as a stimulus on the augmentation. At the simulation model for the dynamic variant, the joint ball is resting eccentrically on the implant. The fixation of the implant was assumed at the end of the conical pins. The eccentric loading and clamping of the pin ends result in a displacement and thus a (rotational) moment at the connection points of the pins with the implant main body. This in turn represents the load situation that occurs when the implant is attached to the bone via the pins and "tilts" the implant surface away due to an eccentric load (with corresponding movement of the shoulder).

Mechanical testing was performed in order to confirm the results from the simulation and to demonstrate the effective strength of the implant variants and the respective fixing methods. For the mechanical tests, a specific test stand was built at the Center for Medical Physics and Biomedical Engineering of the Medical University of Vienna at the AKH Vienna.

The test bench is used to simulate the movements of the shoulder joint. For this purpose, the test stand was primarily designed as an actuator-controlled pendulum system. Thus, it is possible to perform cyclic movements of a simulated humeral head prosthesis in relation to the glenoid protective implant. The contact force (force normal to the implant surface) and the type of movement of the ball of the humeral head (from rolling on the implant surface to a pure friction movement at one point) are adjustable.

The first mechanical test is intended to confirm the mechanical stability of the combination of the implant itself and the fastening elements (screws in the rigid and pins in the dynamic variant) under static loading. The implant itself as well as the fastening elements and the connection of the implant with the fastening elements must be sufficiently strong. In particular, the connection of the implant with the screws or pins represents a critical point. With a static implant, the screw heads must not protrude beyond the implant surface under load.

Following the static load test, a dynamic load test was carried out. For this purpose, the pendulum frame was oscillated with an angular deflection of $\pm 30°$. Due to the dynamics of the system and the centrifugal forces that occur when the pendulum frame is loaded with the additional weight, the test was carried out with a maximum load of 1000 N.

Comparative measurements were carried out to determine differences in the abrasion behavior of different materials. The following test materials were examined in Table 2:

Table 2. Overview of test objects for abrasion measurement.

Test Object No.	Material
Test object 1	Test sample of a glenoid protective prosthesis made of technical PEEK: KETRON PEEK-1000 (not medical grade)
Test object 2	Test plates made of medical grade PEEK No. 1
Test object 3	Test plates made of medical grade PEEK No. 2
Test object 4	Comparison sample of a glenoid component of an anatomical shoulder prosthesis (Global Advantage Keeled Glenoid, DePuy) made of PE (1020 XLK UHMWPE)

To create largely real conditions, the test objects were placed in a shell filled with a physiological solution of H_2O, agar-agar and NaCl during the pendulum motion. Agar-agar was added to increase the viscosity of the liquid, as the synovial liquid is also more viscous.

A high-precision galvo scanner (scanner galvanometer) was used for the tactile measurement of the surface [49] to quantitatively determine the abrasion. The galvo scanner has a resolution of 200 nm. This means that any abrasion in the form of "material shrinkage" can be measured at the relevant point. This method does not determine the abrasion as detached particles, but whether there is abrasion on the implant surface. These depressions in the material can be very precisely measured with the galvo scanner and are thus displayed quantitatively.

The test objects were clamped on a cross table. The galvo scanner was fixed via a 3D articulated tripod so that its lever arm rests on the test object (Figure 7). Now the cross table was moved manually, and the output signal was recorded with a DAQ measuring system (DEWE-43 from DEWESoft, Trobvlje, Slovenia, http://www.dewesoft.com, accessed on 8 July 2022).

Figure 7. Measuring tip of the galvo scanner for tactile surface measurement.

3. Results

In this chapter, the results for the simulation (FEM), mechanical testing results and the abrasion measurements are summarized.

3.1. Simulation

The results of the FEM simulation for both implant variants are presented in Table 3.

Table 3. Summary of FEM simulation results.

Tensile Elongation	Rigid Variant	Dynamic Variant
max. Tensile Elongation PEEK-Optima	20%	20%
Analysis Sample Material	24%	24%
Result FEM (500 N)	2.2%	-
Strain	**Rigid Variant**	**Dynamic Variant**
Tensile Strength PEEK-Optima	100 MPa	100 MPa
Analysis Sample Material	117 MPa	117 MPa
Max. Strain FEM (500 N)	60 MPa	~100 MPa (168 MPa peak)
Max. Strain FEM (1500 N)	90 MPa	>100 MPa
Displacement	**Rigid Variant**	**Dynamic Variant**
Max. Displacement (deflection/sag) at 500 N FEM	0.1 mm	0.95 mm

With the rigid variant at 1500 N load, stresses greater than 90–100 MPa occur selectively in the material (Figure 8).

Since these areas are rather small and surrounded by areas of significantly lower stress, it can be assumed that structural integrity will be maintained. However, it is to be expected that superficial damage may occur (the areas of high tension lie on the surface of the implant in contact with the joint ball).

Peak strain values of up to 168 MPa occur selectively in the material of the dynamic variant (see Figure 9).

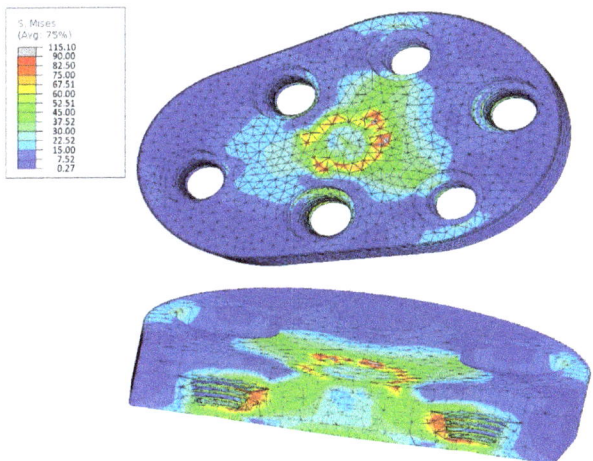

Figure 8. FEM Analysis of the rigid variant: strain at 1500 N, scaling 90 MPa.

Figure 9. FEM Analysis of the Dynamic variant: strain at 500 N, scaling 168 MPa.

However, these areas are very limited and are justified by the fact that the ends of the pins were firmly clamped for the simulation. This is exactly where these high stresses occur. It is more realistic to distribute the forces over a larger surface of the pins. The maximum strain values occurring over a large area are in the range of 100 MPa. However, this is already at the limits of the material.

The simulation with a load of 1500 N resulted in peak strain values well above 100 MPa. This would mean that the implant would no longer be able to withstand the load. The high stresses occur again in isolated areas and could occur due to the modeling (clamping only at the tip of the conical pins would distribute the dissipation of forces along the pins).

Summary:

The results of the FEM analysis of the rigid variant show that there will be no permanent damage (plastic deformation) when the load is 500 N. Under a load of 1500 N, plastic deformations or material damage can occur on the contact surface with the joint ball, but the overall strength of the implant and the screw connections would still be guaranteed.

The results of the FEM analysis of the dynamic implant variant show that loads of 500 N do not lead to any damage to the implant or that the entire structural integrity is preserved. At a load of 1500 N, the maximum permissible stresses in the material would be significantly exceeded. In such a case, modeling becomes very difficult, and it is very likely that the resulting stresses arise due to the model assumptions. However, since a maximum of 500 N can be assumed as a realistic force, the simulation of the dynamic implant variant also shows that it has sufficient mechanical stability.

3.2. Mechanical Testing

3.2.1. Rigid Variant

For the test, the protective glenoid implant was attached at a distance from a test block with three angular stable osteosynthesis screws (Figure 10). This simulates the case where the protective glenoid implant was attached to the scapula above the graft and the graft was already slightly resorbed. This creates a small gap and the forces applied to the implant are transmitted exclusively via the screws into the scapula, which is the worst case from a mechanical point of view. In this case, the force is transmitted in the area of the small thread onto the head of the osteosynthesis screws. The screw heads must remain securely fixed in the implant and not "tear out" so that the screw heads protrude beyond the implant surface.

Figure 10. Fixing the test implant at the test block (represented is only one screw before insertion, the test implant was fixed with three screws).

The test implant was loaded with 1300 N. There was no damage to the implant and the threaded connections also withstood the load. This result corresponds to the behavior expected from the FEM simulation (no permanent deformation, maintenance of overall stability up to 1500 N).

The implant and the threaded connection were also stable in the test with dynamic loads.

3.2.2. Dynamic Variant

The dynamic implant variant with the conical spikes for fixation, which allow the implant to sink, was also fixed in a test block. This test block has a convex surface and parallel holes to accommodate the fixing pins; see Figure 11.

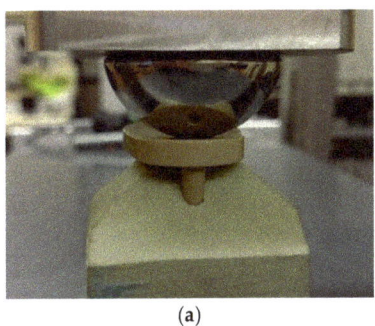
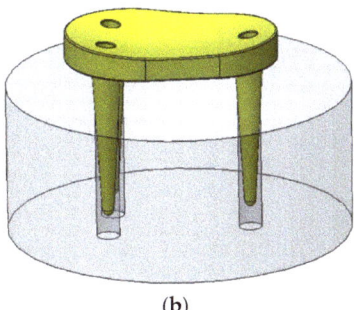

(a) (b)

Figure 11. Test abutment for the dynamic implant variant: (a) view of the actual test setup; (b) schematic view of the test block.

The test implant was loaded in the same way as in the static loading test of the rigid implant variant. The load did not cause any mechanical damage to the test implant.

The dynamic test implant was also tested with dynamic loads, where it was attached to the test block and the pendulum frame was made to vibrate. This resulted in a fracture of the conical pins in the area of the thread transition. However, the fracture did not occur until a forced extreme load was applied, with the force being applied eccentrically and transversely (pivot point of the joint ball outside the central axis). Such loads are not to be expected under real conditions, so this test also confirmed the stability of the dynamic implant variant.

3.3. Abrasion Measurement

The summary of all abrasion measurement results is shown in Figure 12a–d. The different diagrams show the measurement curves of test objects 1–4 (compare Table 2). Figure 12a contains additional descriptive elements.

The position along the implant surface (geometric longitudinal axis) is plotted on the x-axis and the deflection (normal distance) of the galvo scanner measuring tip is plotted on the y-axis. The y-axis thus represents the measured abrasion. A standardized abrasion value in µm/100 cycles is exhibited for each object. Sudden, strongly deviating signals correspond to depressions in the surface and are a direct measure of the abrasion occurring at this point.

The results clearly show that the abrasion to be expected with the glenoid protective prosthesis is much lower than with a standard glenoid component of an anatomical shoulder prosthesis made of PE.

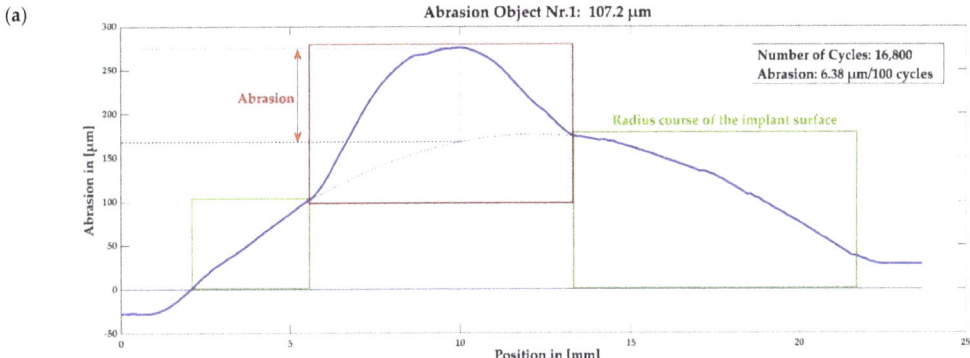

Figure 12. *Cont.*

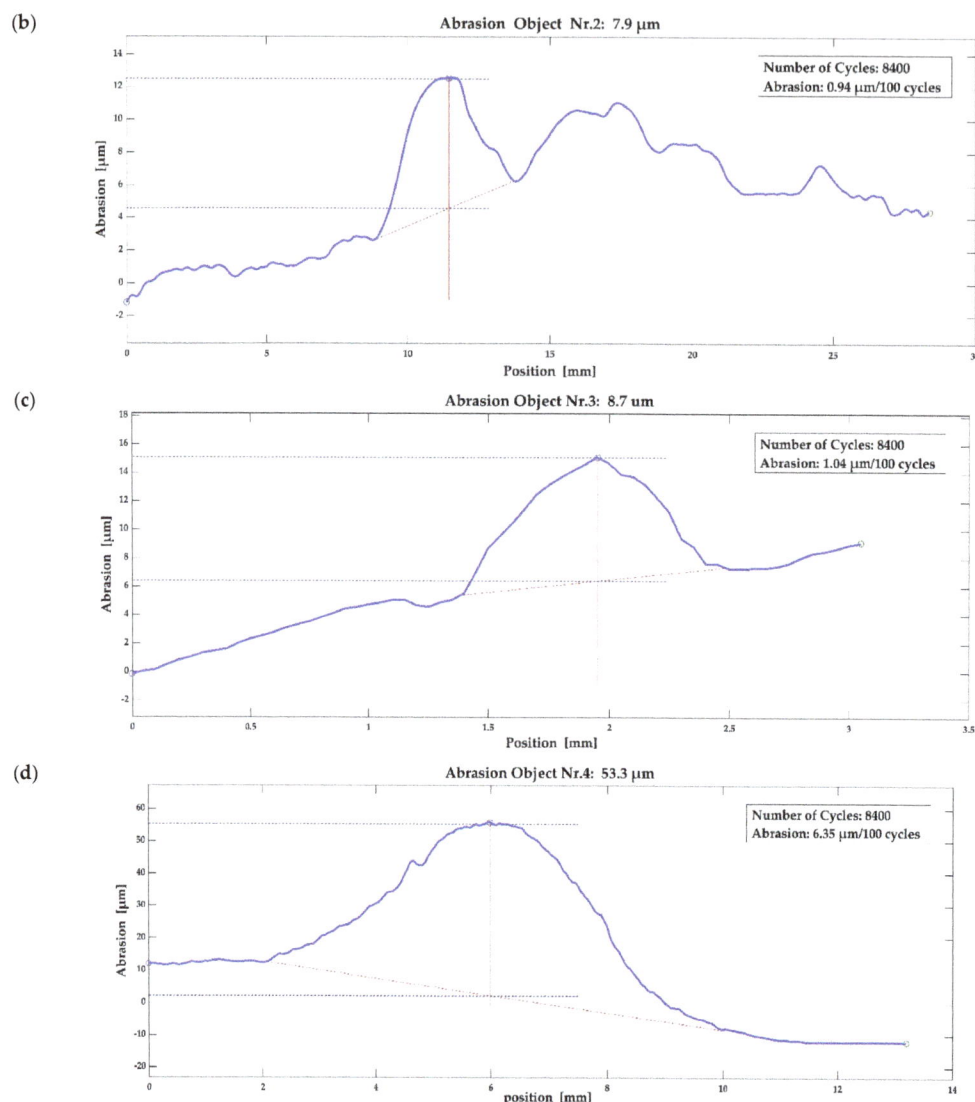

Figure 12. Abrasion measurement curves: (**a**) test object 1, including additional description of the sections of the curve; (**b**) test object 2; (**c**) test object 3; (**d**) test object 4.

This result corresponds to the order of magnitude of a study already carried out, in which a wear factor was determined that is 10 times higher for a friction pairing of UHMWPE with CoCrMo steel than for PEEK with CoCrMo steel [50], as compared in Table 4.

Table 4. Total Wear Factors (values times 10^{-6} mm^3/Nm) for Carbon Fiber-Reinforced PEEK-OPTIMA against different Counterparts and UHMWPE against CoCrMo steel [47].

CFR PEEK-OPTIMA/CFR PEEK-OPTIMA	CFR PEEK-OPTIMA/Alumina	CFR PEEK-OPTIMA/CoCrMo Steel	UHMWPE/ CoCrMo Steel
0.34	0.18	0.1	1.1

The difference between the result with the technical PEEK (test object 1) and the medical grade PEEK (test objects 2 and 3) can be explained by different mechanical properties due to the use of different starting materials for the synthesis. The technical PEEK used for the test has a notched impact strength of 3.5 kJ/m^2, whereas the medical grade PEEK has a notched impact strength of 5.5 kJ/m^2.

4. Discussion

Within the scope of the present paper, two variants of a glenoid protective prosthesis were developed, with the following functions in the foreground:

Protection of the screw heads against direct contact with the joint ball of the metal humeral head prosthesis, prevention of unscrewing or loosening of the screws, formation of a sliding partner during the healing phase, better "holding together" of the bone fragments as well as targeted application of force and uniform pressure on the augmentation material (functional load).

The development was carried out based on specific known problems with the method currently used and potential improvements based on the selected design. The results are prototypes, which were first validated in mechanical tests and then tested in a clinical pilot study. These two variants were developed for the research program for prosthetics, biomechanics and biomaterials research at Paracelsus Medical University, in order to take the second question into account and to enable a direct comparison within the framework of a clinical study. The specific question is whether significantly better results can be achieved with a dynamic system of fixation of the protective prosthesis or the augmentation compared to a rigid fixation. From a biomechanical point of view, it was postulated that the atrophy of the augmentation should be lower with dynamic fixation. This was justified accordingly in the presented research carried out.

Both implant variants were tested for stability and strength by means of FEM simulation. In addition, based on the results of the FEM simulation, it can be assumed that the rigid design of the protective denture also has advantages in terms of osseointegration of the augmentation, since the modulus of elasticity of the implant material used (PEEK) is similar to that of cortical bone tissue and thus exerts a natural load on the augmentation, which in turn is intended to reduce atrophy.

Final mechanical tests confirmed the results previously obtained in the FEM simulations regarding the stability of both implant variants. In addition to checking the stability, the abrasion behavior was also measured during the mechanical tests, since in contrast to the surface loading in a total prosthesis with a form-fitting ball and ball socket, the present protective implant with only a slight concave curvature is subjected to a theoretical point load. Despite the compressibility and elasticity of the material, the contact surface is relatively small, so that an experimental test of the abrasion properties is necessary. Here, too, the result was positive to the extent that the abrasion determined was very low (significantly less than with the material of a commercial glenoid prosthesis measured in comparison).

In addition to the actual development of the glenoid protection prostheses, appropriate documentation was carried out to register and conduct a clinical study to test the implants. The documentation has also been prepared in accordance with applicable standards and guidelines for the development of a medical device in order to facilitate possible approval and marketing.

As mentioned above, a clinical pilot study will be carried out following the discussed activities with the results obtained. The purpose of this study is to demonstrate that the use of a glenoid protective prosthesis can significantly improve the outcome of glenoid revision in shoulder prostheses. Depending on the findings of this study, various further developments would be conceivable. A promising option would be to not solely use a protective prosthesis during the healing period of the augmentation material (still two-stage procedure): The old glenoid component is removed and the glenoid is rebuilt using the protective prosthesis, followed by the insertion of the new glenoid component in a second surgery, but, in addition, a "revision glenoid prosthesis" is inserted which remains permanently implanted.

The theoretical background and the basics for it were determined and corresponding potential suggestions for improvement were implemented in the form of a glenoid protective prosthesis. "In silico" (FEM simulations) investigations and mechanical tests on prototypes served to verify the (bio)mechanical properties of the protective prostheses. A particularly interesting result was the abrasion measurements. It was found that the abrasion of the protective implants made of PEEK is significantly lower than that of the glenoid component of a standard anatomical shoulder prosthesis. This fact opens up the possibility of further developing the proposed protective prosthesis in such a way that it could be used as a permanent revision implant, thus avoiding the need for a second intervention.

Two variants were designed in the development of the protective prosthesis, and the subsequent clinical trial will show whether the dynamic variant has the postulated advantages over the rigid variant.

Author Contributions: Conceptualization, D.S. and P.S.-G.; Investigation, D.S. and D.R.; Methodology, D.S.; Resources, K.E. and D.R.; Supervision, P.S.-G.; Validation, D.R.; Writing—original draft, D.S.; Writing—review & editing, K.E. and P.S.-G. All authors have read and agreed to the published version of the manuscript.

Funding: This research received no external funding.

Institutional Review Board Statement: Not applicable. The presented study did not involve humans or animals. The proposed clinical pilot study was submitted to a competent lead ethics committee (Ethics Committee Salzburg) for evaluation and a positive vote was issued (business number 415-E/1834/8-2014). The study was registered with the competent authority (BASG—Austrian Federal Office for Safety in Heath Care) and the study was granted (proceedings number 9221175). The execution of the study is still pending.

Acknowledgments: We would like to express our sincere thanks to Herbert Resch, acting rector of Paracelsus Medical Private University Salzburg at the time of the study and an experienced specialist in traumatology and shoulder surgery, for his kind support.

Conflicts of Interest: The authors declare no conflict of interest.

References

1. Farley, K.X.; Wilson, J.M.; Kumar, A.; Gottschalk, M.B.; Daly, C.; Sanchez-Sotelo, J.; Wagner, E.R. Prevalence of Shoulder Arthroplasty in the United States and the Increasing Burden of Revision Shoulder Arthroplasty. *JBJS Open Access* **2021**, *6*, e20.00156. [CrossRef] [PubMed]
2. Kim, S.H.; Wise, B.L.; Zhang, Y.; Szabo, R.M. Increasing Incidence of Shoulder Arthroplasty in the United States. *J. Bone Joint Surg. Am.* **2011**, *93*, 2249–2254. [CrossRef] [PubMed]
3. Wiater, J.M.; Fabing, M.H. Shoulder Arthroplasty: Prosthetic Options and Indications. *J. Am. Acad. Orthop. Surg.* **2009**, *17*, 415–425. [CrossRef] [PubMed]
4. Sanchez-Sotelo, J. Glenoid Bone Loss: Etiology, Evaluation, and Classification. *Instr. Course Lect.* **2019**, *68*, 65–78.
5. Hill, J.M.; Norris, T.R. Long-Term Results of Total Shoulder Arthroplasty Following Bone-Grafting of the Glenoid. *J. Bone Joint Surg. Am.* **2001**, *83*, 877–883. [CrossRef]
6. Neyton, L.; Walch, G.; Nové-Josserand, L.; Edwards, T.B. Glenoid Corticocancellous Bone Grafting after Glenoid Component Removal in the Treatment of Glenoid Loosening. *J. Shoulder Elb. Surg.* **2006**, *15*, 173–179. [CrossRef]
7. Steinmann, S.P.; Cofield, R.H. Bone Grafting for Glenoid Deficiency in Total Shoulder Replacement. *J. Shoulder Elb. Surg.* **2000**, *9*, 361–367. [CrossRef]

8. Brown, M.; Eseonu, K.; Rudge, W.; Warren, S.; Majed, A.; Bayley, I.; Higgs, D.; Falworth, M. The Management of Infected Shoulder Arthroplasty by Two-Stage Revision. *Shoulder Elb.* **2020**, *12*, 70–80. [CrossRef]
9. Zhang, B.; Niroopan, G.; Gohal, C.; Alolabi, B.; Leroux, T.; Khan, M. Glenoid Bone Grafting in Primary Anatomic Total Shoulder Arthroplasty: A Systematic Review. *Shoulder Elb.* **2021**, *13*, 509–517. [CrossRef]
10. Sheth, U.; Lee, J.Y.J.; Nam, D.; Henry, P. Early Outcomes of Augmented Glenoid Components in Anatomic Total Shoulder Arthroplasty: A Systematic Review. *Shoulder Elb.* **2021**, *14*, 238–248. [CrossRef]
11. Ghoraishian, M.; Abboud, J.A.; Romeo, A.A.; Williams, G.R.; Namdari, S. Augmented Glenoid Implants in Anatomic Total Shoulder Arthroplasty: Review of Available Implants and Current Literature. *J. Shoulder Elb. Surg.* **2019**, *28*, 387–395. [CrossRef] [PubMed]
12. Gohlke, F.; Werner, B. Humeral and glenoid bone loss in shoulder arthroplasty: Classification and treatment principles. *Orthopade* **2017**, *46*, 1008–1014. [CrossRef] [PubMed]
13. Iannotti, J.P.; Frangiamore, S.J. Fate of Large Structural Allograft for Treatment of Severe Uncontained Glenoid Bone Deficiency. *J. Shoulder Elb. Surg.* **2012**, *21*, 765–771. [CrossRef] [PubMed]
14. Bohsali, K.I.; Wirth, M.A.; Rockwood, C.A.J. Complications of Total Shoulder Arthroplasty. *J. Bone Joint Surg. Am.* **2006**, *88*, 2279–2292. [CrossRef]
15. Chin, P.Y.K.; Sperling, J.W.; Cofield, R.H.; Schleck, C. Complications of Total Shoulder Arthroplasty: Are They Fewer or Different? *J. Shoulder Elb. Surg.* **2006**, *15*, 19–22. [CrossRef]
16. Cofield, R.H.; Edgerton, B.C. Total Shoulder Arthroplasty: Complications and Revision Surgery. *Instr. Course Lect.* **1990**, *39*, 449–462.
17. Hernandez, N.M.; Chalmers, B.P.; Wagner, E.R.; Sperling, J.W.; Cofield, R.H.; Sanchez-Sotelo, J. Revision to Reverse Total Shoulder Arthroplasty Restores Stability for Patients With Unstable Shoulder Prostheses. *Clin. Orthop. Relat. Res.* **2017**, *475*, 2716–2722. [CrossRef]
18. Fox, T.J.; Cil, A.; Sperling, J.W.; Sanchez-Sotelo, J.; Schleck, C.D.; Cofield, R.H. Survival of the Glenoid Component in Shoulder Arthroplasty. *J. Shoulder Elb. Surg.* **2009**, *18*, 859–863. [CrossRef]
19. Kasten, P.; Pape, G.; Raiss, P.; Bruckner, T.; Rickert, M.; Zeifang, F.; Loew, M. Mid-Term Survivorship Analysis of a Shoulder Replacement with a Keeled Glenoid and a Modern Cementing Technique. *J. Bone Joint Surg. Br.* **2010**, *92*, 387–392. [CrossRef]
20. Walch, G.; Young, A.A.; Melis, B.; Gazielly, D.; Loew, M.; Boileau, P. Results of a Convex-Back Cemented Keeled Glenoid Component in Primary Osteoarthritis: Multicenter Study with a Follow-up Greater than 5 Years. *J. Shoulder Elb. Surg.* **2011**, *20*, 385–394. [CrossRef]
21. McLendon, P.B.; Schoch, B.S.; Sperling, J.W.; Sánchez-Sotelo, J.; Schleck, C.D.; Cofield, R.H. Survival of the Pegged Glenoid Component in Shoulder Arthroplasty: Part II. *J. Shoulder Elb. Surg.* **2017**, *26*, 1469–1476. [CrossRef] [PubMed]
22. Franklin, J.L.; Barrett, W.P.; Jackins, S.E.; Matsen, F.A. 3rd Glenoid Loosening in Total Shoulder Arthroplasty. Association with Rotator Cuff Deficiency. *J. Arthroplast.* **1988**, *3*, 39–46. [CrossRef]
23. Karelse, A.; Van Tongel, A.; Verstraeten, T.; Poncet, D.; De Wilde, L.F. Rocking-Horse Phenomenon of the Glenoid Component: The Importance of Inclination. *J. Shoulder Elb. Surg.* **2015**, *24*, 1142–1148. [CrossRef] [PubMed]
24. Grob, A.; Freislederer, F.; Marzel, A.; Audigé, L.; Schwyzer, H.K.; Scheibel, M. Glenoid Component Loosening in Anatomic Total Shoulder Arthroplasty: Association between Radiological Predictors and Clinical Parameters—An Observational Study. *J. Clin. Med.* **2021**, *10*, 234. [CrossRef]
25. Scalise, J.J.; Iannotti, J.P. Bone Grafting Severe Glenoid Defects in Revision Shoulder Arthroplasty. *Clin. Orthop. Relat. Res.* **2008**, *466*, 139–145. [CrossRef]
26. He, L.-H.; Zhang, Z.-Y.; Zhang, X.; Xiao, E.; Liu, M.; Zhang, Y. Osteoclasts May Contribute Bone Substitute Materials Remodeling and Bone Formation in Bone Augmentation. *Med. Hypotheses* **2020**, *135*, 109438. [CrossRef]
27. Hussain, Z.; Ullah, I.; Liu, X.; Shen, W.; Ding, P.; Zhang, Y.; Gao, T.; Mansoorianfar, M.; Gao, T.; Pei, R. Tannin-Reinforced Iron Substituted Hydroxyapatite Nanorods Functionalized Collagen-Based Composite Nanofibrous Coating as a Cell-Instructive Bone-Implant Interface Scaffold. *Chem. Eng. J.* **2022**, *438*, 135611. [CrossRef]
28. Ullah, I.; Hussain, Z.; Zhang, Y.; Liu, X.; Ullah, S.; Zhang, Y.; Zheng, P.; Gao, T.; Liu, Y.; Zhang, Z.; et al. Inorganic Nanomaterial-Reinforced Hydrogel Membrane as an Artificial Periosteum. *Appl. Mater. Today* **2022**, *28*, 101532. [CrossRef]
29. Phipatanakul, W.P.; Norris, T.R. Treatment of Glenoid Loosening and Bone Loss Due to Osteolysis with Glenoid Bone Grafting. *J. Shoulder Elb. Surg.* **2006**, *15*, 84–87. [CrossRef]
30. Antuna, S.A.; Sperling, J.W.; Cofield, R.H.; Rowland, C.M. Glenoid Revision Surgery after Total Shoulder Arthroplasty. *J. Shoulder Elb. Surg.* **2001**, *10*, 217–224. [CrossRef]
31. Elhassan, B.; Ozbaydar, M.; Higgins, L.D.; Warner, J.J.P. Glenoid Reconstruction in Revision Shoulder Arthroplasty. *Clin. Orthop. Relat. Res.* **2008**, *466*, 599–607. [CrossRef] [PubMed]
32. Bonnevialle, N.; Melis, B.; Neyton, L.; Favard, L.; Mólé, D.; Walch, G.; Boileau, P. Aseptic Glenoid Loosening or Failure in Total Shoulder Arthroplasty: Revision with Glenoid Reimplantation. *J. Shoulder Elb. Surg.* **2013**, *22*, 745–751. [CrossRef] [PubMed]
33. Wolfson, M.; Curtin, P.; Curry, E.J.; Cerda, S.; Li, X. Giant Cell Tumor Formation Due to Metallosis after Open Latarjet and Partial Shoulder Resurfacing. *Orthop. Rev.* **2020**, *12*, 60–63. [CrossRef]
34. Lederman, E.S.; Nugent, M.T.; Chhabra, A. Metallosis after Hemiarthroplasty as a Result of Glenoid Erosion Causing Contact with Retained Metallic Suture Anchors: A Case Series. *J. Shoulder Elb. Surg.* **2011**, *20*, e12–e15. [CrossRef] [PubMed]

35. Sedrakyan, A. Metal-on-Metal Failures–in Science, Regulation, and Policy. *Lancet* **2012**, *379*, 1174–1176. [CrossRef]
36. Smith, A.J.; Dieppe, P.; Vernon, K.; Porter, M.; Blom, A.W. Failure Rates of Stemmed Metal-on-Metal Hip Replacements: Analysis of Data from the National Joint Registry of England and Wales. *Lancet* **2012**, *379*, 1199–1204. [CrossRef]
37. Seaman, S.; Kerezoudis, P.; Bydon, M.; Torner, J.C.; Hitchon, P.W. Titanium vs. Polyetheretherketone (PEEK) Interbody Fusion: Meta-Analysis and Review of the Literature. *J. Clin. Neurosci. Off. J. Neurosurg. Soc. Aust.* **2017**, *44*, 23–29. [CrossRef]
38. Honigmann, P.; Sharma, N.; Schumacher, R.; Rueegg, J.; Haefeli, M.; Thieringer, F. In-Hospital 3D Printed Scaphoid Prosthesis Using Medical-Grade Polyetheretherketone (PEEK) Biomaterial. *Biomed Res. Int.* **2021**, *2021*, 1301028. [CrossRef]
39. de Ruiter, L.; Rankin, K.; Browne, M.; Briscoe, A.; Janssen, D.; Verdonschot, N. Decreased Stress Shielding with a PEEK Femoral Total Knee Prosthesis Measured in Validated Computational Models. *J. Biomech.* **2021**, *118*, 110270. [CrossRef]
40. Morrison, C.; Macnair, R.; MacDonald, C.; Wykman, A.; Goldie, I.; Grant, M.H. In Vitro Biocompatibility Testing of Polymers for Orthopaedic Implants Using Cultured Fibroblasts and Osteoblasts. *Biomaterials* **1995**, *16*, 987–992. [CrossRef]
41. Skirbutis, G.; Dzingutė, A.; Masiliūnaitė, V.; Šulcaitė, G.; Žilinskas, J. PEEK Polymer's Properties and Its Use in Prosthodontics. A Review. *Stomatologija* **2018**, *20*, 54–58. [PubMed]
42. Horák, Z.; Pokorný, D.; Fulín, P.; Šlouf, M.; Jahoda, D.; Sosna, A. Polyetheretherketon (PEEK)—I. Část: Perspektivní Materiál pro Ortopedickou a Traumatologickou Praxi. *Acta Chir. Orthop. Traumatol. Cech.* **2010**, *77*, 463–469. [PubMed]
43. Pokorný, D.; Fulín, P.; Slouf, M.; Jahoda, D.; Landor, I.; Sosna, A. [Polyetheretherketone (PEEK). Part II: Application in Clinical Practice]. *Acta Chir. Orthop. Traumatol. Cech.* **2010**, *77*, 470–478.
44. Williams, D. Polyetheretherketone for Long-Term Implantable Devices. *Med. Device Technol.* **2008**, *19*, 10–11.
45. Abdullah, M.R.; Goharian, A.; Abdul Kadir, M.R.; Wahit, M.U. Biomechanical and Bioactivity Concepts of Polyetheretherketone Composites for Use in Orthopedic Implants-a Review. *J. Biomed. Mater. Res. A* **2015**, *103*, 3689–3702. [CrossRef]
46. Modjarrd, K.; Ebnesajjad, S. (Eds.) *Handbook of Polymer Applications in Medicine and Medical Devices*, 1st ed.; Elsevier Science: Oxford, UK, 2013; ISBN 9780323221696.
47. Bola, M.; Simões, J.A.; Ramos, A. Finite Element Modelling and Experimental Validation of a Total Implanted Shoulder Joint. *Comput. Methods Programs Biomed.* **2021**, *207*, 106158. [CrossRef] [PubMed]
48. Bergmann, G.; Graichen, F.; Bender, A.; Rohlmann, A.; Halder, A.; Beier, A.; Westerhoff, P. In Vivo Gleno-Humeral Joint Loads during Forward Flexion and Abduction. *J. Biomech.* **2011**, *44*, 1543–1552. [CrossRef]
49. Rafolt, D.; Gallasch, E. Surface myomechanical responses recorded on a scanner galvanometer. *Med Biol Eng Comput.* **2002**, *40*, 594–599. [CrossRef]
50. Scholes, S.C.; Unsworth, A. Investigating the potential of implantable grade PEEK as a bearing material against various counterfaces. In Proceedings of the Europen Society for Biomaterials Conference, Nantes, France, 27 September–1 October 2006.

Article

Numerical Modeling of a New Type of Prosthetic Restoration for Non-Carious Cervical Lesions

Anna A. Kamenskikh [1,*], Lyaysan Sakhabutdinova [1], Nataliya Astashina [2], Artem Petrachev [2] and Yuriy Nosov [1]

[1] Department of Computational Mathematics, Mechanics and Biomechanics, Perm National Research Polytechnic University, 614990 Perm, Russia; lyaysans@list.ru (L.S.); ura.4132@yandex.ru (Y.N.)

[2] Department of Orthopedic Dentistry, Perm State Medical University Named after Academician E.A. Wagner, 26 Petropavlovskaya St., 614990 Perm, Russia; caddis@mail.ru (N.A.); artem@petrachev.ru (A.P.)

* Correspondence: anna_kamenskih@mail.ru; Tel.: +7-(342)-239-15-64

Abstract: The paper considers a new technology for the treatment of non-carious cervical lesions (NCCLs). The three parameterized numerical models of teeth are constructed: without defect, with a V-shaped defect, and after treatment. A new treatment for NCCL has been proposed. Tooth tissues near the NCCLs are subject to degradation. The main idea of the technology is to increase the cavity for the restoration of NCCLs with removal of the affected tissues. The new treatment method also allows the creation of a playground for attaching the gingival margin. The impact of three biomaterials as restorations is studied: CEREC Blocs; Herculite XRV; and Charisma. The models are deformed by a vertical load from the antagonist tooth from 100 to 1000 N. The tooth-inlay system is considered, taking into account the contact interaction. Qualitative patterns of tooth deformation before and after restoration were established for three variants of the inlay material.

Keywords: tooth; NCCL; contact; modeling; finite element method (FEM); biomaterials; strain

1. Introduction

1.1. Research Objectives

The object of the study is new prosthetic inlays in non-carious cervical lesions (NCCLs). They suggest the expansion of the cavity of the NCCLs.

Research objectives:

- the creation of parametrized models of teeth with and without the NCCL;
- the creation of parametrized models with the restoration of the NCCL in the form of a new prosthetic inlay;
- the implementation of a series of numerical experiments on strain of the tooth before and after restoration;
- the analysis of the impact of prosthetic inlay materials on strained teeth.

The development of a new method for replacing NCCL and a preliminary assessment of the effect of restoration materials on the tooth-inlay system deformation are the main goals of this study. This method can increase the restorations' service life and their aesthetics. An NCCL is often combined with gingival margin recession. The new treatment method allows the creation of a playground for attaching the gingival margin. The evaluation of the restoration materials' performance is also an important factor.

The work includes only computational experiments at this stage. All studies are in silico. The analysis of therapy parameters and the preliminary selection of materials occur on numerical models.

1.2. Problem Context

Computer modeling and the accumulated experience of describing the behavior of various materials allow them to be used in the field of medical research. Today there are many examples of successful applications of numerical models in dentistry [1,2]. Research of the finite element method (FEM) allows us to expand our understanding of the causes of the occurrence and development of mechanical damage to teeth. Analysis of NCCLs in terms of mechanical behavior seems interesting and promising [3,4].

NCCLs are a fairly common dentition disease in the world [5–7]. This is a non-carious disease of the teeth with tissue degradation near the cervical area. There are four types of NCCLs according to the nature of development in the tooth tissues, as well as two types according to the geometry of the "wedge" section (V and U shaped) [8,9]. It is scientifically substantiated that the shape of the NCCLs affects the load distribution in the tooth. In the area of development of a V-shaped NCCL, the stress concentration is four times higher [9]. Tissue degradation in the defect zone develops without treatment. It has a significant impact on the patient's life quality. Such NCCLs cause a number of inconveniences to the patient: violation of the tooth aesthetics, pain, overload of healthy teeth, etc.

There are many works aimed at studying the causes of the occurrence and development of NCCLs. The hard tooth tissue properties of NCCLs affected are being investigated [10–14]. The biomechanical reasons for the occurrence of abfraction are analyzed [12,15–17]; Occlusal loads and resulting stress-strain states of teeth are being studied [16,18–20]. It is found that the enamel strength decreases in the direction from the outer surface to the dentine-enamel junction [21]. The axial load is 30% more strongly distributed in the enamel. The response to axial load is 30% higher in the enamel than in the rest of the tooth tissues. Multiple multidirectional loads cause reactions that are five times higher than the reactions to the axial load of the same level [17,18]. At the same time, the enamel-cement border and the cervical part of the vestibular surface experience the maximum load, mainly in the incisors and premolars. Abfraction is considered the main cause of developing tooth NCCLs [12,22,23]. However, Grippo J.O. and Masai J.V. [24] found that the stresses occurring in the cervical region on the vestibular surface are similar to those on the oral one, where NCCLs are extremely rare. The combination of acids and internal stresses on tooth enamel is the reason for this effect according to scientists [24,25]. A relationship is also established between parafunctions, such as bruxism, and the occurrence of tooth NCCLs [26,27].

The loss of restoration of NCCL is a serious problem that dentists face daily. This is due to many factors, for example, high loads on the tooth crown part [11,12,28]. Part of modern research is aimed at analyzing the effect of the material on the strain of the tooth-restoration system [29–31]. Other authors consider the influence of different mechanisms of inelastic strain on the restorations' performance [32,33]. The modification of the restorations' geometric configuration and new treatment technologies is another research area [34,35]. Science-intensive approaches in dentistry have made it possible to study the processes occurring in the tooth tissues [10,15,19,36–39]. The influence of occlusal and parafunctional loads is researched. An analysis of change in the properties of hard tooth tissues is performed. The pattern of the stress distribution in healthy and affected teeth is investigated. However, to date, the information obtained on the etiology and pathogenesis of abfractions is still insufficient to provide quality care to patients with NCCLs.

This problem requires new effective solutions. The modern level of providing highly qualified medical care makes high demands in the treatment of dental system pathologies. The methods of biomechanical modeling and mathematical analysis acquire a special role in the planning stage of dental treatment [11]. FEM is used for the modeling and analysis of complex systems, including biomechanical ones [9,18,32–35]. The numerical model will allow planning options for the formation of a cavity for optimal long-term restoration. Changing the defect geometry during preparation can change or transfer the load vector to stronger areas. The prediction of the impact of loading for various restoration options will help to assess the possible effectiveness of orthopedic treatment. A reasonable choice of treatment tactics will prevent or reduce the rate of progression of the disease.

1.3. Problem Description

An NCCL is a fairly common tooth lesion. The disease causes increased tooth sensitivity and violation of dentition aesthetics. NCCLs are located near the cervical area. The tooth perceives the worst loads from the antagonist's. Further development of the defect is possible. With the disease progression, movement of the gingival margin in the apical direction is possible, which leads to developing the gingival recession.

An NCCL and its development cause a change in the tissues surrounding the lesion [40]. According to research, a local detachment of enamel from dentin was found in 30% of clinical cases. The separation causes a gap. The gap leads to the breaking off of the enamel area and the development of the defect in the future [25]. According to existing data, enamel changes are of a different nature. Focal enamel demineralization occurs in the areas around the defect [25]. Demineralization is aggravated when exposed to an acidic environment [41]. Various lesions are fixed in the enamel. The nature of the damage depends on the defect form and the main cause of its development. The damage can take the form of microscopic furrows, cracks, and craters [42,43]. Microdamages can lead to enamel chipping and defect expansion in the future [25,44]. The directions of the cracks do not have a definite dependence. Cracks propagate deep into the pulp chamber in the direction of the lesion, also in different directions from the defect [41,44]. Dentin changes occur in the NCCL zone and depend on the shape and depth of the defect [44,45]. Replacement dentin appears with deep damage and prevents the expansion of the cavity with gradual development NCCLs [25,46].

At the moment, the most widespread are direct restorations. However, they have a number of disadvantages:

- polymerization shrinkage;
- dependence on the manual skills of a particular specialist;
- short service life;
- pigmentation;
- the inability to ensure reconstruction of the periodontal attachment between the artificial material and the gum.

The necessity to create new technologies and methods for NCCL restoration has arisen. The new prosthetic inlays with an additional expansion of the wedge cavity are one of the solutions. The proposed method involves a significant amount of tissue preparation, both to expand the defect and to create zones of additional retention. It is important to note that mainly tooth tissues with accumulated macro- and micro-damages are removed.

Deformation of the tooth before and after restoration by a new method is considered in the work. Modeling is performed in the ANSYS Mechanical APDL application package (ANSYS Inc., Canonsburg, PA, USA). The tooth geometry is modeled as a first approximation and is more rounded. Crown geometry can be changed. In the first approximation, the crown geometry is not symmetrical and has different heights, i.e., an attempt is endeavored to bring the geometry of the crown closer to the individual case.

2. Materials and Methods

2.1. Model

The central section of the tooth, with and without taking into account the NCCL, is shown in Figure 1. The tooth model includes the volume of enamel (1) and dentine (2). The tooth pulp (3) is not modeled but is taken into account when parameterizing the NCCL. When deepening the NCCL in the tooth tissue, it is taken into account that it should not penetrate into the pulp.

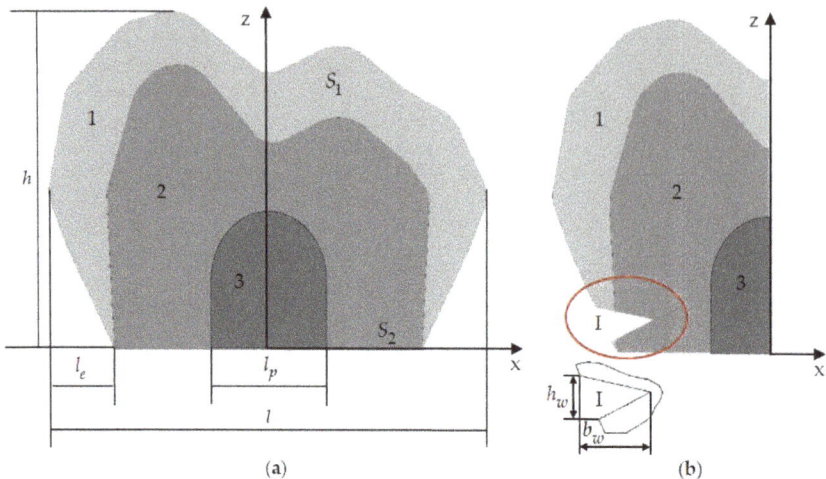

Figure 1. The central section of the tooth, taking into account (**a**) and without taking into account (**b**) an NCCL: 1 is enamel; 2 is dentine; 3 is pulp; I is defect.

The tooth geometry without taking into account the root system is often used in practice [47]. Such models make it possible to quickly obtain qualitative results on the unit deformation. The tooth models do not take into account the root system. This decision was made to qualitative assessment of tooth deformation in the first approximation.

Parameterization of the tooth geometry is performed according to its main parameters: height h, length l, and thickness of the enamel l_e. l_e is the parameter of the enamel's maximum possible thickness. The actual enamel thickness in the model can be more or less by 10–15%. Figure 1 shows the geometrical configuration of a tooth with $h \approx 7.2$, $l \approx 9.44$, and $l_e \approx 1.5$ mm. Overall dimensions of the tooth correspond to premolars and molars. Static boundary conditions are set on the tooth surface S_1 and kinematic ones on the surface S_2. The boundary conditions for all models are the same. The load varies from 100 to 1000 N. The NCCL is parameterized. Parameterization is based on the position and coordinates of the defect. As a result of modeling, an NCCL (I) $2l_w \times h_w \times b_w$ is obtained. Figure 1 shows the NCCL (I) with parameters $5.38 \times 0.73 \times 1.3$ mm.

A cavity is created in the tooth for a prosthetic inlay when creating a new type of restoration. Figure 2 shows the geometry of the cavity central section of the tooth model prosthetic inlay.

An original new method of treatment is proposed. The formed cavity to fix the inlay (II) includes the main part obtained by expanding the lesion, with an additional retention point in the form of a cavity passing to the proximal surface of the tooth, an additional platform at the top of the cavity for fixing the veneer part of the inlay, and a gingival fold (III) located more apically than the lesion.

Element (III) was introduced into the construct to recreate the dentogingival attachment. Often the NCCL is combined with a gingival margin recession. The reconstruction of the periodontal attachment between the artificial material and the gum is not possible.

The inlay cavity was parameterized using 3 parameters:

- h_1 is to create a gingival fold;
- h_2 is to create the main cavity;
- h_3 is to create an additional retention point with an additional platform at the top of the cavity for fixing the veneer part of the inlay.

A geometric configuration of the inlay cavity is shown in Figure 3 for parameter values $h_1 = 0.3$, $h_2 = 0.2$, and $h_3 = 0.5$ mm. The central section of the cavity extends over the entire length of the NCCL. The inlay's final appearance is formed with the geometry of the veneer part.

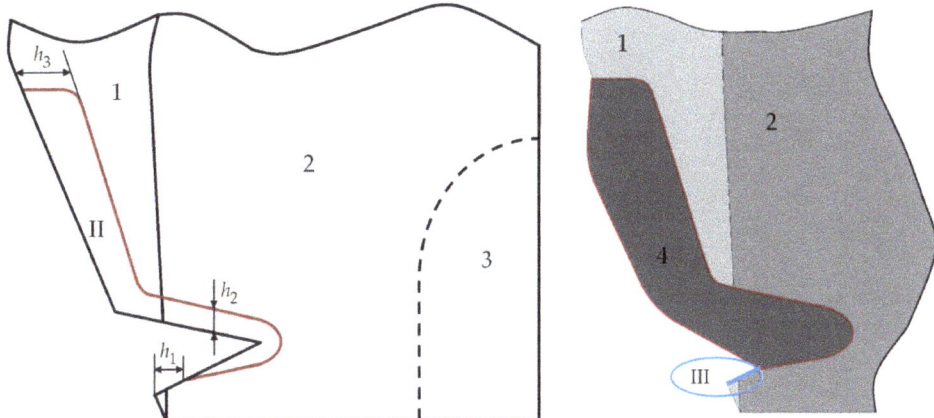

Figure 2. Modeling a cavity for a prosthetic inlay: 1 is enamel; 2 is dentine; 3 is pulp; 4 is inlay; II is cavity to fix the inlay; III is gingival fold.

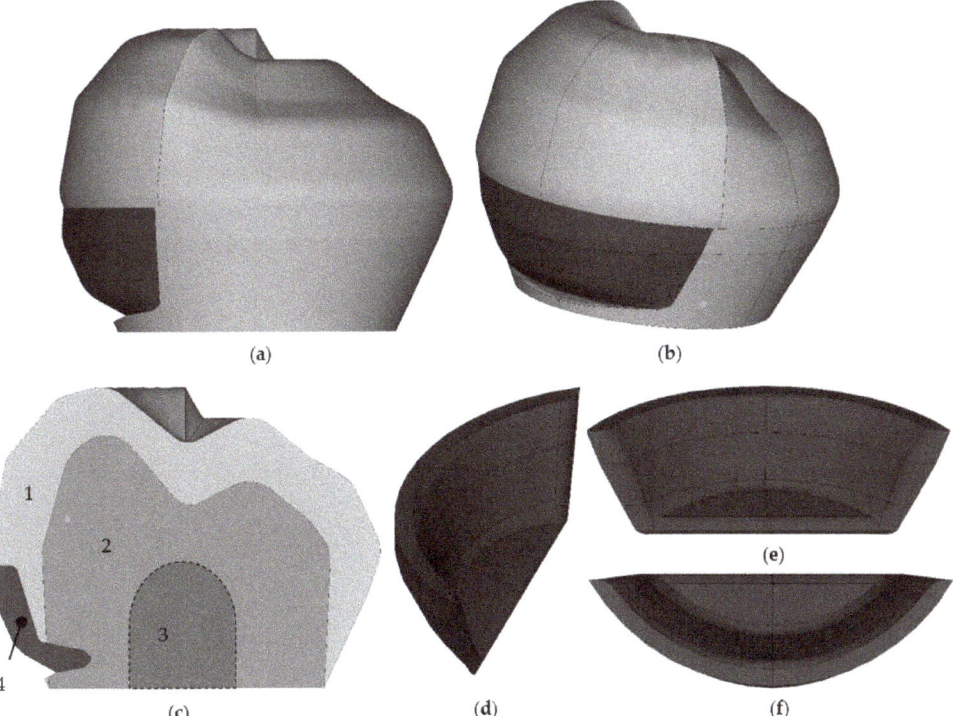

Figure 3. Tooth with a prosthetic inlay: (**a**,**b**) are isometry; (**c**) is the central section; (**d**) is inlay isometry; (**e**) is inside view inlay; (**f**) is top view inlay; 1 is enamel; 2 is dentine; 3 is pulp; 4 is inlay.

Figure 3 shows the tooth geometry with the NCCL restoration using a new prosthetic inlay, as well as the view of the inlay in the tooth cavity. Model 1 is enamel, 2 is dentin, and 4 is a new restoration of an NCCL using different materials. The pulp of the tooth (3) is not modeled but is taken into account when creating a cavity for a prosthetic inlay.

The inlay is quite streamlined on all sides. The inlay model is parameterized. The thickness, veneer part, and cavity area can change. The figure shows one of the options for the veneer part geometry of the inlay. Given inlay geometry will be used in a numerical experiment series.

The prosthetic inlay maintains the aesthetics of the dentition. The inlay restores the aesthetics of the tooth and increases the contact area of the tooth and the prosthetic structure.

The main limitations of the model at the moment:

- roots and gingiva are not taken into account in the model to save computational resources and a detailed study of the tooth-inlay contact zone;
- only the vertical load from the antagonist tooth in a wide range is considered;
- only the case of complete adhesion of the inlay and tooth is considered, although in reality sliding is possible;
- the elastic deformation behavior of materials is considered at this stage. It is planned to study the effect of heat shrinkage on the stress state of the biomechanical unit and refine the behavior model of the system materials in the future;
- the degradation of materials is not taken into account, as well as the formation of cracks due to the complexity of such mechanical models.

2.2. Mechanical Properties of the Mouthguard Components

The materials of the model are considered in elastic formulation. The elastic compression modulus (E) and Poisson's ratio (v) of enamel and dentine are shown in Table 1. Properties of dentine and enamel are taken from reference literature.

Table 1. Physical-mechanical properties of dental tissues.

Parameter	Enamel	Dentine
E, GPa	72.7	18.6
v	0.33	0.31

The materials for the prosthetic inlay: CEREC Blocs (Sirona, Bensheim, Germany) is material 1; Herculite XRV (Kerr Corp, Orange, CA, USA) is material 2; Charisma (Heraeus Kulzer GmbH, Hanau, Germany) is material 3. The physical-mechanical properties of the inlay materials are presented in Table 2. The material properties for restoration are taken from reference literature [48,49].

Table 2. Physical-mechanical properties of the prosthetic inlay materials.

Parameter	Material 1	Material 2	Material 3
E, GPa	45.0	9.5	14.1
v	0.3	0.24	0.24

The most promising material for creating an inlay is considered to be fine-structured feldspar ceramic blocks of industrial production CEREC (material 1). The material is used to make inlays, onlays, crowns, and veneers. The minimum values of the cavity parameters for the inlay are selected according to the restrictions imposed on the material when milling.

For comparison, inlays from two different composite materials are considered. Herculite XRV (material 2) is a versatile microhybrid composite material. Charisma (material 3) is a radiopaque glass-based composite material.

2.3. Numerical Finite Element (FE) Solution and Convergence

The simulation is implemented in the applied ANSYS Mechanical APDL engineering analysis package (ANSYS Inc., Canonsburg, PA, USA). Volume finite elements SOLID185 (four-node tetrahedra) with Lagrangian approximation and three degrees of freedom at each node are used. Contact gluing is modeled in the inlay-tooth interface zones, taking into account friction. The model eliminates the divergence of contact surfaces and appearance of sliding zones. The contact interaction is modeled using a contact pair of elements (CONTA173, TARGE170). The surface-surface contact is considered. The contact algorithm is the extended Lagrange method.

The finite element partition of the model is chosen within the assessment of the influence of the system discretization degree on the numerical solution of the problem. The minimum overall dimension of the finite element near the tooth-inlay contact area is 0.05 mm. When moving away from the contact zone, the size of the finite elements increases in a gradient. The maximum overall dimension of the final element reaches 0.15 mm.

3. Results

The crown part of the tooth is deformed together, i.e., there is no change in the crown geometry. The position and number of the points of load application from the antagonist tooth has little effect on the stress-strain state of the tooth-inlay system.

The stress-strain state of a tooth without defect was considered in advance (Figure 4). The nature of the distribution of the intensity of stresses and strains is shown on the example of the load of 500 N from the antagonist tooth.

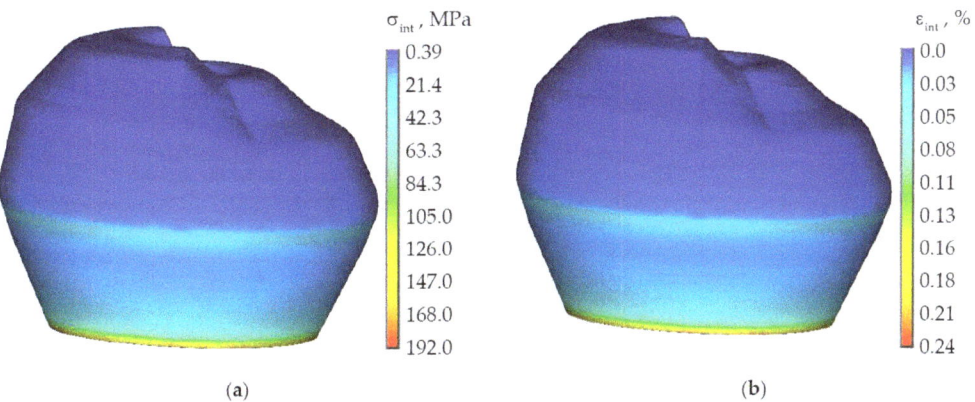

Figure 4. Stress and strain intensity of the tooth without NCCL: (**a**) is stress intensity; (**b**) is strain intensity.

The maximum stress and strain intensity is observed in the zone of kinematic boundary conditions. The maximum intensity of stresses and strains is observed in the tooth enamel near the cervical area. The maximum stress level in dentine is 85% lower than in enamel and reaches 28.8 MPa. The maximum level of strains in dentine does not exceed 0.2%. The intensity of stresses and strains in the zone where the NCCL will be modeled reaches the level of approximately 70 MPa and 0.1%, respectively.

The stress and strain intensity of the tooth with NCCL at a load of 500 N are shown in Figure 5.

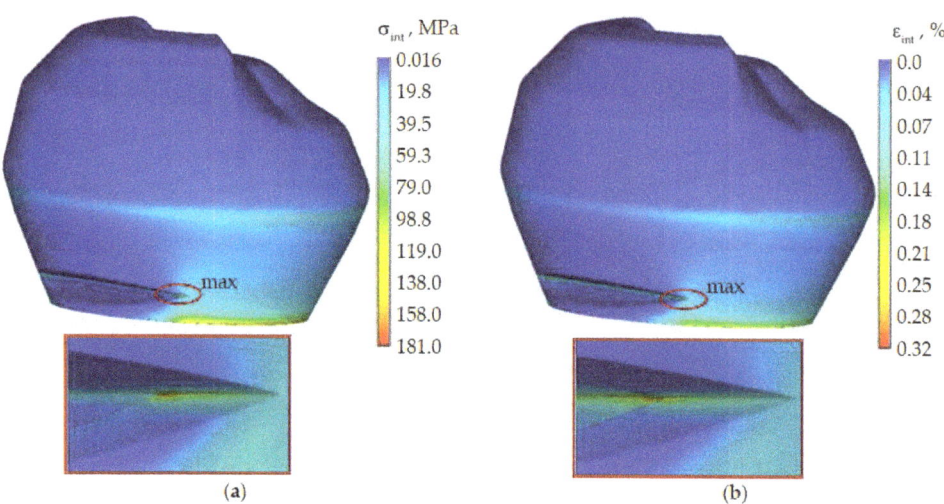

Figure 5. Stress and strain intensity of the tooth with NCCL: (**a**) is stress intensity; (**b**) is strain intensity.

The maximum level of stress and strain in the tooth model with NCCL is observed in the "wedge" zone. The maximum stress intensity is observed at the edge of the NCCL in the enamel and reaches 181 MPa. The maximum stress intensity is higher more than 2.5 times than in the tooth without NCCL. The maximum strain intensity is also observed in the NCCL in the dentine and reaches 0.32%, which is 1.6 times higher than in the model without defects.

The next stage of the study is analyzing the effect of the prosthetic inlay in the NCCL by changing the "wedge" geometry. Figures 6 and 7 show the stress intensity distribution in the biomechanical tooth-inlay system under the load of 500 N from the antagonist tooth. The qualitative view of the distribution of the deformation behavior parameters of the tooth-inlay system does not depend on the inlay material. The main difference between the solutions is in quantitative values. Stress and strain intensities are shown in the example of a model with an inlay from material 1.

The level of the stress intensity in the area of the prosthetic inlay is comparable to the stresses in the tooth without defects. The maximum stress intensity in the tooth-inlay system has shifted to the cervical area of the tooth.

The stress intensity in the inlay is 60.2% lower than in the tooth. At the lower boundary from the outside, local stress concentrators are observed at the level of 77 MPa. The distribution of stress fields in dentine corresponds to the loading conditions. The main stresses from the antagonist tooth action are realized in the enamel and in the inlay. The stress intensity on the outer surface inlay is lower than on the inner one.

The dependences of the maximum values of stress intensities in the biomechanical system elements on the applied load value are shown in Figure 8.

The inlay material does not significantly affect the values of maximum stresses in the enamel. The more uniform distribution of the stress intensity in the biomechanical tooth-inlay system is observed by the use of material 1. A significant increase of stress intensity in the tooth dentin is observed in this case. The maxσ_{int} in dentine when inlay material 1 is 2.5 and 1.2 times higher than in the tooth model without and with an NCCL, respectively. A decrease maxσ_{int} in enamel and an increase in dentine were also observed when using prosthetic inlays from materials 2 and 3. A comparison of the stress level with models without and with NCCL is shown in Table 3.

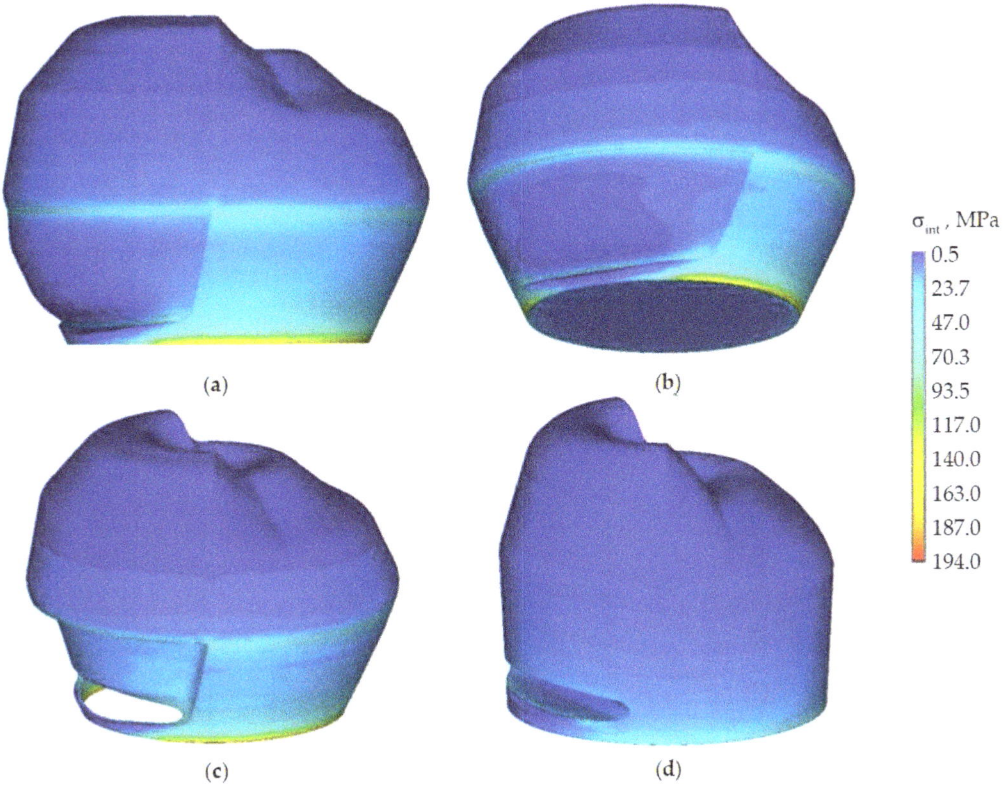

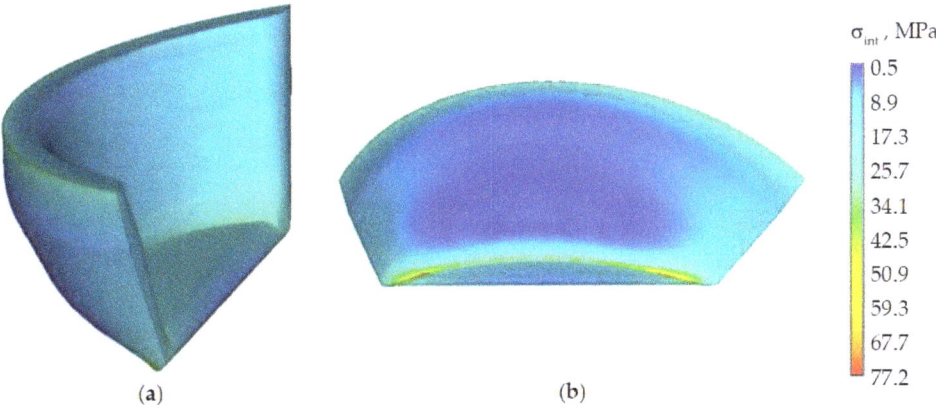

Figure 6. Stress intensity of the biomechanical assembly: (**a**,**b**) are tooth-inlay system; (**c**) is enamel; (**d**) is dentine.

Figure 7. Stress intensity of the inlay: (**a**) is general view, (**b**) is outer side.

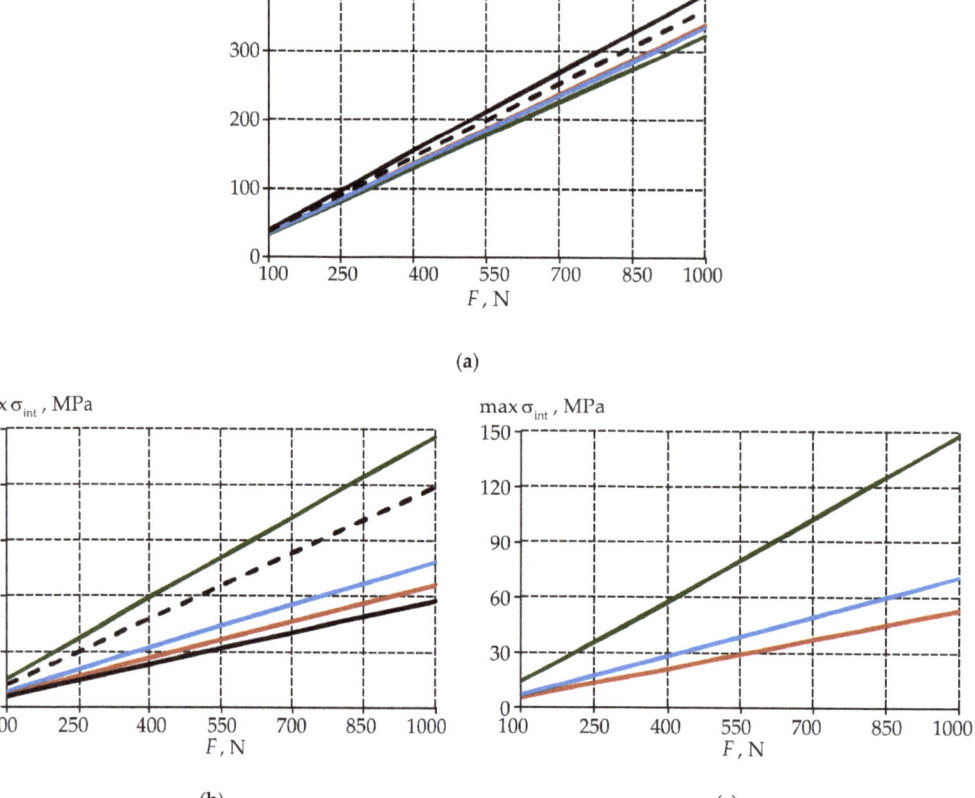

Figure 8. Dependence of the maximum values of the biomechanical assembly stress intensity on the load: (**a**) is enamel; (**b**) is dentine; (**c**) is inlay; black (solid line) is model without defect; black (dotted line) is model with defect; green is model with material 1 inlay, red is model with material 2 inlay, blue is model with material 3 inlay.

Table 3. Comparison maxσ_{int} (%) in the biomechanical assembly elements of different models.

Model	Element	Model Accounting for Prosthetic Inlay		
		Material 1	Material 2	Material 3
Not taking into account the NCCL	Enamel Dentine	<by 16.00% >by 154.62%	<by 11.86% >by 15.17%	<by 12.93% >by 38.83%
Taking into account the NCCL	Enamel Dentine	<by 10.65% >by 23.39%	<by 6.24% <by 44.19%	<by 7.38% <by 32.72%

A decrease in the stress intensity in the tooth enamel when using a restoration in the form of a prosthetic inlay by 12–16% can be noted. An increase in the maximum intensity of stresses in the dentine and the inlay near of the gingival fold is observed due to the contact gluing. The stress intensity in the model with a prosthetic inlay made of material 1 is more than two times lower on the main volume of materials. The influence of the geometry of the cavity for the inlay and veneer part on the deformation behavior of the biomechanical assembly must be studied.

Let us consider the nature of the strain intensity distribution in the biomechanical tooth-inlay assembly (Figure 9).

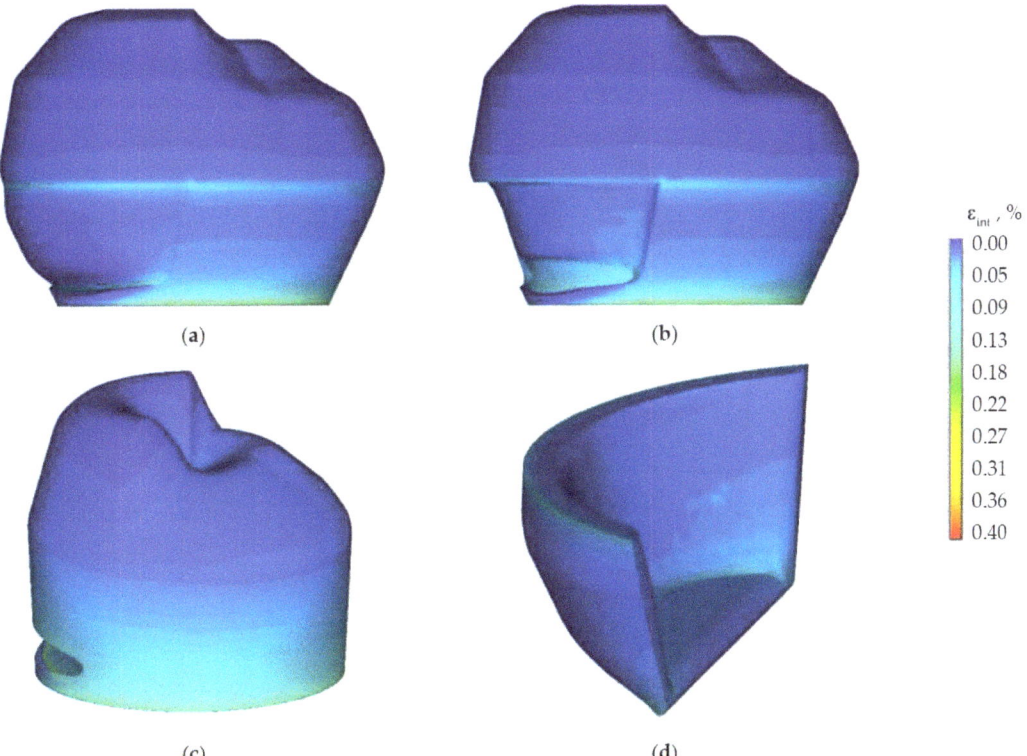

Figure 9. Strain: (**a**) is tooth-inlay system; (**b**) is enamel; (**c**) is dentine; (**d**) is inlay.

The maximum strain intensity is observed in the dentine near the edge of the contact interaction zone with the prosthetic inlay. The strain intensity is lower by two or more times on the rest of the dentine volume. The maxε_{int} in the enamel is observed near the cervical area of the tooth. The maxε_{int} in the prosthetic inlay is located on the lower surface near the edge of the contact zone. This effect can be eliminated by: changing the geometry of the prosthetic inlay; the selection of material and refinement of the finite element model.

A dependence of the strain maximum intensity on the load level is shown in Figure 10.

The strain intensity in the enamel is lower in the models with a prosthetic inlay in the NCCL. The maximum influence on the nature of the distribution and the level of strain intensity is observed in the tooth dentine. The strain increase in the dentine and a shift of the maximum level ε_{int} to the "wedge" area in model with a defect can be noted. The maxε_{int} dentine in the model with an NCCL is 1.7 times more than in the tooth without a defect.

The use of a new NCCL restoration in the form of a prosthetic inlay makes it possible to reduce the strain intensity in the dentine when using materials 2 and 3. An increase maxε_{int} in dentine is observed in the model with an inlay made of material 1 by 2.1 times than in the tooth without defect. The maximum level zone ε_{int} is localized near the inlay-tooth contact border near the area of the gingival fold. The level of strain intensity is comparable to the strains of the tooth without defect on the main volume of dentine.

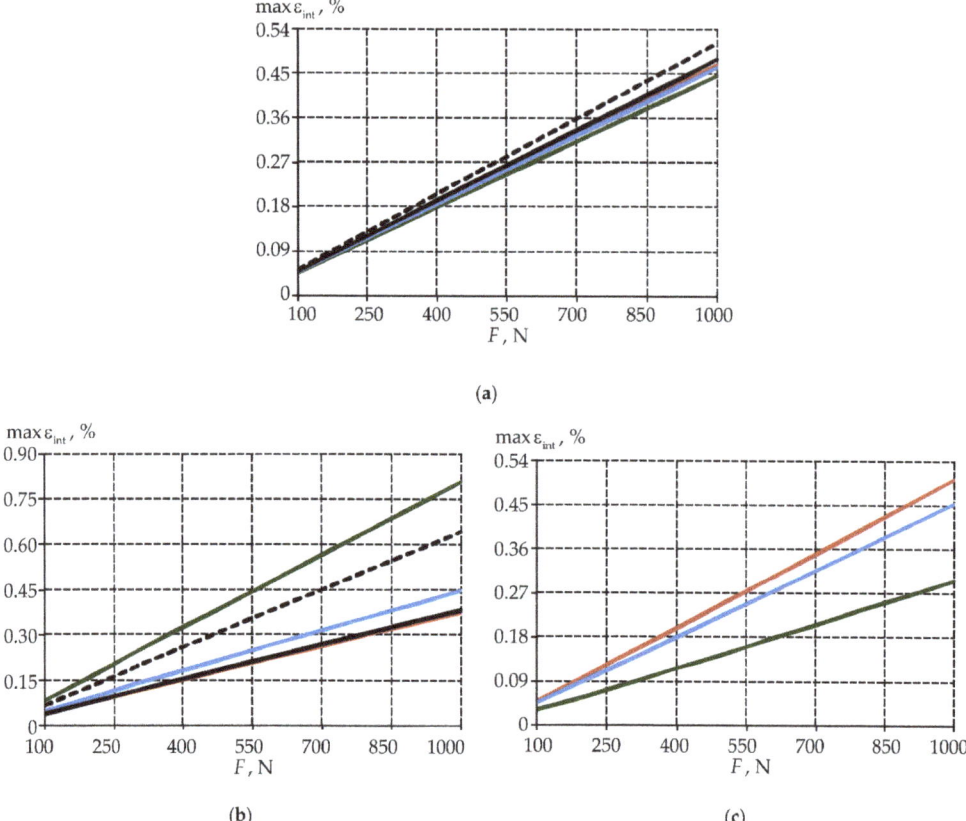

Figure 10. Dependence of the maximum values of the strain intensity of the biomechanical assembly on the load: (**a**) is enamel; (**b**) is dentine; (**c**) is inlay; black (solid line) is model without defect; black (dotted line) is model with defect; green is model with material 1 inlay, red is model with material 2 inlay, blue is model with material 3 inlay.

The minimum value maxε_{int} is observed in the inlay from material 1. The maximum strain level of the inlay from materials 2 and 3 is comparable to the tooth enamel. This can adversely affect the service life of the prosthetic structure.

It is important to evaluate the dependence of the contact parameters because of the problem statement. The interface zone parameters are indicators of the strain behavior of the tooth-inlay system: contact pressure P_K and contact tangential stress τ_K. A dependence of the maximum (max) and average (Δ) levels of contact parameters on the tooth-inlay mating surface for three inlay materials are shown in Figure 11.

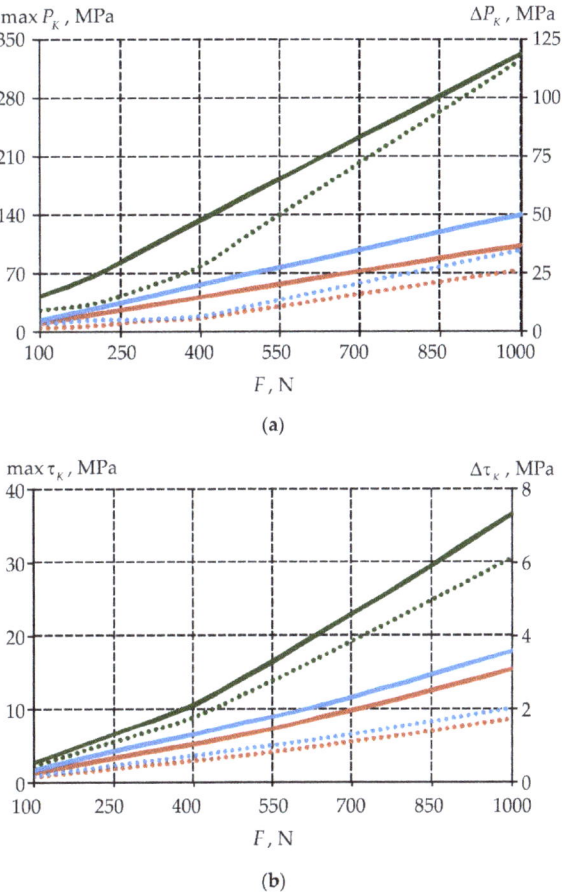

Figure 11. Dependence of contact parameters in the inlay on the load: (**a**) is contact pressure; (**b**) is tangential contact stress; solid lines is the maximum; points is the average; green is model with material 1 inlay, red is model with material 2 inlay, blue is model with material 3 inlay.

The maximum contact parameters are observed near the edge of the tooth-inlay contact zone near the gingival fold, similar to stresses and strains. The average level of contact pressure and contact tangential stress is 3–4 and 7–8.9 times lower than the maximum values of the parameters respectively. The maxτ_K and $\Delta\tau_K$ are 7–9 and 15–19 times lower than maxP_K and ΔP_K. The study of the influence values of the friction coefficient on the biomechanical assembly deformation is required.

The obtained estimates give an idea of the qualitative patterns of the influence of the new restoration type and its materials on tooth deformation.

4. Discussion
4.1. Limitation Statement

This article is the result of preliminary research for a new type of restoration. The object of study has the following limitations:

- tooth root system in the model is discarded;
- interaction with gums is not taken into account;
- dental pulp is not modeled;

- enamel and dentin are deformed together;
- contact interaction is modeled as a complete adhesion of mating surfaces.

Material models have the following limitations:
- the behavior is described as isotropic elastic;
- materials shrinkage is not taken into account.

The main task of the work was to evaluate the effect of the material and geometric configuration of a new type of NCCL restoration on tooth deformation. The patterns of change in the deformation behavior of the restored tooth from the level of external load are also revealed. The accepted limitations are acceptable, but the geometric model and description of the materials behavior needs to be improved. The researchers face a number of tasks:

- the refinement of design schemes and finite element modeling of a biomechanical assembly;
- the analysis of the effect of the polymerization shrinkage of restorative composites on the total deformation of the tooth;
- the analysis of the influence of the cavity geometry for the prosthetic inlay in the NCCL;
- the analysis of the influence of materials for the restoration of an NCCL with different tooth configurations and cavities according to the prosthetic design;
- the analysis of the influence of loads from the antagonist tooth acting at an angle to the biomechanical system;
- the analysis of the influence of the nature of the tooth elements conjugation and the tooth-inlay system.

4.2. Materials

Researchers [50–52] note that one of the main criteria for the performance of NCCL restorations is adhesion between the restoration material and tooth elements. Battancs et al. concluded that the performance of the tooth-restoration system does not depend on the material. Thus, any material that already exists and is used in the treatment of an NCCL is suitable for prosthetics. Ichim et al. [3] come to the opposite conclusion and recommend the use of materials with a Young's modulus of less than 1 GPa in restorations. The results of this work confirm that a material with a lower elastic compression modulus delivers a favorable strain behavior to the tooth-restoration system. At the same time, it should be noted that in the restoration of a harder material, a lower level of deformation is observed. Machado et al. come to the conclusion that non-straight ceramic inlays in the NCCL have less roughness, which favors subsequent periodontal treatment [53]. The authors of [53] also obtain information that composite restoration materials can reduce the stress level in the tooth-inlay system, but at the same time they have significant thermal shrinkage. Ceramic inlays and crowns have become widespread in the last decade [54,55]. Ceramic inlays make it possible to obtain good adhesion with the tooth elements, which is considered an important factor, especially in the prosthetics of an NCCL [55]. The maximum values of stress and strain intensity are observed near the contact between the lower part of the inlay and the tooth, which does not contradict the study data [3]. Additionally, there is a significant excision of the size of the tooth tissues, a significant number of defects, which probably does not only affect the increase in load, but also creates additional retention losses.

Today, there is a lot of research on the applicability of innovative biomaterials in dentistry. An interesting idea is to use oral-derived stem cells together with biomaterials or scaffold-free techniques to obtain strategic tools for regenerative and translational dentistry [56]. The authors [57] have developed a synthetic P26 peptide that demonstrates a remarkable dual mineralization potential to repair incipient enamel decay and mineralization defects localized in peripheral dentin below the dentin-enamel junction. The research in [58] indicated the prospect of using black phosphorene (BP) for pharmacological applications as scaffolds and prosthetic coverings. In our opinion, the combined use of innovative materials and new restorative inlay has great potential in the treatment of NCCLs and the prevention of their development. In the implementation of new clinical practices, it will be important to confirm them with the help of reliable computer models, which we are creating.

4.3. Influence of Taking into Account the Root System on the Tooth Deformation

Creating complete parameterized tooth geometry is a complex process. The modeling of teeth with a truncated root system [4,30,59] and without taking into account the root system [47,60,61] is often encountered in practice. Such limitations can have a significant impact on the numerical simulation results. Edge effects appear in the rootless model.

The root system of a tooth is often modeled with simplified geometry to pilot studies of new treatment methodologies [62,63]. The canonical geometry in the form of a single root is modeled [62,63]. The simulation of the simple geometry of two root canals is also encountered [35,64].

Influence estimates of the tooth root system on the modeling results are of interest. It was decided to consider the tooth deformation (Figures 1 and 3), taking into account the root system in order to clarify quantitative patterns. The tooth root system is modeled in a simplified setting (Figure 12).

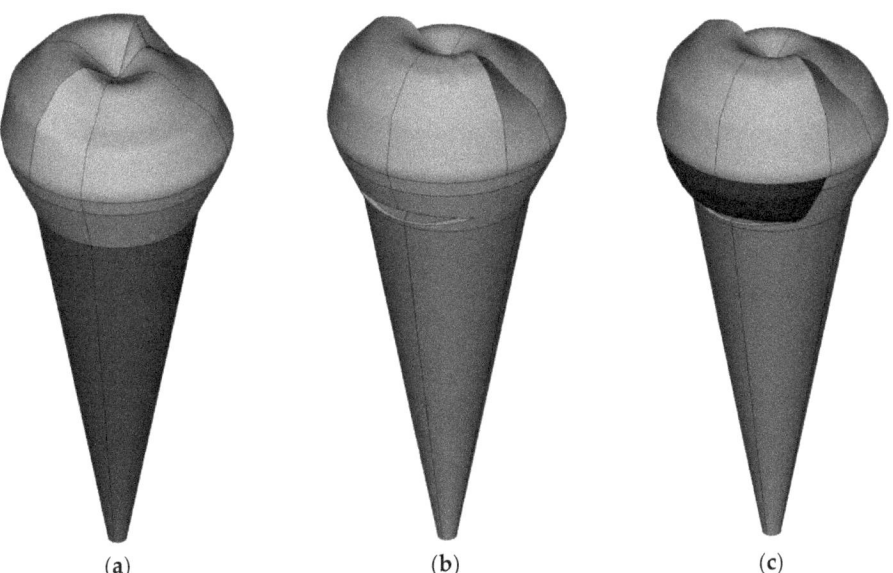

Figure 12. Models of teeth taking into account the root system simple geometry: (**a**) is without defect; (**b**) is with a defect; (**c**) is with the restoration new type.

The tooth root is modeled as a truncated cone. The boundary conditions are: the prohibition of the normal displacement of the side surfaces; the prohibition of all coordinates displacements of the root system lower part.

It has been established that taking into account the root system has little effect on the tooth deformation qualitative picture. A significant influence on the quantitative values of the deformation state parameters and the contact characteristics of the tooth-inlay system can be noted in this case.

The analysis of quantitative differences in the deformation behavior parameters of teeth with and without taking into account the root system will be performed according to the formula, where we take the model taking into account the root as reference values:

$$\Delta A = \sum_{F=100}^{1000} [(A|_{\text{without root}} - A|_{\text{with root}}) / A|_{\text{with root}} \cdot 100] / 10\%, \quad (1)$$

where A is maximum parameters σ_{int} (Figure 13), ε_{int} (Figure 14), P_K and τ_K (Figure 15). The difference percentage in deformation parameters slightly depends on the load. ΔA is the arithmetic mean of the deviation. The parameters were compared in terms of materials volumes in the crown area. The root volume was not taken into account when determining the parameters maxima.

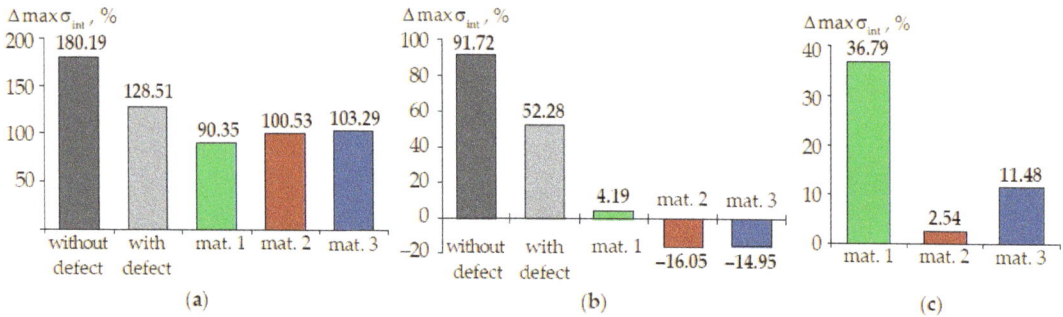

Figure 13. $\Delta \max \sigma_{\text{int}}$: (**a**) is enamel; (**b**) is dentine; (**c**) is inlay; dark-grey is model without defect; light-grey is model with defect; green is model with material 1 inlay, red is model with material 2 inlay, blue is model with material 3 inlay.

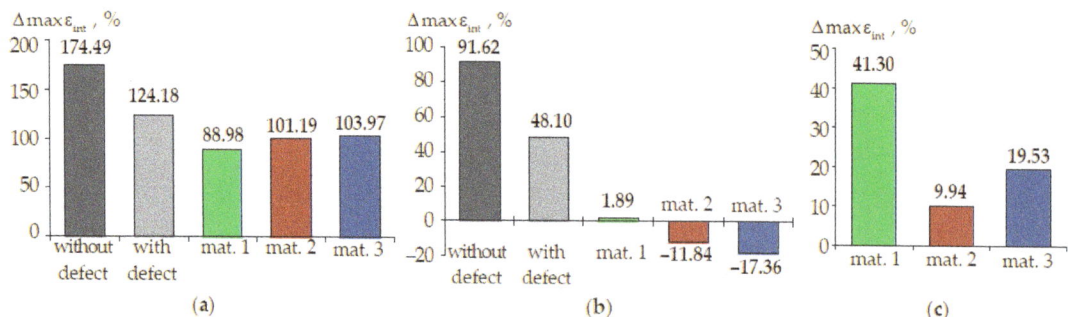

Figure 14. $\Delta \max \varepsilon_{\text{int}}$: (**a**) is enamel; (**b**) is dentine; (**c**) is inlay; dark-grey is model without defect; light-grey is model with defect; green is model with material 1 inlay, red is model with material 2 inlay, blue is model with material 3 inlay.

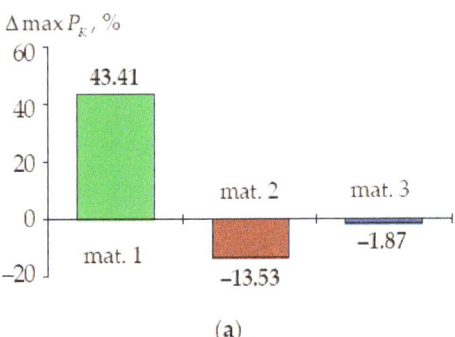 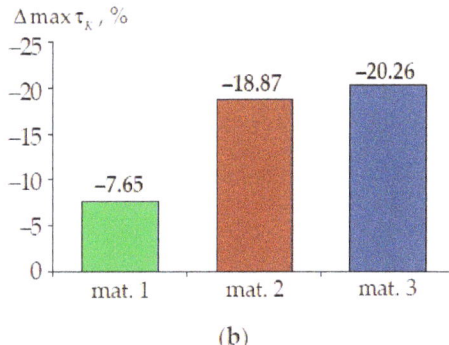

Figure 15. $\Delta\max P_K$ (**a**) and $\Delta\max\tau_K$ (**b**): green is model with material 1 inlay, red is model with material 2 inlay, blue is model with material 3 inlay.

Accounting for the root has the maximum effect on the behavior of the tooth without and with the defect. The maximum intensity of stresses and strains has a more pronounced localization in the defect area in this case.

The intensity of stresses and strain in the enamel is approximately two times lower in the tooth-inlay system when the root is taken into account. Accounting for the root system has an insignificant effect on the dentin deformation parameters (less than 20%).

The effect of taking into account the root on the contact parameters depends on the inlay material. For contact pressure: in material 1, there is a decrease in the parameter maximum level by approximately 40%; in materials 2 and 3, there is a slight increase in the parameter maximum level by 14 and 2%, respectively. The contact tangential stress in the model taking into account the root is higher.

The model without taking into account the root system gives only a qualitative idea of the tooth deformation. The refinement of models taking into account the root system close to real geometry is required. An analysis of the possibility of tooth root truncation to increase the computational procedures speed is also necessary.

This research direction is a priority for the scientific group. The rationalization of calculation schemes is necessary due to the wide scope of future research: influence analysis of the geometry cutout under inlay, inlay materials, occlusal load, the conjugation patterns tooth-inlay, etc.

4.4. Main Results

The distribution nature and the stress intensity level in a tooth with an NCCL are comparable to the results obtained by Jakupović et al. [9]. According to [3], the maximum stresses in the interface between the restoration and the tooth are observed near the edge of the lower surface of the inlay. The data obtained in this work show a similar result. An important result of the article is the investigation of a new technology for the restoration of an NCCL that allows the use of ceramic inlays that contribute to the creation of the required adhesion with the tooth elements. The use of a contact strain mechanism of the biomechanical system tooth restoration brings the nature of the strain closer to the real case. Many works on numerical analysis consider the strain of a tooth with the restoration of an NCCL within the framework of joint strain of the elements, which does not reflect all the features of the strain of a biomechanical assembly [3,9,32–35]. At the moment, a significant increase in stress in the dentine has been established, with a decrease in the stress level in the enamel at the inlay from the CEREC Blocs material. This effect can be avoided by the rational selection of the prosthetic inlay geometry and analyzing the influence of the tooth-restoration interface nature.

5. Conclusions

In general, the new restoration of the NCCL with the expansion of the cavity under the prosthetic inlay and the formation of the area of the gingival fold for the gum growth shows its viability. On the main volume of tooth materials, there are levels of strain parameters comparable to the tooth without taking into account the lesion. It is required to study the influence of the cavity geometry for the prosthetic construction and the veneer part of the inlay to eliminate local zones of the maximum level of stresses and strains in the biomechanical assembly. The problem of frictional contact interaction in the region of the tooth-inlay interface with the refinement of the finite element model should be extended.

It is established that the material of the Herculite XRV inlay (Kerr Corp, Orange, CA, USA) in this formulation of the problem delivers a favorable strain state to the biomechanical system. The nature and levels of the distribution of the stress values of the maximum and average level of the contact parameters of the interface zones and strain intensity are close to those of a tooth without taking into account the lesion. At the same time, the level of strain of the Herculite XRV inlay is comparable to the strains in the tooth enamel, which can adversely affect the service life of the structure.

The results obtained in this article reflect the qualitative patterns of tooth deformation behavior. Accounting for the root system and its rationalization is required for quantification.

6. Patents

A patent of the Russian Federation "Method of treating a wedge-shaped tooth lesion and a device for its implementation" No. 2 719 898, registration date 23 April 2020. The authors of the patent are Astashina Natalia, Petrachev Artem, Kazakov Sergei, Rogozhnikova Evgenia. The patent holder is Perm State Medical University named after E.A. Wagner.

Author Contributions: Conceptualization, A.A.K. and N.A.; methodology, A.A.K. and N.A.; software, A.A.K., L.S. and Y.N.; validation, A.A.K. and N.A.; writing—original draft preparation, A.A.K., L.S. and A.P.; writing—review and editing, A.A.K., N.A., L.S. and A.P.; visualization, A.A.K., L.S. and Y.N.; funding acquisition, A.A.K. All authors have read and agreed to the published version of the manuscript.

Funding: The study was funded by Perm National Research Polytechnic University in the framework of the Federal Academic Leadership Program "Priority-2030".

Institutional Review Board Statement: Not applicable.

Informed Consent Statement: Not applicable.

Data Availability Statement: Not applicable.

Conflicts of Interest: The authors declare no conflict of interest.

References

1. Olesova, V.N.; Ivanov, A.S.; Olesov, E.E.; Romanov, A.S.; Zaslavskiy, R.S. Biomechanical comparison of ceramic, titanium and chrome cobalt post inlays in post-traumatic dental defects repair. *Disaster Med.* **2022**, *1*, 53–58. [CrossRef]
2. Krupnin, A.E.; Kharakh, Y.N.; Gribov, D.A.; Arutyunov, S.D. Biomechanical analysis of new constructions of adhesive bridge prostheses. *Russ. J. Biomech.* **2019**, *23*, 423–434. [CrossRef]
3. Ichim, I.P.; Schmidlin, P.R.; Li, Q.; Kieser, J.A.; Swain, M.V. Restoration of non-carious cervical lesions: Part II. Restorative material selection to minimise fracture. *Dent. Mater.* **2007**, *23*, 1562–1569. [CrossRef]
4. Dikova, T.; Vasilev, T.; Hristova, V.; Panov, V. Finite element analysis of V-shaped tooth defects filled with universal nanohybrid composite using incremental technique. *J. Mech. Behav. Biomed. Mater.* **2021**, *118*, 104425. [CrossRef]
5. Ceruti, P.; Menicucci, G.; Mariani, G.; Pittoni, D.; Gassino, G. Non carious cervical lesions. A review. *Minerva Stomatol.* **2006**, *55*, 43–57.
6. Sarode, G.; Sarode, S. Abfraction: A review. *J. Oral Maxillofac. Pathol.* **2013**, *17*, 222–227. [CrossRef]
7. Elmarakby, A.; Sabri, F.; Alharbi, S.; Halawani, S. Noncarious Cervical Lesions as Abfraction: Etiology, Diagnosis, and Treatment Modalities of Lesions: A Review Article. *Dentistry* **2017**, *7*, 1000438. [CrossRef]
8. Noda, N.-A.; Chen, K.-K.; Tajima, K.; Takase, Y.; Yamaguchi, K.; Nagano, H. Intensity of Singular Stress Field due to Wedge-Shaped Defect in Human Tooth after Restored with Composite Resins. *Trans. Jpn. Soc. Mech. Eng. Part A* **2009**, *75*, 1209–1216. [CrossRef]

9. Jakupović, S.; Anić, I.; Ajanović, M.; Korać, S.; Konjhodžić, A.; Džanković, A.; Vuković, A. Biomechanics of cervical tooth region and noncarious cervical lesions of different morphology; three-dimensional finite element analysis. *Eur. J. Dent.* **2016**, *10*, 413–418. [CrossRef]
10. Soares, P.V.; Souza, L.V.; Veríssimo, C.; Zeola, L.F.; Pereira, A.G.; Santos-Filho, P.C.; Fernandes-Neto, A.J. Effect of root morphology on biomechanical behaviour of premolars associated with abfraction lesions and different loading types. *J. Oral Rehabil.* **2014**, *41*, 108–114. [CrossRef]
11. Srirekha, A.; Bashetty, K. A comparative analysis of restorative materials used in abfraction lesions in tooth with and without occlusal restoration: Three-dimensional finite element analysis. *J. Conserv. Dent.* **2013**, *16*, 157–161. [CrossRef]
12. Vandana, K.L.; Deepti, M.; Shaimaa, M.; Naveen, K.; Rajendra, D. A finite element study to determine the occurrence of abfraction and displacement due to various occlusal forces and with different alveolar bone height. *J. Indian Soc. Periodontol.* **2016**, *20*, 12–16. [CrossRef]
13. Rees, J.S.; Hammadeh, M.; Jagger, D.C. Abfraction lesion formation in maxillary incisors, canines and premolars: A finite element study. *Eur. J. Oral Sci.* **2003**, *111*, 149–154. [CrossRef]
14. Du, J.K.; Wu, J.H.; Chen, P.H.; Ho, P.S.; Chen, K.K. Influence of cavity depth and restoration of non-carious cervical root lesions on strain distribution from various loading sites. *BMC Oral Health* **2020**, *20*, 98. [CrossRef]
15. Jakupovic, S.; Cerjakovic, E.; Topcic, A.; Ajanovic, M.; Prcic, A.K.; Vukovic, A. Analysis of the abfraction lesions formation mechanism by the finite element method. *Acta Inform. Med.* **2014**, *22*, 241–245. [CrossRef]
16. Haralur, S.B.; Alqahtani, A.S.; AlMazni, M.S.; Alqahtani, M.K. Association of non-carious cervical lesions with oral hygiene habits and dynamic occlusal parameters. *Diagnostics* **2019**, *9*, 43. [CrossRef]
17. Romeed, S.A.; Malik, R.; Dunne, S.M. Stress analysis of occlusal forces in canine teeth and their role in the development of non-carious cervical lesions: Abfraction. *Int. J. Dent.* **2012**, *2012*, 234845. [CrossRef]
18. Ahmić Vuković, A.; Jakupović, S.; Zukić, S.; Bajsman, A.; Gavranović Glamoč, A.; Šečić, S. Occlusal stress distribution on the mandibular first premolar—FEM analysis. *Acta Med. Acad.* **2019**, *48*, 255–261. [CrossRef]
19. Rogozhnikov, G.I.; Nemenatov, I.G.; Astashina, N.B.; Shulyatnikova, O.A.; Rogozhnikov, A.G.; Olshanskii, E.V. Orthopedic treatment of patients with clinoid defects of premolars and molars hard tissues. *Actual Probl. Dent.* **2010**, *3*, 24–27.
20. Zeola, L.F.; Pereira, F.A.; Machado, A.C.; Reis, B.R.; Kaidonis, J.; Xie, Z.; Townsend, G.C.; Ranjitkar, S.; Soares, P.V. Effects of non-carious cervical lesion size, occlusal loading and restoration on biomechanical behaviour of premolar teeth. *Aust. Dent. J.* **2016**, *61*, 408–417. [CrossRef]
21. Meredith, N.; Sherriff, M.; Setchell, D.J.; Swanson, S.A.V. Measurement of the microhardness and Young's modulus of human enamel and dentine using an indentation technique. *Arch. Oral Biol.* **1996**, *41*, 539–545. [CrossRef]
22. Benazzi, S.; Grosse, I.R.; Gruppioni, G.; Weber, G.W.; Kullmer, O. Comparison of occlusal loading conditions in a lower second premolar using three-dimensional finite element analysis. *Clin. Oral Investig.* **2014**, *18*, 369–375. [CrossRef]
23. Lee, H.E.; Lin, C.L.; Wang, C.H.; Cheng, C.H.; Chang, C.H. Stresses at the cervical lesion of maxillary premolar—A finite element investigation. *J. Dent.* **2002**, *30*, 283–290. [CrossRef]
24. Grippo, J.O.; Masai, J.V. Role of biodental engineering factors (BEF) in the etiology of root caries. *J. Esthet. Dent.* **1991**, *3*, 71–76. [CrossRef]
25. Leal, N.M.S.; Silva, J.L.; Benigno, M.I.M.; Bemerguy, E.A.; Meira, J.B.C.; Ballester, R.Y. How mechanical stresses modulate enamel demineralization in non-carious cervical lesions? *J. Mech. Behav. Biomed. Mater.* **2017**, *66*, 50–57. [CrossRef]
26. Xhonga, F.A. Bruxism and its effect on the teeth. *J. Oral Rehabil.* **1977**, *4*, 65–76. [CrossRef]
27. Telles, D.; Pegoraro, L.F.; Pereira, J.C. Prevalence of noncarious cervicai lesions and their relation to occiusal aspects: A clinical study. *J. Esthet. Restor. Dent.* **2000**, *12*, 10–15. [CrossRef]
28. Geramy, A.; Sharafoddin, F. Abfraction: 3D analysis by means of the finite element method. *Quintessence Int.* **2003**, *34*, 526–533.
29. Vasudeva, G.; Bogra, P. The effect of occlusal restoration and loading on the development of abfraction lesions: A finite element study. *J. Conserv. Dent.* **2008**, *11*, 117–120. [CrossRef]
30. Dikova, T.; Vasilev, T.; Hristova, V.; Panov, V. Finite Element Analysis in Setting of Fillings of V-Shaped Tooth Defects Made with Glass-Ionomer Cement and Flowable Composite. *Processes* **2020**, *8*, 363. [CrossRef]
31. Beresescu, G.; Ormenisan, A.; Raluca Monica, C.; Veliscu, A.; Manea, M.; Ion, R. FEM Analysis of Stress in Non-carious Cervical Lesion Restoration with Four Different Restorative Materials. *Mater. Plast.* **2018**, *55*, 42–45. [CrossRef]
32. Czerwiński, M.; Żmudzki, J.; Kwieciński, K.; Kowalczyk, M. Finite element analysis of the impact of the properties of dental wedge materials on functional features. *Arch. Mater. Sci. Eng.* **2021**, *112/1*, 32–41. [CrossRef]
33. Ichim, I.; Li, Q.; Loughran, J.; Swain, M.; Kieser, J. Restoration of non-carious cervical lesions. Part, I. Modelling of restorative fracture. *Dent. Mater.* **2007**, *23*, 1553–1561. [CrossRef]
34. Brailko, N.N.; Tkachenko, I.M.; Kovalenko, V.V.; Lemeshko, A.V.; Fenko, A.G.; Kozak, R.V.; Kalashnikov, D.V. Investigation of stress-strain state of "restoration & tooth" system in wedge-shaped defects by computed modeling method. *Wiadomości Lek.* **2021**, *74*, 2112–2117.
35. Kichenko, A.A.; Kiryukhin, V.Y.; Rogozhnikov, G.I.; Neminatov, I.G.; Astashina, N.B.; Shulyatnikova, O.A. Finite element analysis of wedge-shaped defect restoration at the first molar of a patient. *Russ. J. Biomech.* **2008**, *2*, 80–96.

36. Kasuya, A.V.B.; Favarão, I.N.; Machado, A.C.; Spini, P.H.R.; Soares, P.V.; Fonseca, R.B. Development of a fiber-reinforced material for fiber posts: Evaluation of stress distribution, fracture load, and failure mode of restored roots. *J. Prosthet. Dent.* **2020**, *123*, 829–838. [CrossRef]
37. Tajima, K.; Chen, K.K.; Takahashi, N.; Noda, N.; Nagamatsu, Y.; Kakigawa, H. Three-dimensional finite element modeling from CT images of tooth and its validation. *Dent. Mater. J.* **2009**, *28*, 219–226. [CrossRef]
38. Magne, P.; Oganesyan, T. CT scan-based finite element analysis of premolar cuspal deflection following operative procedures. *Int. J. Periodontics Restor. Dent.* **2009**, *29*, 361–369.
39. Shimada, Y.; Yoshiyama, M.; Tagami, J.; Sumi, Y. Evaluation of dental caries, tooth crack, and age-related changes in tooth structure using optical coherence tomography. *Jpn. Dent. Sci. Rev.* **2020**, *56*, 109–118. [CrossRef]
40. Sarycheva, I.; Yanushevich, O.; Minakov, D. Diagnostics of non-carious lesions of dental hard tissues with the methods of optical spectroscopy and radiography. *Braz. Dent. Sci.* **2020**, *23*, 1–8. [CrossRef]
41. Aw, T.C.; Lepe, X.; Johnson, G.H.; Mancl, L. Characteristics of noncarious cervical lesions: A clinical investigation. *J. Am. Dent. Assoc.* **2002**, *133*, 725–733. [CrossRef]
42. Mercuț, V.; Popescu, Ț.S.M.; Scrieciu, M.; Amărăscu, M.O.; Vătu, M.; Diaconu, O.A.; Osiac, E.; Ghelase, Ș.M. Optical coherence tomography applications in tooth wear diagnosis. *Rom. J. Morphol. Embryol.* **2017**, *58*, 99–106.
43. Abdalla, R.; Mitchell, R.J.; Ren, Y. Non-carious cervical lesions imaged by focus variation microscopy. *J. Dent.* **2017**, *63*, 14–20. [CrossRef]
44. Karan, K.; Yao, X.; Xu, C.; Wang, Y. Chemical profile of the dentin substrate in non-carious cervical lesions. *Dent. Mater.* **2009**, *25*, 1205–1212. [CrossRef]
45. Sakoolnamarka, R.; Burrow, M.F.; Prawer, S.; Tyas, M.J. Raman spectroscopic study of noncarious cervical lesions. *Odontology* **2005**, *93*, 35–40. [CrossRef]
46. Kantardžić, I.; Vasiljević, D.; Blažić, L.; Luzanin, O. Influence of cavity design preparation on stress values in maxillary premolar: A finite element analysis. *Croat. Med. J.* **2012**, *53*, 568–576. [CrossRef]
47. Babaei, B.; Cella, S.; Farrar, P.; Prentice, L.; Prusty, B.G. The influence of dental restoration depth, internal cavity angle, and material properties on biomechanical resistance of a treated molar tooth. *J. Mech. Behav. Biomed. Mater.* **2022**, *133*, 105305. [CrossRef]
48. *CEREC Blocs—For CEREC/InLab: Operating Instructions*; Imprime en Allemagne: Bensheim, Germany, 2006.
49. Hussein, L.A. 3-D finite element analysis of different composite resin MOD inlays. *Am. J. Sci.* **2013**, *9*, 422–428.
50. Oneț, D.B.; Tudoran, L.B.; Delean, A.G.; Șurlin, P.; Ciurea, A.; Roman, A.; Bolboacă, S.D.; Gasparik, C.; Muntean, A.; Soancă, A. Adhesion of flowable resin composites in simulated wedge-shaped cervical lesions: An in vitro pilot study. *Appl. Sci.* **2021**, *11*, 3173. [CrossRef]
51. Battancs, E.; Fráter, M.; Sáry, T.; Gál, E.; Braunitzer, G.; Szabó, P.B.; Garoushi, S. Fracture behavior and integrity of different direct restorative materials to restore noncarious cervical lesions. *Polymers* **2021**, *13*, 4170. [CrossRef]
52. MacHado, A.C.; Soares, C.J.; Reis, B.R.; Bicalho, A.A.; Raposo, L.H.A.; Soares, P.V. Stress-strain analysis of premolars with non-carious cervical lesions: Influence of restorative material, loading direction and mechanical fatigue. *Oper. Dent.* **2017**, *42*, 253–265. [CrossRef] [PubMed]
53. Siegel, S. *The Dental Reference Manual: A Daily Guide for Students and Practitioners*; Springer: Cham, Switzerland, 2017. [CrossRef]
54. Tamimi, F.; Hirayama, H. *Digital Restorative Dentistry A Guide to Materials, Equipment, and Clinical Procedures: A Guide to Materials, Equipment, and Clinical Procedures*; Springer: Cham, Switzerland, 2019. [CrossRef]
55. Colombo, M.; Gallo, S.; Padovan, S.; Chiesa, M.; Poggio, C.; Scribante, A. Influence of different surface pretreatments on shear bond strength of an adhesive resin cement to various zirconia ceramics. *Materials* **2020**, *13*, 652. [CrossRef] [PubMed]
56. Tatullo, M.; Codispoti, B.; Paduano, F.; Nuzzolese, M.; Makeeva, I. Strategic tools in regenerative and translational dentistry. *Int. J. Mol. Sci.* **2019**, *20*, 1879. [CrossRef] [PubMed]
57. Mukherjee, K.; Visakan, G.; Phark, J.H.; Moradian-Oldak, J. Enhancing collagen mineralization with amelogenin peptide: Towards the restoration of dentin. *ACS Biomater. Sci. Eng.* **2020**, *6*, 2251–2262. [CrossRef] [PubMed]
58. Tatullo, M.; Genovese, F.; Aiello, E.; Amantea, M.; Makeeva, I.; Zavan, B.; Rengo, S.; Fortunato, L. Phosphorene is the new graphene in biomedical applications. *Materials* **2019**, *12*, 2301. [CrossRef]
59. Kamenskikh, A.; Kuchumov, A.G.; Baradina, I. Modeling the contact interaction of a pair of antagonist teeth through individual protective mouthguards of different geometric configuration. *Materials* **2021**, *14*, 7331. [CrossRef]
60. Baraka, M.M.; Geigerb, S.; Lev-Tov Chattaha, N.; Shaharc, R.; Weiner, S. Enamel dictates whole tooth deformation: A finite element model study validated by a metrology method. *J. Struct. Biol.* **2009**, *168*, 511–520. [CrossRef]
61. Ausielloa, P.; Ciaramellab, S.; Fabianellic, A.; Gloriad, A.; Martorellib, M.; Lanzottib, A.; Wattse, D.C. Mechanical behavior of bulk direct composite versus block composite and lithium disilicate indirect Class II restorations by CAD-FEM modeling. *Dent. Mater.* **2017**, *33*, 690–701. [CrossRef]
62. Rivera, J.L.V.; Gonçalves, E.; Soares, P.V.; Milito, G.; Perez, J.O.R.; Roque, G.F.P.; Fernández, M.V.; Losada, H.F.; Pereira, F.A.; del Pino, G.G.; et al. The Restored Premolars Biomechanical Behavior: FEM and Experimental Moiré Analyses. *Appl. Sci.* **2022**, *12*, 6768. [CrossRef]

63. Laohachaiaroon, P.; Samruajbenjakun, B.; Chaichanasiri, E. Initial Displacement and Stress Distribution of Upper Central Incisor Extrusion with Clear Aligners and Various Shapes of Composite Attachments Using the Finite Element Method. *Dent. J.* **2022**, *10*, 114. [CrossRef]
64. Bonaba, M.F.; Mojraa, A.; Shirazi, M. A numerical-experimental study on thermal evaluation of orthodontic tooth movement during initial phase of treatment. *J. Therm. Biol.* **2019**, *80*, 45–55. [CrossRef] [PubMed]

Article

Design and Finite Element Analysis of Patient-Specific Total Temporomandibular Joint Implants

Shirish M. Ingawale [1] and Tarun Goswami [1,2,*]

[1] Department of Biomedical, Industrial & Human Factors Engineering, Wright State University, 3640 Col Glen Hwy, Dayton, OH 45435, USA; shirishingawale@gmail.com
[2] Department of Orthopaedic Surgery and Sports Medicine, Wright State University, Dayton, OH 45435, USA
* Correspondence: tarun.goswami@wright.edu; Tel.: +1-(937)-775-5120

Abstract: In this manuscript, we discuss our approach to developing novel patient-specific total TMJ prostheses. Our unique patient-fitted designs based on medical images of the patient's TMJ offer accurate anatomical fit, and better fixation to host bone. Special features of the prostheses have potential to offer improved osseo-integration and durability of the devices. The design process is based on surgeon's requirements, feedback, and pre-surgical planning to ensure anatomically accurate and clinically viable device design. We use the validated methodology of FE modeling and analysis to evaluate the device design by investigating stress and strain profiles under functional/normal and para-functional/worst-case TMJ loading scenarios.

Keywords: TMJ; biomaterials; custom devices; finite element analysis; 3D models; mandibles

Citation: Ingawale, S.M.; Goswami, T. Design and Finite Element Analysis of Patient-Specific Total Temporomandibular Joint Implants. Materials 2022, 15, 4342. https://doi.org/10.3390/ma15124342

Academic Editor: Oskar Sachenkov

Received: 17 May 2022
Accepted: 16 June 2022
Published: 20 June 2022

Publisher's Note: MDPI stays neutral with regard to jurisdictional claims in published maps and institutional affiliations.

Copyright: © 2022 by the authors. Licensee MDPI, Basel, Switzerland. This article is an open access article distributed under the terms and conditions of the Creative Commons Attribution (CC BY) license (https://creativecommons.org/licenses/by/4.0/).

1. Introduction

In treating the TMJ dysfunction, all nonsurgical approaches should be exhausted. In some select patients, the end-stage TMJ pathology resulting in distortion of anatomical architectural form and physiological dysfunction dictates the need for total joint replacement (TJR) [1–4]. The goal of TMJ TJR is the restoration of mandibular function and form; any pain relief attained is considered of secondary benefit [1,2]. The TMD patients with serious osteoarthritis, rheumatoid arthritis, psoriatic arthritis, and ankylosis might be good candidates for receiving TMJ prosthesis [1–8].

TMJ resections have been carried out for about 150 years [4,9,10]. Before 1945, the technique of alloplastic reconstruction of TMJ was mainly limited to replacement of condyle [9]. Interposition of alloplastic implants, resection dressings and prostheses were the dominant techniques [9]. Sterilization, biocompatibility and fixation of the alloplastic implants were main concerns in early days [9]. No evidence-based data on outcomes are available from that time. By 1945 reconstruction of the TMJ involved the close cooperation of surgeons and dentists [5,9]. In view of the rare application of TMJ prostheses, their relatively wide variety described over past six decades emphasizes that alloplastic TMJ reconstruction is still evolving.

TMJ implants can be differentiated into fossa-eminence prostheses, ramus prostheses and condylar reconstruction plates, and total joint prostheses. Although singular replacement of the fossa or condyle is preferred as a temporary solution, the partial TMJ reconstruction finds comparatively declining usage by surgeons for clinical reasons. Total TMJ implants are recommended when the glenoid fossa is exposed due to excessive stress in conditions such as degenerative disorders, arthritis ankylosis, and multiply operated pain patients [1–7]. Table 1 lists indications for alloplastic reconstruction of the TMJ.

Table 1. Indications for the alloplastic reconstruction of the temporomandibular joint (TMJ).

Sr. No.	Indications for Alloplastic TMJ Reconstruction
1.	Ankylosis or reankylosis [1,3,4], degeneration, or resorption [3,4] of joints with severe anatomic abnormalities.
2.	Failed autogenous grafts in multiply operated patients [1,3,4].
3.	Destruction of autogenous graft tissue by pathology [1,3,4].
4.	Failed Proplast–Teflon that results in severe anatomic joint mutilation [1,3,4].
5.	Failed Vitek–Kent total or partial joint reconstruction [1].
6.	Severe inflammatory joint disease, such as rheumatoid arthritis which results in anatomic mutilation of the joint components and functional disability [1,3,4].

Relative contraindications to the use of alloplast in reconstruction of the TMJ are age of the patient, mental status of the patient, uncontrolled systemic disease such as diabetes mellitus or myelodysplasia, active or chronic infection at the implantation site, and allergy to materials that are used in the devices to be implanted [1,3,4]. The perceived potential disadvantages of the alloplastic TMJ TJR are cost of the device, need for two-stage procedure in ankylosis cases, material wear debris with associated pathologic responses failure of the prostheses secondary to loosening of the screw fixation or fracture from metal fatigue, lack of long-term stability, inability of alloplastic implant to follow physical growth of the younger patients, and unpredictable need for revision surgery. Long-term studies comparing functional and aesthetic outcomes of various TMJ prostheses are not available (with an exception of one study by [11] with up to 14-year follow-up), which leaves the choice of prosthesis to surgeon's personal preference.

We performed a comprehensive review of published literature [1–36] regarding TMJ reconstruction, and based our TMJ prostheses design approach on the knowledge gained from clinical, biomechanical and scientific reports about the history, designs, efficacy, and clinical outcomes of TMJ prostheses. There are two categories of the TMJ TJR devices approved for implantation by the United States Food and Drug Administration (FDA); the stock or off-the-shelf devices, and the custom or patient-fitted devices. At the time of implantation, the surgeon has to 'make fit' the stock (off-the-shelf) device. In contrast, the custom (patient-fitted) devices are 'made to fit' each specific case. To date, there is only one study [13], reported in the literature that compares a stock and a custom TMJ TJR system. This study concluded that patients implanted with the custom TMJ TJR system had statistically significant better outcomes in both subjective and objective domains than did those implanted with the stock system devices studied [13].

The history of alloplastic TMJ reconstruction has, unfortunately, been characterized by multiple highly publicized failures based on inappropriate design, lack of attention to biomechanical principles, and ignorance of what already had been documented in the orthopedic literature [3,4,12,14]. In addition, because TMJ is the only ginglymoarthrodial joint in human body, and because its function is intimately related to occlusal harmony, a prosthetic TMJ necessitates characteristics not considered in orthopedic implant design [4]. The use of inappropriate materials and designs has resulted in success rate of many TMJ implants being lower than those for total hip and knee prostheses [14]. Most of the published literature regarding TMJ implants has been clinical and case reports, with much less studies investigating the design and biomechanics of the TMJ implants. In view of paucity of this information, and the need for more efficient and durable total TMJ implants, we undertook a study aimed at designing and evaluating customized total TMJ prostheses.

Design Requirements for Total TMJ Prosthesis

van Loon, et al. [15] indicated that there are three major requirements in TMJ TJR; (i) to imitate the functional movement, (ii) to realize a close fit to the skull, and (iii) to achieve a long lifetime. Table 2 lists summary of requirements for successful total reconstruction of the TMJ. Stability of alloplastic joint replacements depends not only on fixation, but also on adaptation of the implant to the bone to which it is to be fixed [1,2,11]. The orthopedic experience with implantation of alloplastic joints has shown that better adaptation of the device to the host bone results in more stability and functional longevity of the implant [1,2,16,17]. Stability of TMJ prosthesis at the time of implantation is equally important for its success. Motion of the implanted prosthesis under a load can cause the surrounding bone to degenerate, leading to further device loosening and consequent failure [1]. Currently, screw fixation of TMJ implants is the most predictable and stable form of stabilization developed [1]. Screws may loosen with time and function, requiring replacement. To assure long-term success of the TMJ implants, primary stability of prosthetic components must be ensured by biointegration of the screws [2].

Table 2. Criteria for the successful alloplastic total reconstruction of the TMJ.

Sr. No.	Requirements/Criteria for Success of Alloplastic Total Joint Replacement Devices
1.	The materials from which the devices are made must be biocompatible [1,2,15,16].
2.	The devices must be designed with sufficient mechanical strength to withstand the loads delivered over the full range of function of the joint [1,2,15,16].
3.	The devices must be stable in-situ [1,2,15,16].
4.	The surgery to implant the prosthesis must be performed for the proper indications, and it must be performed aseptically [1,2,16].
5.	The prostheses should imitate the condylar translation during mouth opening, and without restricting movements of non-replaced TMJ [15].
6.	The prostheses should be fitted correctly to the mandible and the skull [15].
7.	Expected lifetime of more than 20 years [15].
8.	Low wear rate; and wear particles must be tolerated by the body [15].
9.	Simple and reliable implantation procedures [15].

Most patients requiring TMJ replacement have deformed local bony anatomy. During implantation of the stock TMJ prosthesis, the surgeon confronts with a difficult challenge of making 'off-the-shelf' components fit and remain stable, and often the precious host bone needs to be sacrificed to make the stock TMJ components to create stable component-to-host-bone contact [2]. Surgeons attempt to make stock devices fit by bending or shimming may lead to component or shim material fatigue and/or overload fostering early failure under repeated cyclic functional loading. Potential micromotion of any altered or shimmed component adversely affects the screw fixation biointegration. Micromotion leads to the formation of a fibrous connective tissue interface between the altered component and the host bone, and can cause early loosening of the screws leading to device failure. Our patient-specific TMJ implants are designed to accurately fit each patient's specific anatomical condition. They conform to any unique or complex anatomical host bone condition. These designs do not require any alteration or shimming of either the device or the host bone to achieve initial fixation and stability. The screws secure implant components intimately to the host bone mitigating possibility of micromotion and maximizing the opportunity for biointegration [2].

2. Materials and Methods

In this section, we discuss the methodology of designing condylar and fossa components of the custom-designed total TMJ prostheses. Also discussed, are unique design features such as accurate fit of the prosthetic surface to the host bone in contact, perforated notches of implant which protrude and fit into the custom-cut slots in native bone, customized surgical guides, and screws with locking mechanism.

2.1. Design of Patient-Specific Total TMJ Prosthesis

The schematic in Figure 1 outlines our approach to developing a novel patient-specific total TMJ implant system. Our unique patient-fitted designs based on computed tomography (CT) images of the patient's TMJ and associated anatomic structures offer accurate anatomical fit and better fixation to the host bone. The novel/unique features of the prostheses promise an improved osseo-integration and durability. Our design process is based on surgeon's requirements, feedback, and pre-surgical planning to ensure anatomically accurate and clinically viable device design. Pre-planning of the surgery is an integral part of the proposed design and development methodology, and is intended to reduce intra-operative adjustments of the device components, complexity of the already challenging operating procedure, and the overall time spent in the operating room.

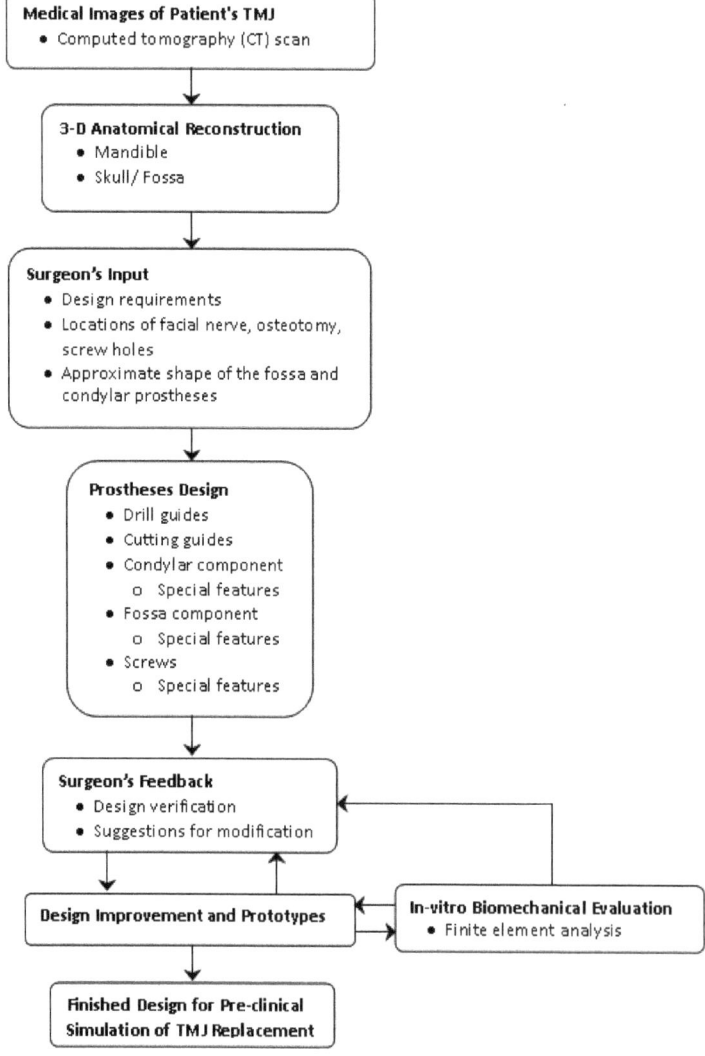

Figure 1. Methodology followed for design and preliminary analysis of the patient-specific total temporomandibular joint (TMJ) prostheses.

Subject-specific 3D anatomical reconstruction of the patient's mandible and skull/fossa/articular eminence is performed using commercial software Mimics v14.12 (Materialise, Plymouth, MI, USA) from computed tomography (CT) scans (see Figure 2). Upon importing the patient's CT images in Mimics, anatomical model comprising of the patient's mandible and fossa eminence is developed by performing a series of operations such as image processing, segmentation, region growing, mask formation for the anatomic region of interest (i.e., bone and teeth), and calculation of 3D equivalent similar to the 3D reconstruction method described elsewhere [18]. The prostheses and accessories are designed using commercial software packages 3-matic v6.0 (Materialise, Plymouth, MI, USA) and SolidWorks v2010 (SIMULIA, Providence, RI, USA) as discussed in following sections.

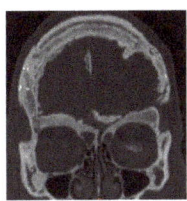
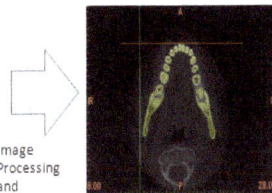

Patient's CT Data — Image Processing and Segmentation — Mask For Region of Interest — 3D Reconstruction — Patient-specific Anatomical Model — Mandible and Fossa

Figure 2. Subject-specific 3D anatomical reconstruction of the patient's mandible and fossa eminence performed from computed tomography (CT) data using Mimics software.

2.1.1. Surgical Pre-Planning and Surgeon Input

Close collaboration between device designer and surgeon (treating the given TMJ patient) is an important aspect of the proposed design and development approach. Mutual sharing of knowledge and expertise, clinical and design requirements and constraints is vital in ensuring the optimal design and performance of TMJ devices. We utilize the computerized anatomical model and its 3D printed equivalent to acquire surgeon's design requirements such as location of the facial nerve (to keep it from any damage or injury during surgery), location of the condylar osteotomy (i.e., removal of the degenerated or damaged condylar bone), outline of shape for the planned condylar and fossa prostheses, location of screws to secure the condylar and fossa components to host bone, number and dimension of screws, etc.

To help the surgeon accurately remove the damaged part of condylar neck/head, a surgical guide is custom designed for each reconstruction case as shown in Figure 3. During surgery, after putting the patient in intermaxillary fixation (IMF) and gaining access to TMJ capsule, the surgical guide can be fixated to mandible using screws located superior and inferior to the line of osteotomy/condylectomy. In other words, the surgical guide is secured using screws at the condylar head and condylar neck/ramus depending on osteotomy location and surgeon's preference. After completing the osteotomy, surgical guide is detached from the bone by removing the screws. The location of condylectomy guide screw hole inferior to the anterio-posterior excision line can be selected (and custom designed) such that the same screw hole can also be used later by one of the screws used to secure condylar/ramus implant to the host mandible.

Our design approach and pre-surgical planning enables appropriate design of screws and pre-drilled screw holes in implants to avoid unintentional injury to facial nerve, soft tissue, and other delicate structures in the vicinity of the complex surgical site. Following the similar design approach used for osteotomy guide, the screw-drill-guide is custom designed each for the condylar/ramus component and the fossa-eminence part of the TMJ prostheses. These drill guides are intended to create a hole of preferred dimension (diameter and depth) at the accurate location and orientation for each screw as prescribed by the surgeon. For a given screw, a drill of smaller diameter than that of the particular screw is selected so ensure less bone damage, optimal purchase, and rigid interface between the screw and host bone during and after implantation.

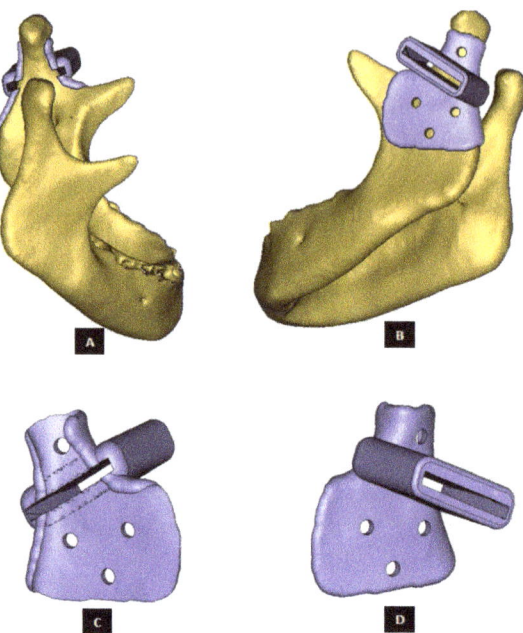

Figure 3. Custom-designed surgical guide for condylectomy (i.e., removal of damaged part of the condylar bone). (**A,B**) Show the medial and lateral view, respectively, of surgical guide placed at the location on mandible where osteotomy is to be performed. (**C,D**) Show medial and lateral–anterior view, respectively, of the surgical guide alone. The visuals demonstrate that custom-design of the device enables it to accurately adapt to the native bone. This methodology allows the designer to control size and shape of the device, and location of its fixation screws as prescribed by the surgeon.

Based on a surgeon's initial design requirements, the prostheses, drill guides, osteotomy guide, and templates are designed. In response to surgeon's feedback about the initial designs, suggested changes are incorporated to improve the device design. This feedback loop is kept open, and the designs are fine-tuned, till the surgeon approves the designs. In-vitro biomechanical assessment of patient's host bone and TMJ prostheses is incorporated in the design validation and improvement loop as described later in this paper. After sufficiently improving the designs, the prostheses graduate to the next stage where finished implants and accessories are ready for prototyping and pre-surgical simulation of the operating procedure using anatomical models and finished prototypes. In real-world scenario, before applying our methodology to actual clinical application, it has to be verified and validated through cadaver studies.

2.1.2. Design of Condylar Prosthesis

Anatomically accurate fit of TMJ prosthesis to the host bone is essential for stable fixation leading to efficacy and longevity of the device. Our custom-designed condylar components follow the anatomical geometry and contours on the lateral surface of ramus and condylar part of host anatomy to which the prosthesis is to be fixated. Custom shape of the prosthesis maximizes the possibility of precise fit and secure fixation. Different shapes of the condylar and ramal parts can be designed per surgeon's recommendations to conform to the patient's unique/complex anatomical situation. Figures 4–12 show various such shapes of the condylar component of our TMJ prostheses. Since these components are custom designed to meet the unique requirements of each individual patient's situation, the characteristic length, width, and thickness of condylar component; the number and locations of screws; and dimensions of condylar neck and head vary from patient to patient. The minimal level of the condylar thickness, width, head diameter, and number and location of screws are maintained (based on orthopaedic experience listed in the literature, surgeon's prescription, and pre-clinical biomechanical evaluation of the device designs) to ensure that the device provides sufficient mechanical strength and stability during functional and para-functional loading after implantation.

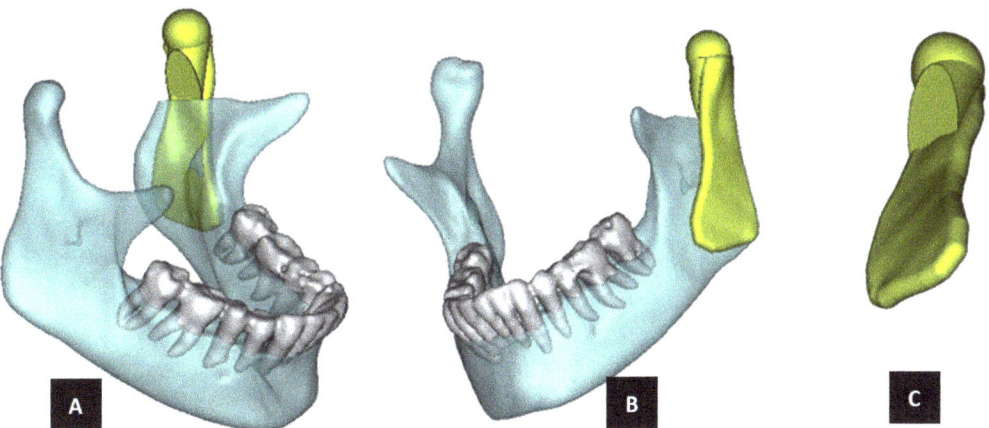

Figure 4. Shape outline of a custom-designed condylar/ramus prosthesis. (**A**,**B**) Show medial–anterior view and lateral–anterior view, respectively, of the prosthesis accurately adapting to the host bone. (**C**) Shows medial-inferior view of the prosthesis shape outline.

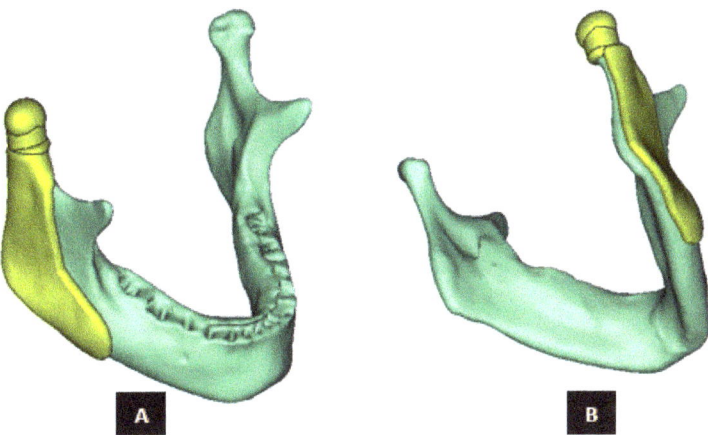

Figure 5. Shape outline of a custom-designed condylar/ramus prosthesis for the replacement of right TMJ of a patient. (**A**,**B**) Show lateral–anterior view and lateral–posterior view, respectively, of the prosthesis accurately conforming to geometric shape of patient's mandible.

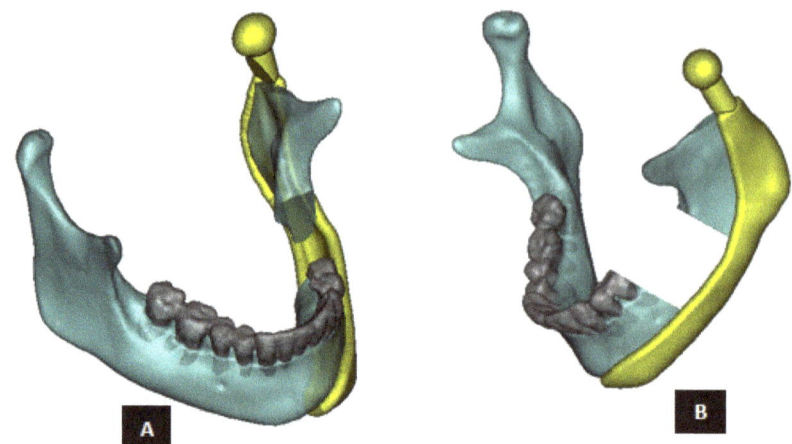

Figure 6. Shape outline of a custom-designed condylar/ramus/mandibular component of the TMJ prosthesis for reconstruction of left TMJ. (**A**,**B**) Show medial–anterior view and lateral–anterior view, respectively, of the prosthesis along with 3D anatomical model of the patient's mandible after condylectomy. The osteotomy gap seen in the left mandibular body is due to removal of a tumor in that region. This osteotomy gap can be filled with a graft, and the mandibular component of this TMJ prosthesis is designed to provide mechanical support to the host bone and graft.

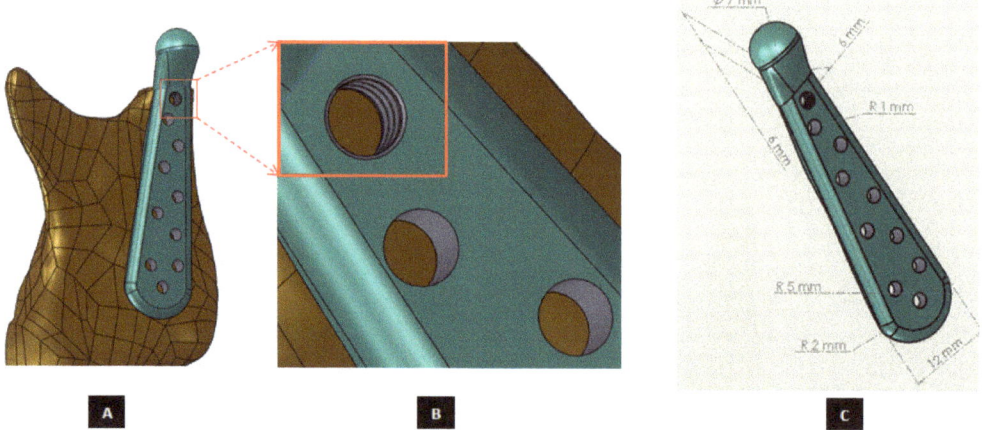

Figure 7. Custom-designed condylar/ramus component of the TMJ total joint replacement prosthesis for left TMJ of a patient. (**A**) Shows lateral view of the implant with screw holes. (**B**) Shows an enlarged view of the screw holes, where the first superiorly located screw hole has threads to incorporate locking-plate-screw mechanism by engaging the threads on the head of a locking screw described in the text. (**C**) Shows engineering dimensions of this patient-specific implant.

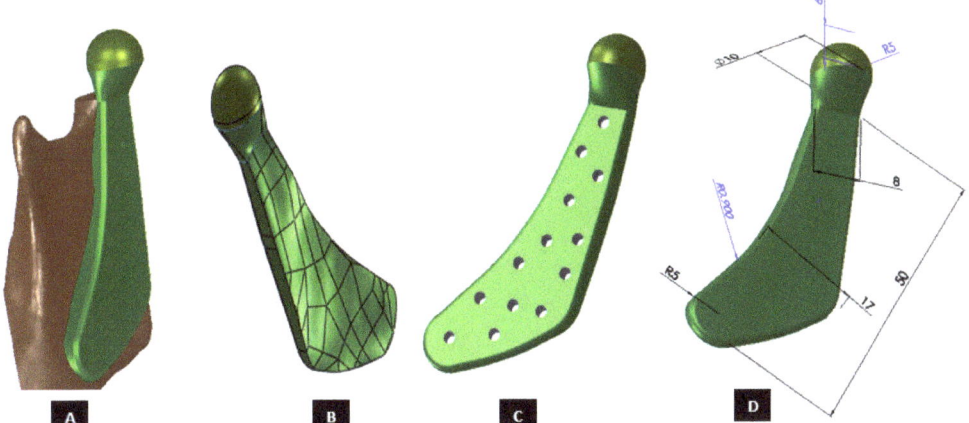

Figure 8. Shape outline of a custom-designed condylar/ramus component of the TMJ total joint replacement prosthesis for left TMJ of a patient. (**A**) Shows anterior–lateral view of the implant with host bone after condylectomy. The posterior–medial view in (**B**) shows that the medial surface of prosthesis is shaped to accurately follow geometric contours of the lateral surface of mandibular host bone for optimal geometrical match between the implant and host bone. (**C**) Shows lateral view of the prosthesis with screw holes. Dimensions of various parts of this patient-specific implant are shown in (**D**).

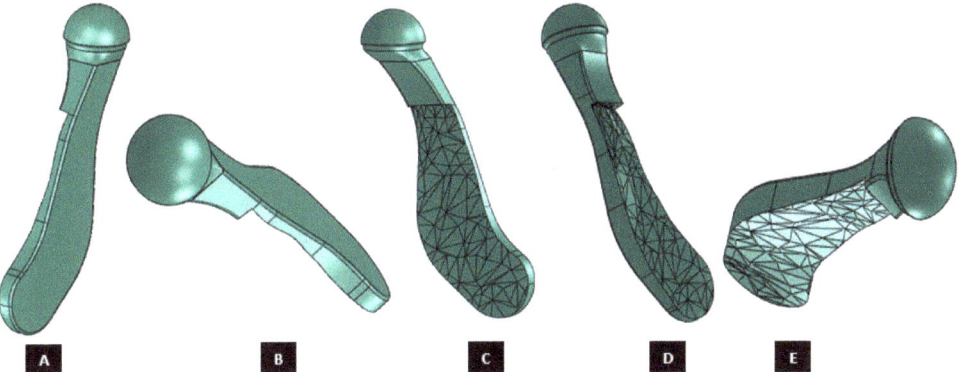

Figure 9. Shape outline of a custom-designed condylar/ramus component of the TMJ total joint replacement prosthesis for left TMJ of a patient. Visuals in (**A**–**E**) demonstrate that shape of medial surface of the prosthesis accurately follows the geometric contours of the lateral surface of the mandibular host bone, and maximizes the opportunity for optimal adaptation of implant to the host bone. The lateral surface of the implant is flat, condylar head is spherical, and the condylar neck has a curvature to avoid problems seen in most right-angled designs of orthopaedic implants.

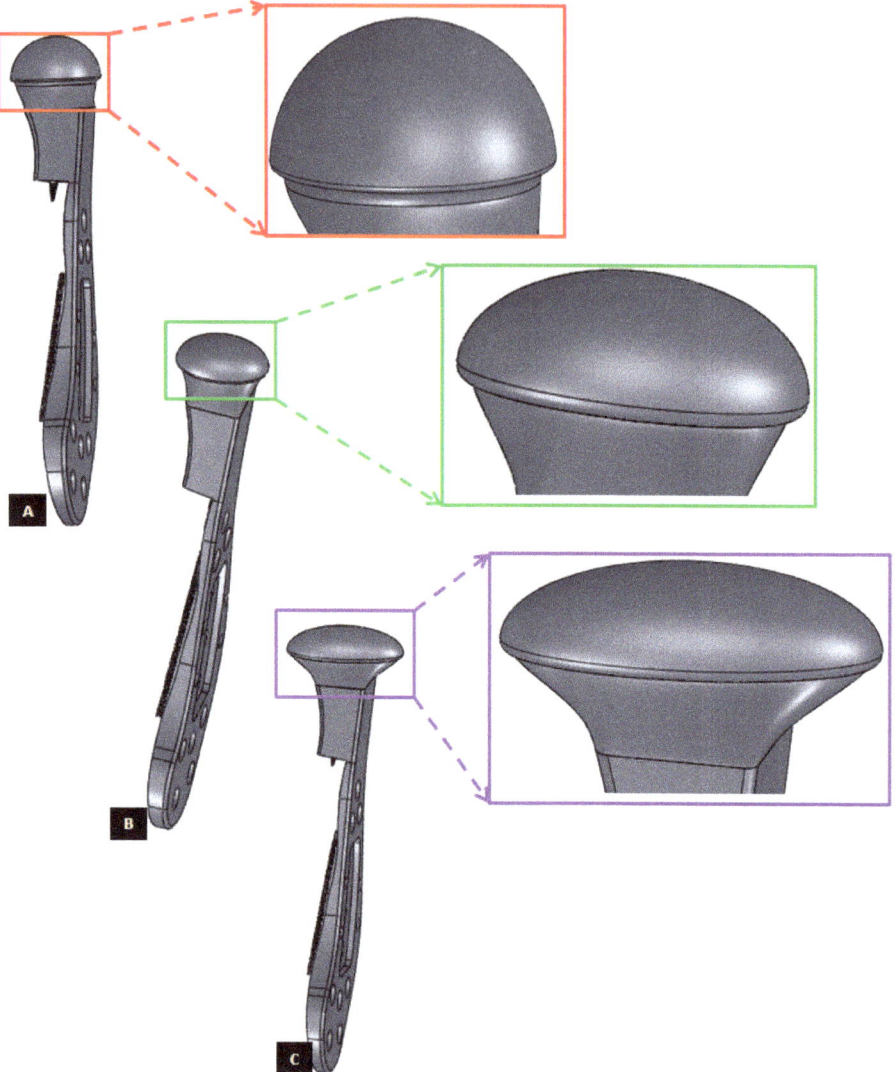

Figure 10. Different shapes of the condylar head of the custom-designed condylar/ramus component of the TMJ prosthesis. (**A**) Shows a prosthesis with spherical condylar head. (**B**,**C**) Show prostheses with elliptical head of different dimensions. The condylar heads are designed to offer larger articulating surface area to avoid stress concentration at small area which may lead to more wear of the articulating surfaces of reconstructed TMJ.

An important advantage of our patient-specific design approach is that the components can be precisely designed to withstand the loads encountered by unique anatomic condition. For the custom-designed condylar/ramus component, the center of rotation of its head can be moved vertically to correct the open bite deformity. The prosthetic condylar head can be placed such that its center of location in located inferior to that of the natural condyle it replaced, thereby allowing low-wear articulation of the reconstructed total TMJ and natural movements of the non-replaced contra-lateral TMJ. Ramal component can be shaped to accommodate the amount of available mandibular host bone. The condylar heads can be designed in different shapes to offer larger articulating surface area to avoid stress concentration in small area of the articulating condylar head and fossa as illustrated in Figure 10. Figures 4–12 show custom-designed condylar/ramus prostheses of varying shape and size. These models demonstrate that our methodology of custom design enables the condylar component to conform to the anatomic situation of damaged and/or complex mandibular host bone. Though the shown designs of condylar component vary in size and shape per anatomic demands and surgeon's prescription, an important common design feature among all these models is that each device provides accurate adaptation to the host bone.

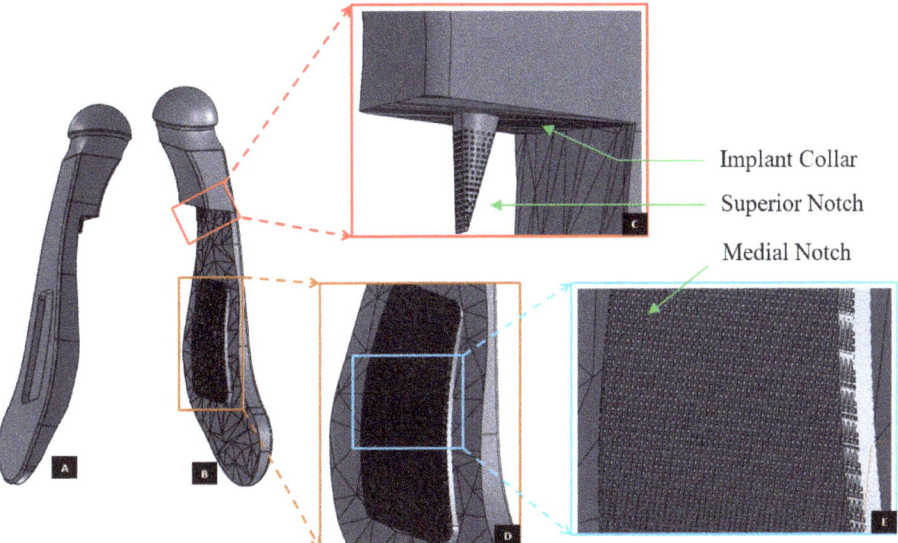

Figure 11. Modification of the custom-designed condylar/ramus component, shown in Figure 9, to include a novel feature; perforated notches protruding into host bone at implantation. (**A**) Shows a grove in the flat lateral surface of the condylar implant. The opposite side of this grove, as shown in (**B**), protrudes out of the medial surface as a notch with perforations. The enlarged views of medial notch and its perforations are shown in (**D,E**). The device also has pointed and perforated notch protruding from inferior surface of the implant's collar/neck. Perforated surfaces of these notches are designed to permit bone in-growth into the prosthesis after implantation to provide added stability. Dimensions of these notches can be customized to fit the size and shape of patient's native bone. Protrudes out of the medial surface as a notch with perforations (**C**).

One novel feature of our TMJ prostheses is the perforated notches protruding into the host bone. Figures 11 and 12 show a condylar/ramus component with its medial surface accurately following the geometric shape of patient's mandibular bone. Also seen protruding out of the medial surface of this device is a rectangular notch with perforations on its surface. This notch is intended to be placed in a custom-cut grove to be created on the lateral surface of mandibular ramus by the surgeon during implantation. Custom-designed cutting guides and templates can be provided to the surgeon to accurately create a small grove in the host bone. This intentional removal of native bone is performed in exchange of the opportunity for maximizing implant stability through bony ingrowth into perforated surfaces of the notch. Figures 11 and 12 show a perforated notch protruding from the inferior surface or collar of condylar neck. This notch is intended to be placed into a custom-cut grove in the superior surface of mandibular condyle/ramus resulting from osteotomy (performed to remove damaged condylar head/neck). In addition to providing better stabilization, the notches also provide an avenue for load transfer between the prosthesis and host bone. This will reduce forces and resultant stress experienced by fixation screws which act as the only mode of load transfer between most TMJ prostheses, especially for the condylar devices in which the collar of condylar prosthesis does not adequately contact the host bone or the medial surface of the implant does not adapt accurately to the complex geometry of patient's mandible. Though having both medial and superior notches in the condylar prosthesis is likely to be advantageous from biomechanical viewpoint, this may make surgical implantation of the device more challenging for the surgeon. Therefore, it will be the surgeon's choice to have either one or both notches for condylar implant.

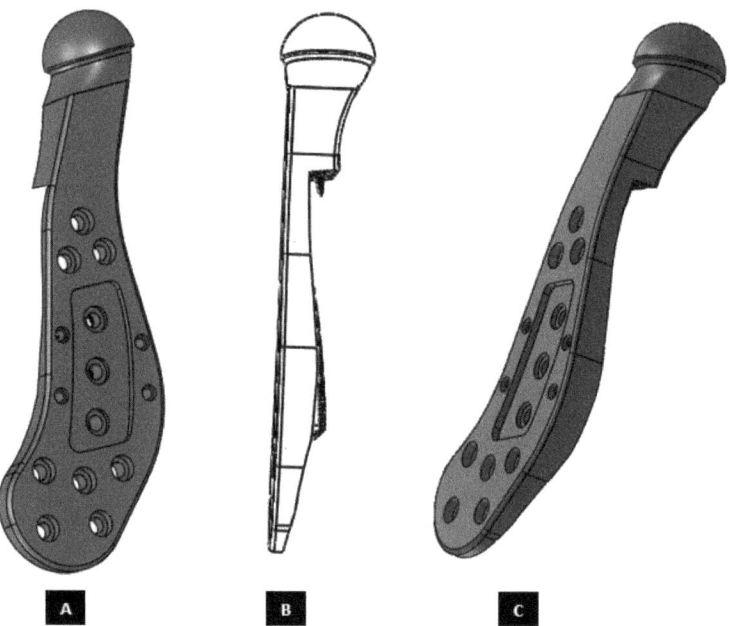

Figure 12. Modification and refinement of custom-designed condylar/ramus component shown in Figures 9 and 11. (**A**,**C**) Show pre-drilled screw holes and a grove in the lateral surface of implant. As shown in (**C**), lateral surface of the device is flat and medial surface is shaped to match the host bone geometry. (**B**) Shows a perforated notch each protruding from the medial surface of the ramus and inferior surface of the implant collar/neck.

2.1.3. Design of Fossa Prosthesis

Fitting the skull is a major problem in TMJ reconstruction patients because of the irregular shape of their TMJs [14,15] The patient-specific design approach enables developing accurately fitting models for the complex shape of patient's fossa-eminence anatomy. Using a similar design approach discussed earlier for the condylar implants, patient-fitted custom designs of fossa prosthesis can be developed such that the device fits accurately to the available host bone. Such custom designed fossa implants can correctly adapt to the natural components of patient's TMJ, and provide improved stability through locking screws and perforated notches fitting into patient's skull. Figures 13–21 show different shapes and features of our custom-designed fossa prostheses.

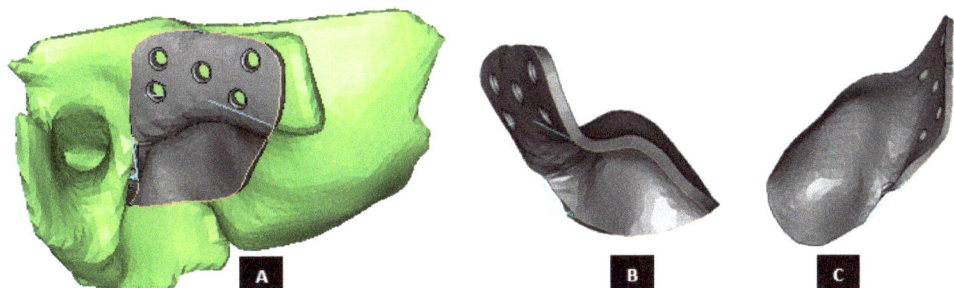

Figure 13. A simple custom-design of the fossa prosthesis with screw holes. (**A**) Demonstrates that the implant is designed for optimal usage of natural fossa eminence for fixation using screws. (**B,C**) Show different views of the implant illustrating the custom shape accurately conforms to the contours of host anatomy. The implant has constant thickness throughout its body, and the shape of articulating surface is same as that of the natural articular surface.

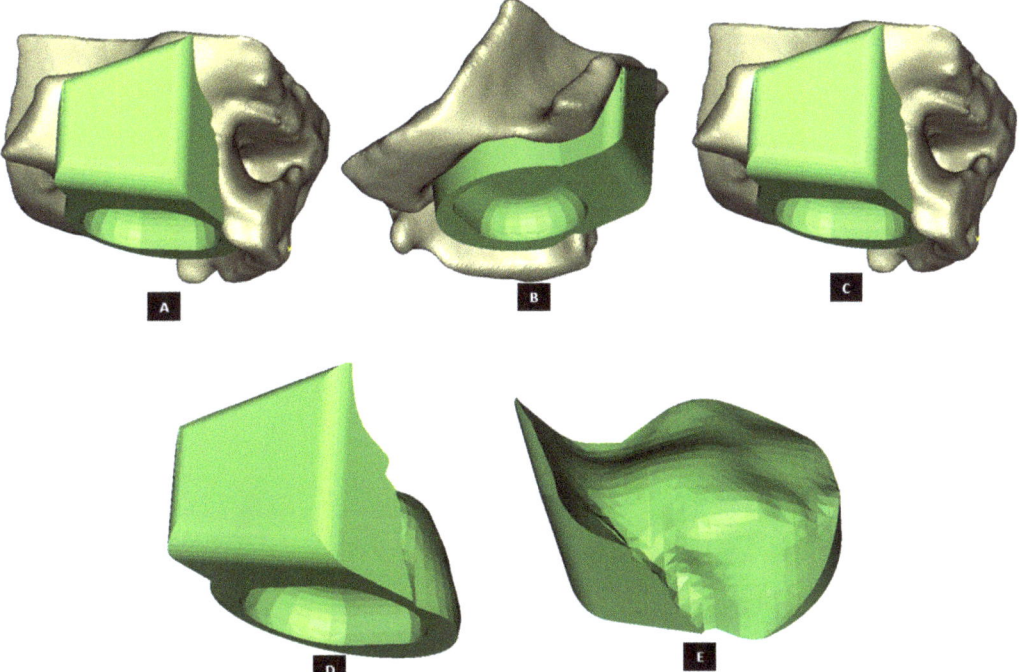

Figure 14. Patient-fitted design of a fossa prosthesis. (**A–C**) Illustrate accurate fit of the device to the patient's natural fossa and eminence. The rectangular slot (with curved anterior and posterior edges) in inferior surface of the implant is designed to provide sufficient rotation and opportunity for anterior-posterior and medio–lateral translation of the matching prosthetic condylar head. The articular grove is designed such that it would prevent dislocation of the prosthetic condylar head during functional movements of the jaw. Visuals in (**D,E**) show that the superior surface of the implant is designed to accurately match the shape of natural fossa. Sufficient thickness is maintained for the lateral portion of implant to pre-drill screw holes which can host locking screws for better fixation and stability.

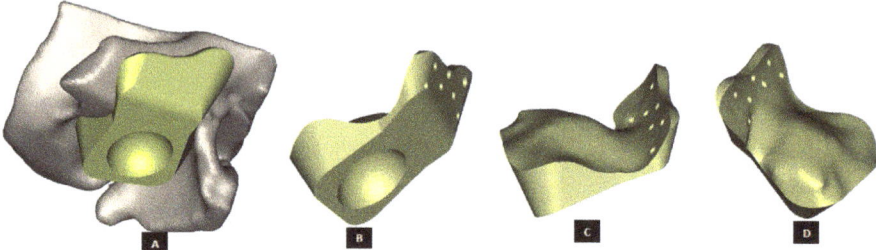

Figure 15. Patient-specific design of fossa prosthesis. Inferior rectangular surface of the device has a circular grove for articulation with condylar head (**A,B**). Visuals illustrate customized size and shape of the device for accurate fit and fixation (**C,D**) to native anatomical structure. Superior edge of the lateral surface (which hosts screw holes) is custom cut to follow the curvature of native eminence and bone situation.

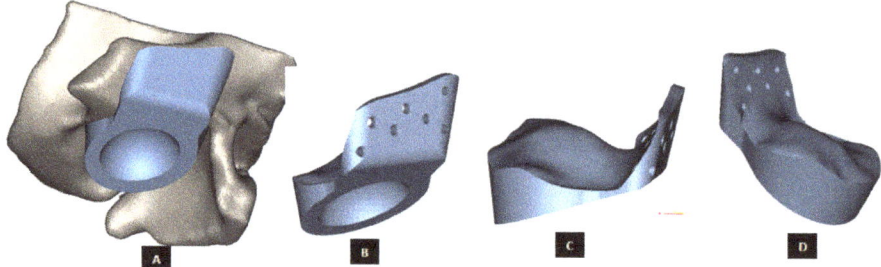

Figure 16. Patient-fitted fossa implant with circular inferior surface which also has a circular grove for articulation with condylar head. Visuals in (**A–D**) demonstrate the customized size and shape of the implant.

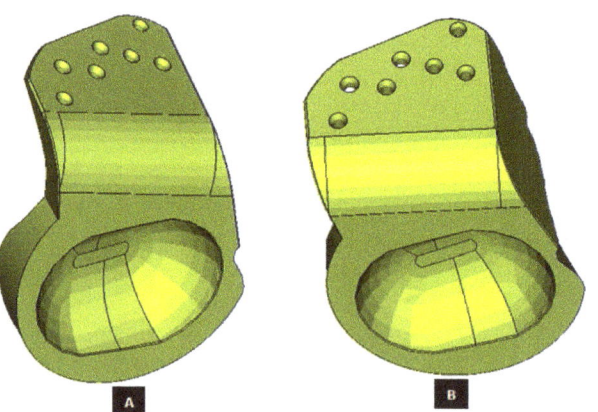

Figure 17. Patient-specific design of fossa prosthesis (**A,B**). The device has a rectangular grove (with curved anterior and posterior edges) in its inferior surface for articulation with condylar head. The uniquely designed articulating surface/hole is slanted in anterior direction. This anterior slope of articulating surface is intended to provide opportunity for anterior translation of the condylar head during movements of mandible. This feature of our fossa prostheses provides an advantage over currently available total TMJ implants which, when implanted, only rotate but do not translate during functional movements of the patient's jaw [19].

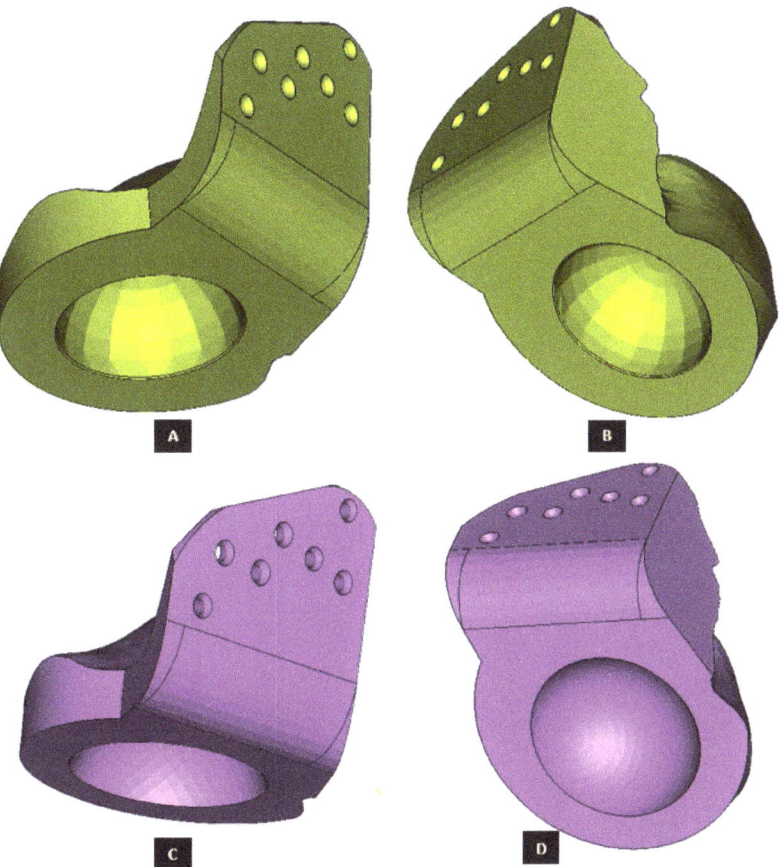

Figure 18. Custom-designed fossa prosthesis with circular articular surface. The device shown in (**A,B**) has relatively smaller articulating circular hole compared to the one shown in (**C,D**). Additionally, articulating surface of the device shown in (**C,D**) is slanted anteriorly to augment anterior translation of condylar head during mastication.

2.1.4. Design of Screws

Unlike hip or knee joints, the bony anatomy of mandibular ramus and temporal glenoid fossa do not afford the use of modular stock components for TMJ TJR that can be stabilized initially with press-fitting or cementation [2]. Therefore, TMJ devices have to rely only on screws for initial fixation and stabilization of their components. Clinicians have underlined the need for improved methods of internal fixation of prosthetic TMJ devices to minimize or eliminate implant loosening and joint failure [1]. The position of inserted screws was more important than the number of screws for stable fixation of the condylar TMJ prosthesis [20]. Our methodology of designing patient-specific total TMJ prostheses based on anatomically accurate 3D models provides realistic and accurate options in deciding positions of fixation screws for the prostheses. The positions of pre-drilled screw holes in the prostheses can be selected to avoid unintentional injury to delicate structures in the vicinity while ensuring stable fixation of the devices. Moreover, unlike stock TMJ implants, the custom-designed TMJ prostheses do not have any unused screw holes which may act as stress-risers under functional in-vivo loading post-implantation.

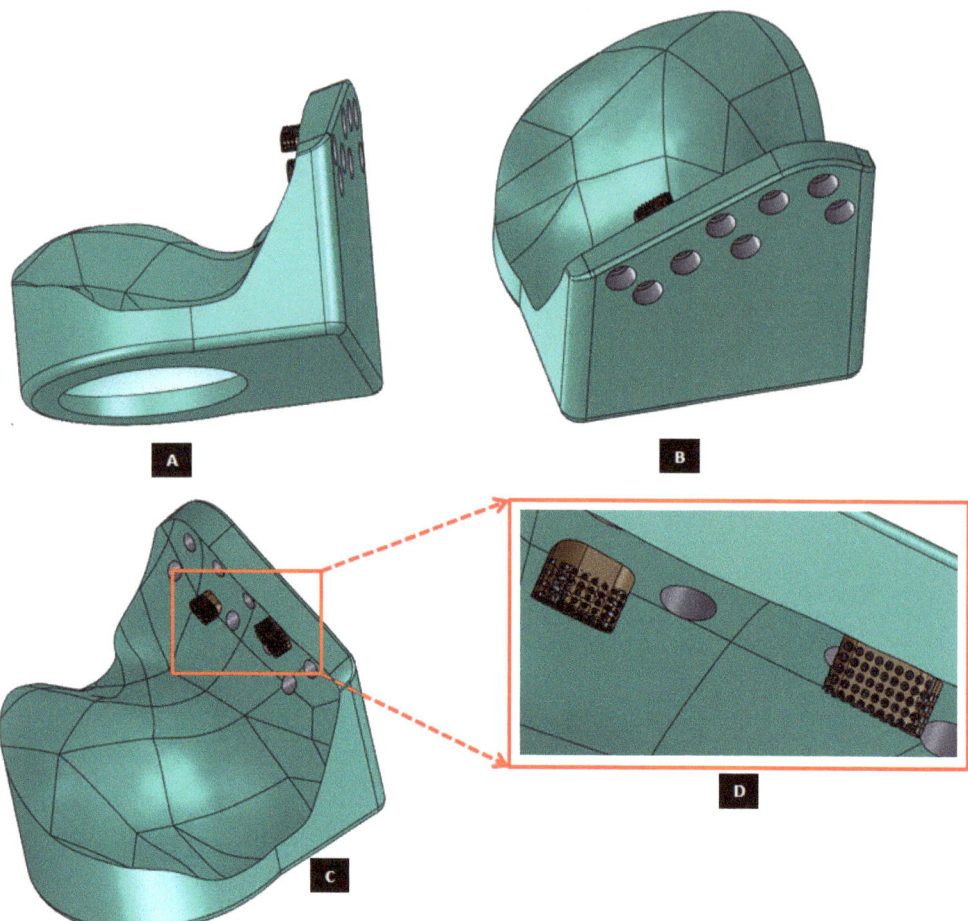

Figure 19. Patient-specific design of a fossa implant with circular articular surface/hole in the inferior face of the device (**A**). The device has a novel feature; perforated medial notches (**B**) protruding into host bone at implantation. Each perforated notch is designed to fit into surgically created mating grove in the host bone, thereby maximizing device stability by allowing ingrowth of bone into the prosthesis after implantation (**C,D**). The notches also provide a mode for load transfer between the prosthesis and native bone, thereby reducing the amount of load and resultant stress acting on the fixation screws. The surgeons can be provided with custom-designed templates and cutting guides to accurately cut the slots in native bone to accommodate perforated notches.

Motion of implanted TMJ prosthesis under load can cause degeneration of the surrounding bone, which may lead to further device loosening and possible failure [1]. Screws may loosen with time and function, requiring replacement. For long-term success of the TMJ implants, forces from the implant to the bone and vice versa must occur without relative motion or without intermittent loading [2,21]. The use of bone screws with sharp threads in TMJ implants prevents movement between the screw head and prosthesis [15]. The custom-designed screws of our TMJ prostheses system provide optimal fixation through locking mechanism—a unique feature not currently offered by any of the US FDA-approved TMJ TJR devices (see Figures 22 and 23). Threads on the screw-head surface provide high grade fixation by firmly engaging with the matching threads in the screw hole of either condylar/ramal or fossa implants. The possibility of movement between the screws and prosthesis can be eliminated by using such locked screws. In addition to the surgical condylectomy guides, our methodology also provides the surgeons with customized screw-drill-guides and templates for the TMJ prostheses. These drill guides are intended to create a hole of preferred dimension (diameter and depth), and at the accurate location and orientation for each screw as prescribed by the surgeon.

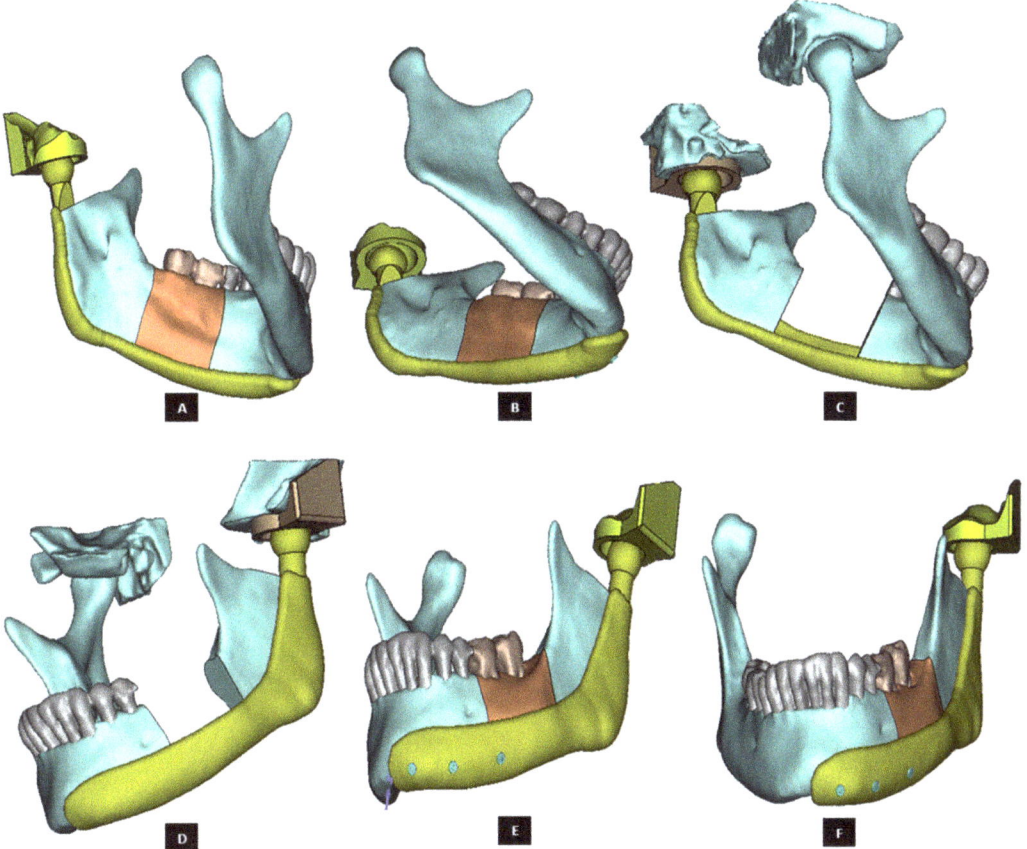

Figure 20. Shape outline of the patient-specific total TMJ prosthesis. Ramal component of the prosthesis is extended anteriorly up to the chin (**A–F**) to support mandibular host bone and graft (with aesthetic dental implant) after removal of the imaginary tumor (shown in red) in the left mandibular body/molar region.

2.1.5. Implant Materials

Using advantageous physical characteristics of biocompatible materials is an essential aspect in the design and manufacture of a successful prosthetic device. Some of the important characteristics of materials used to manufacture the TMJ prostheses from are biocompatibility, mechanical strength, low wear-rate, and harmless wear particles. Advancement in materials research has led to materials [22] such as medical grade pure titanium (Ti), titanium alloy (Ti-6Al-4V), cobalt-chromium-molybdenum (Co-Cr-Mo), and ultrahigh molecular weight polyethylene (UHMWPE) becoming gold standard for low friction orthopaedic joint replacement, Table 3.

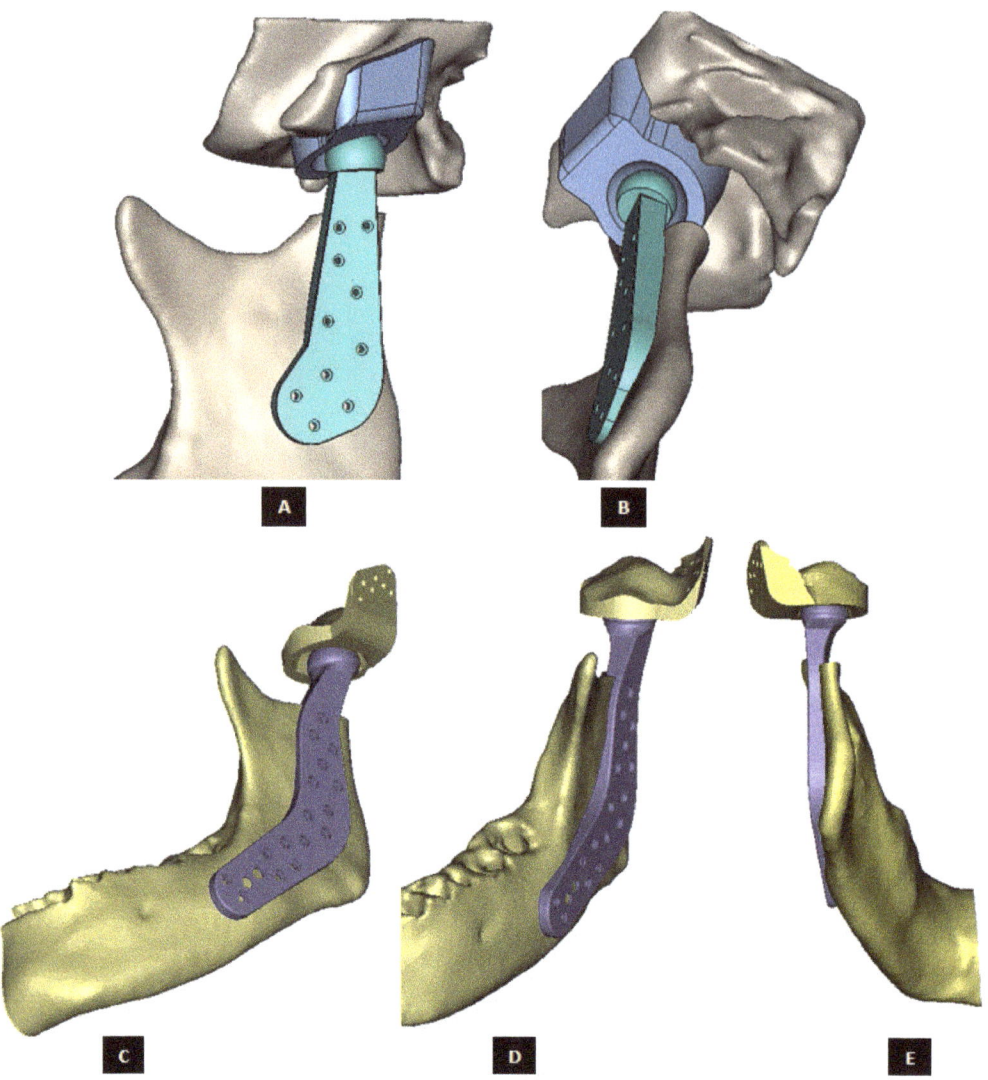

Figure 21. Patient-specific total TMJ prostheses with different articulations and fixation (A–E).

Wrought Co-Cr-Mo is reported to have excellent wear resistance when articulated against UHMWPE in the non-movable articulating surface of most orthopaedic TJR devices [23]. However, TMJ is a highly mobile joint in which articulating surfaces of the reconstructed joint undergo repeated mechanical stresses resulting from movement of the jaw. Metallurgical flaws, such as porosity, found in cast Cr-Co are suggested to cause the fatigue failure of Cr-Co TJR components resulting in noxious metallic debris in the patient's body [23].

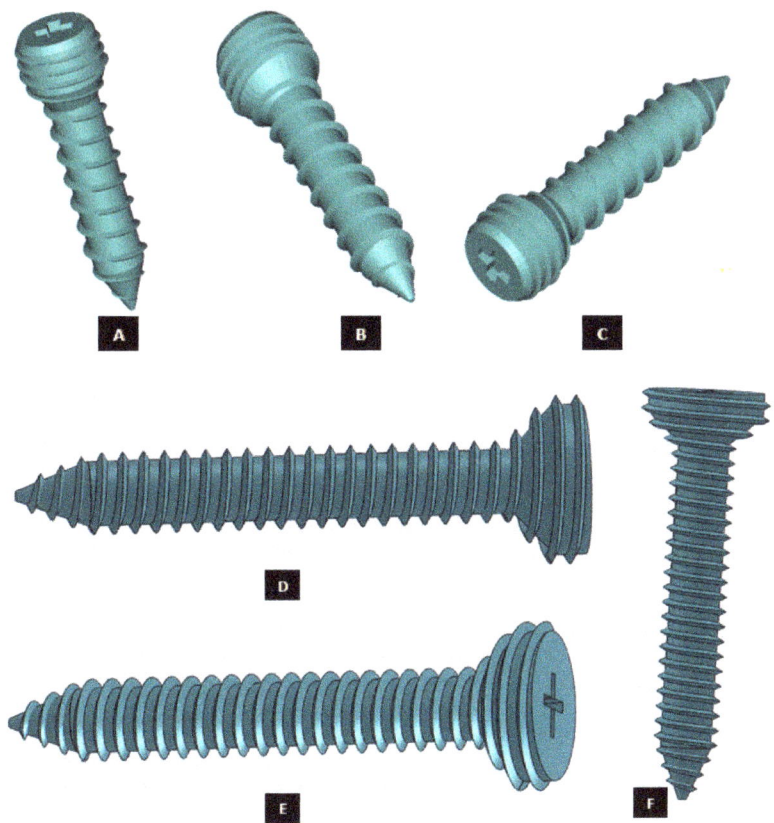

Figure 22. Custom-designed screws with locking mechanism. The threads on the screw-head surface provide improved/optimal fixation by firmly engaging in the matching threads in the screws holes of either condylar/ramal or fossa component of the total TMJ prosthesis (**A–F**).

Titanium alloy (Ti-6Al-4V) for condylar/ramus component and bone screws, and UHMPE for fossa component is based on their favorable characteristics and successful long-term applications reported in scientific and clinical literature. Unalloyed titanium reacts rapidly with oxygen in the air to form a thin (<10 μm) layer of chemically inert titanium oxide which provides a favorable surface for biointegration of prosthesis with bone [2]. In addition to its biocompatibility and biointegration, Titanium also offers properties of strength, corrosion resistance, ductility, and machinability [23]. UHMWPE is a linear un-branched polyethylene chain with a molecular weight of more than one million. UHMWPE is shown to have excellent wear and fatigue resistance for a polymeric material [24]. Untill year 2011, no cases of UHMWPE particulation-related osteolysis have been reported in the TMJ prostheses literature [2,25].

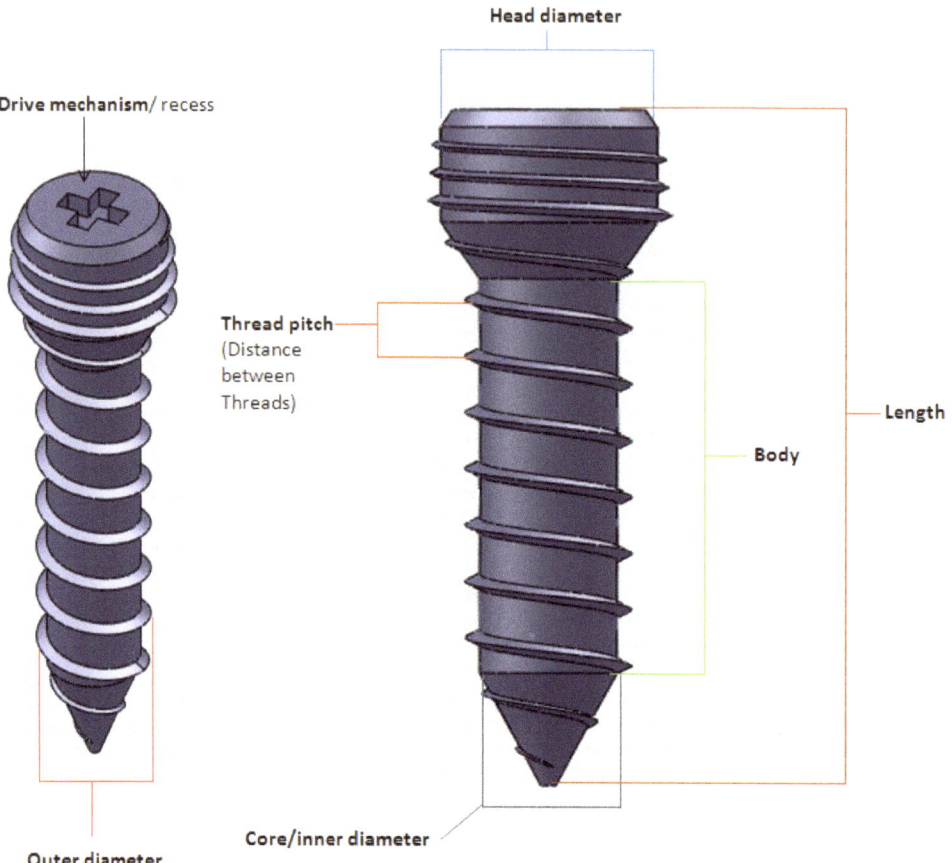

Figure 23. A custom-designed locking screw for TMJ prosthesis. Visuals show different features of the screw. Total length of the screw depends on the size of prosthesis and native bone. The body/shaft of screw is designed long enough to utilize maximum amount of host bone (condyle/ramus or fossa eminence) but avoid protrusion of screws from medial surface of the bone. Length of the screw head varies depending on the thickness of condylar or fossa prosthesis in the particular screw-hole location. The outer diameter of screw is kept in the range of 1.5 mm–3.00 mm as this range is reported to be optimal for the screws of TMJ implants. The screw has varying pitch, with more threads per unit length of screw-head than the body/shaft.

2.2. FEA of Total TMJ Implant

Methods for biomechanical assessment of prosthetic TMJ must be developed to make the implantation outcomes more predictable and reliable, and to evaluate the methods of device fixation to minimize or eliminate implant loosening and joint failure [1]. We performed FE simulations of two patient-specific total TMJ prostheses—one device with medial notches fitting into the groves created in the host bone (see Figure 24) and another 'simple implant' without such notches in the condylar and fossa components (see Figure 25)—using our validated methodology described elsewhere [18]. The objective of this study was to investigate stress and strain distribution in prosthetic components and host bone surrounding the screws under normal and worst-case/over–loading conditions. To account for the user-induced errors due to variations in selecting the nodes of FE mesh for applying boundary conditions and loads, we performed three repetitions of FE simulation under each loading condition for both total TMJ prostheses systems. Results reported in Tables 4 and 5 are average of the values obtained from three runs of each FE simulation.

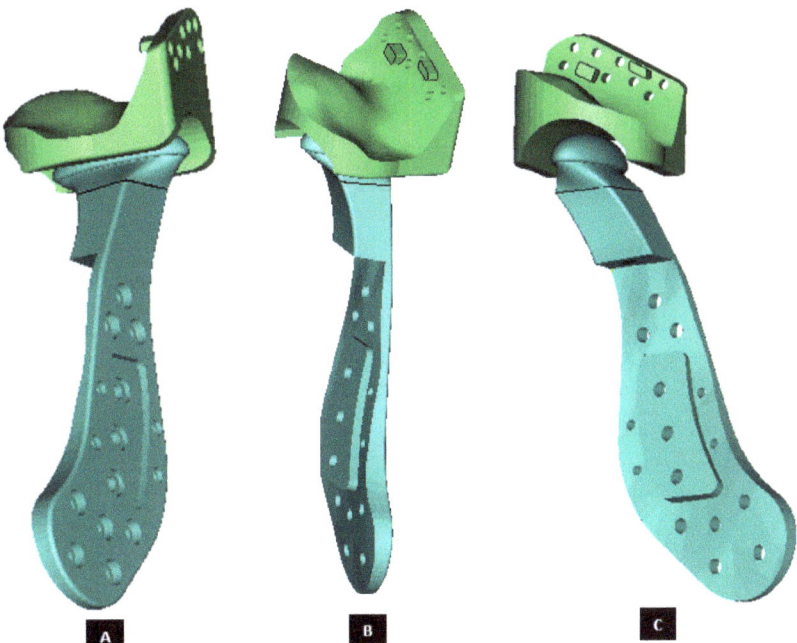

Figure 24. A patient-specific total TMJ prosthesis with medial notches in fossa and condylar components. (**A**) Shows anterior–lateral view of the 'notched implants' with screw holes. Fossa prosthesis has two medial notches to be fit into host bone (**B**,**C**). The articular surface of fossa implant has medio–lateral openings, and is designed to allow optimal anterior and medial translation along with rotation of the prosthetic condylar head along the medio–lateral axis.

2.2.1. FE Modeling and Mesh Generation

Subject-specific 3D anatomical reconstruction of patient's mandible and skull/articular eminence was performed using commercial software Mimics v14.12 (Materialise, Plymouth, MI, USA) from computed tomography (CT) scans of patient's TMJ. Upon importing the patients CT images in Mimics, anatomical model comprising of patient's mandible and fossa eminence was developed from the CT scan by performing a series of operations such as image processing, segmentation, mask formation for bone and teeth, region growing, and calculation of 3D equivalent similar to the 3D reconstruction method described elsewhere [18]. Using the design methodology discussed in previous sections, two patient-specific total TMJ prostheses systems—a 'simple implant' without notches, and another 'implant with notches'—were designed for total reconstruction of the patient's left TMJ (see Figures 24 and 25). For FE simulations, volume bound within surfaces of anatomical components (cortical bone, cancellous bone, and teeth) and prosthetic components (condyle, fossa, and screws) were meshed. Three-D volume mesh was generated for each of these components with ten-node quadratic tetrahedral elements of type C3D10 (see Figures 26 and 27). Mesh convergence was achieved using the technique discussed previously [18]. The finite element analyses were performed using a commercial FE package ABAQUS v6.10 (SIMULIA, Providence, RI, USA).

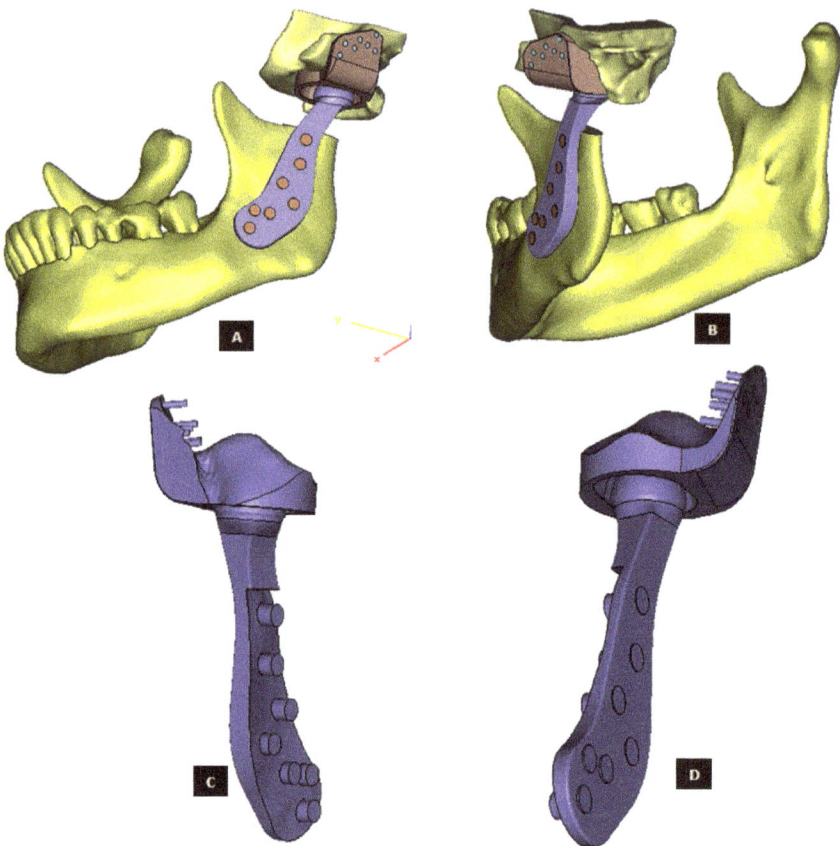

Figure 25. A patient-specific total TMJ prosthesis. (**A**,**B**) Show two views of the 'simple' total TMJ prosthesis along with left fossa bone and mandible after removal of left condyle. (**C**,**D**) Show two views of the total TMJ along with screws.

2.2.2. Model Constraints and Loads

As illustrated in Figure 28, the condylar head of the prosthetic TMJ was allowed to rotate along the medio–lateral axis, and translate in anterior-posterior direction. Similarly, for the TMJ on contralateral side, the natural condylar head was allowed to only rotate along the medio–lateral axis, and translate in anterior-posterior direction. The incisor teeth were fixed so that they could not translate in three directions, but could rotate. The entire fossa host bone was constrained in all directions.

The interface between prosthetic condylar head and articulating surface of fossa prosthesis was modeled as sliding contact with a coefficient of friction of 0.3. The interface between the prostheses and bone in contact was modeled with contact elements having a coefficient of friction of 0.42 as reported in literature [26]. The screw-to-prosthesis and screw-to-bone interfacial conditions were assumed to be bonded. Since use of locking screws eliminates the possibility of movement between screw and prosthesis, the interface condition between screw heads and TMJ prostheses (condylar and fossa) was assumed to be perfect bonding. Two oblique bite forces, each 200 N for normal loading condition and 400 N for over–loading/worst-case scenario, were applied to the mandibular model in the angulus area as shown in Figure 28.

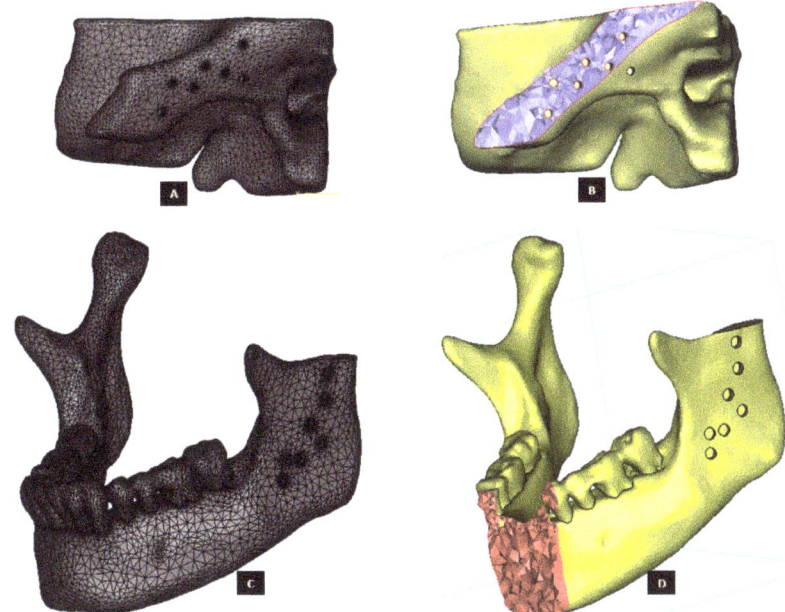

Figure 26. Three-D finite element mesh of the host bone components prepared for total prosthetic replacement of the left TMJ. (**A**) Shows FE surface mesh of left fossa, and (**B**) shows a lateral cross-section of the 3D volume mesh of left fossa bone with screw holes. Similarly, (**C**,**D**) show surface mesh and anterior cross-section of volume mesh, respectively, of the mandible with screw holes and removal of damaged left condyle.

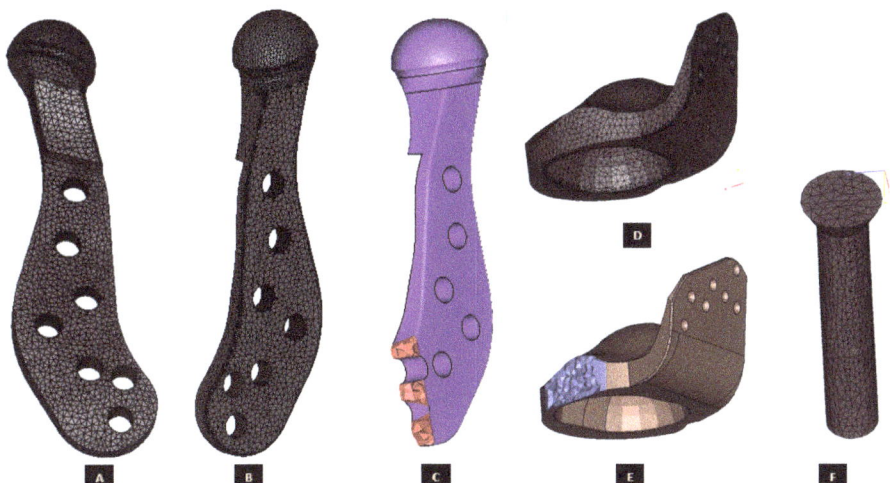

Figure 27. Three-D finite element mesh of the components of patient-specific total TMJ prostheses. (**A**–**C**) Show FE mesh of the condylar/ramal component of the 'simple' TMJ implant (without notches). (**D**,**E**) Show FE mesh of the fossa component, and (**F**) demonstrates FE mesh of a screw for device fixation.

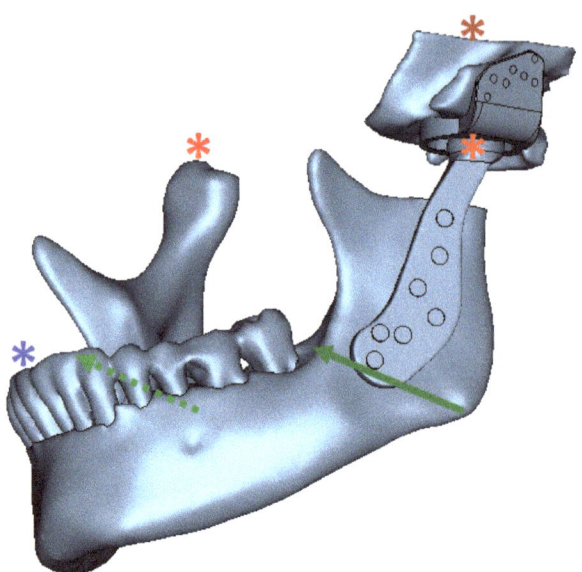

Figure 28. Assembly of all parts of the FE model (including anatomic and prosthetic components), and schematic representation of model constraints and load application for FE simulation of total TMJ prostheses and anatomical components. Green arrows depict the location and direction of bite forces applied in the angulus region on both sides of the mandibular mesh. The asterisks indicate constrained nodes at condyle, fossa, and incisor teeth. Left prosthetic condylar head and right natural condylar head were constrained such that they could only rotate along the medio–lateral axis and translate in anterior-posterior direction. The nodes at incisor teeth were so constrained such that they could only rotate. The entire fossa host bone was constrained in all directions. The interface between prosthetic condylar head and articulating surface of prosthetic fossa was modeled as sliding contact. The prosthesis-to-bone, screw-to-prosthesis, and screw-to-bone interfaces were assumed to be bonded. The interfacial and boundary conditions were kept similar for normal and over–load configurations; and only magnitude of applied forces was changed across the two loading configurations.

2.2.3. Material Properties

Young's modulus and Poissson's ratio of anatomical components (fossa and mandible bone with teeth), titanium alloy (for condylar component and all screws), and UHMWPE (for fossa component) were selected as listed in Table 3. All anatomical components of the model (i.e., cortical bone, cancellous bone, and teeth) were assigned properties of the cortical bone similar to other researchers [27,28] who have previously followed this practice since variation in material properties of these components have negligible influence on biomechanics of the mandible. All materials used in FE models were assumed to be isotropic, homogeneous, and linearly elastic [26]. Static FE simulations were performed using ABAQUS software. Three repetitions/runs of FE simulation under each loading condition were performed for both types of total TMJ implants to account for any user-induced errors such as variation in selecting exactly the same nodes of FE mesh across different simulations. The results summarized in Tables 4 and 5 are average of three simulations for each loading condition for both types of total TMJ prosthesis.

Table 3. Material properties for anatomical and prosthetic TMJ components.

Part	Young's Modulus (MPa)	Poisson's Ratio	References
Host bone	1.47×10^4	0.3	[25,29]
Titanium alloy (Ti-6Al-4V)	1.10×10^5	0.3	[28]
UHMWPE	830	0.317	[30]

Table 4. Peak von Mises stresses developed in condyle/ramus and fossa components, and fixation screws of the simple and notched designs of patient-specific total TMJ prostheses during FE simulations under normal and worst-case/over–loading configurations.

Implant Type	Loading Type	Peak von Mises Stress in Implant Components (MPa) *			
		Condyle/Ramus	Condylar Screws	Fossa	Fossa Screws
Simple	Normal	44.3	61.4	11.3	28
	Over–load	56.7	106.7	14.2	43.1
With Notches	Normal	42.6	59.6	10.5	23.4
	Over–load	59.1	108.3	13.4	38.6

* Average of three simulations performed under similar constraints and loading at three different times (to account for variations induced by the user/operator).

Table 5. Peak stress and strain developed in host bone surrounding the fixation screws of the simple and notched designs of patient-specific total TMJ prostheses during FE simulations under normal and worst-case/over-loading configurations.

Implant Type	Loading Type	Peak von Mises Stress in Host Bone Adjacent to Screw Holes (MPa) *		Peak von Mises Strain in Host Bone Adjacent to Screw Holes (μStrain) *	
		Condyle/Ramus	Fossa	Condyle/Ramus	Fossa
Simple	Normal	4.7	3.5	1983	1253
	Over–load	13.6	7.4	3586	1711
With Notches	Normal	4.3	3.2	1893	1210
	Over–load	12.5	7.1	3374	1564

* Average of three simulations performed under similar constraints and loading at three different times (to account for variations induced by the user/operator).

3. Results

The von Mises stress and micro-strain in the TMJ prostheses (fossa and condylar), screws, and native bone in regions adjacent to screws were measured. Figures 29 and 30 show visuals of stress profiles in the anatomical components and simple TMJ prostheses, respectively. Figures 31 and 32 show visuals of stress profiles in the corresponding anatomical components and 'notched' TMJ prostheses, respectively. Table 4 summarizes peak von Mises stress occurred in prosthetic components and screws. Peak von Mises stress and strain developed in host bone surrounding the fixation screws of simple and notched TMJ prostheses under normal and worst-case/over-loading configuration are summarized in Table 5.

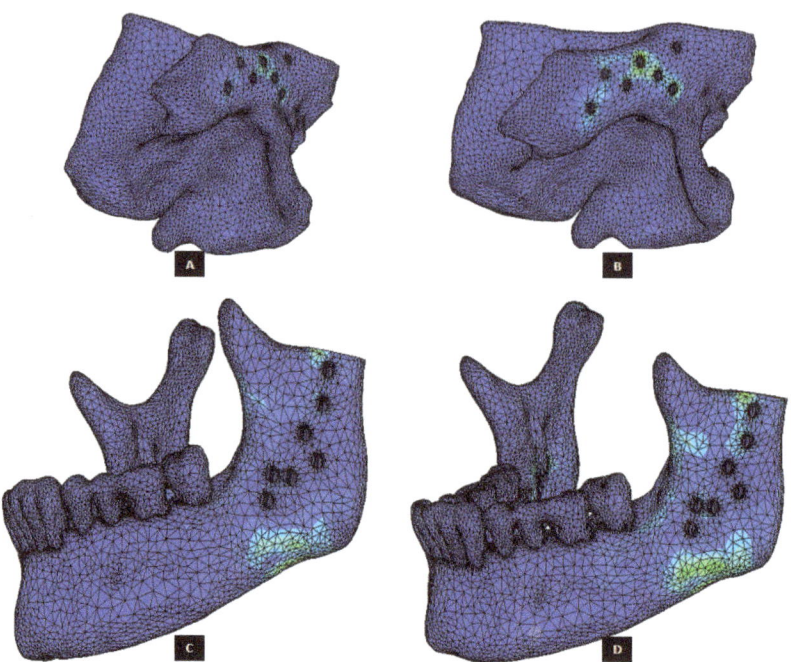

Figure 29. Stress distribution in the host bone components during FE simulations of the total TMJ replacement with custom designed simple TMJ prostheses. (**A,B**) Show von Mises stress in the fossa bone under normal and worst-case/over–load configurations, respectively. (**C,D**) Show von Mises stress in the mandibular bone under normal and worst-case/over–load configurations, respectively.

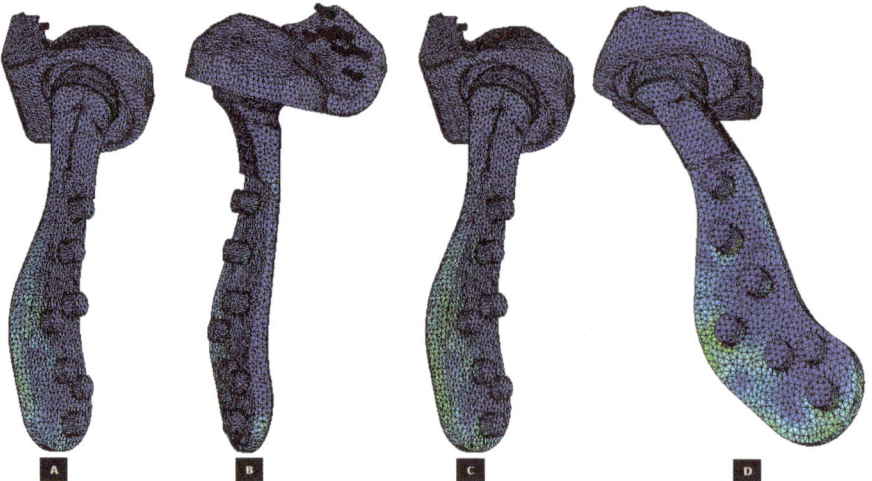

Figure 30. Peak von Mises stress in patient-specific 'simple' TMJ prosthesis (without notches) during FE simulations of two different loading scenarios. (**A,B**) Show von Mises total TMJ prosthesis under normal loading configuration. (**C,D**) Show von Mises stress profile in the prosthesis during FE simulation of worst-case/over-loading scenario.

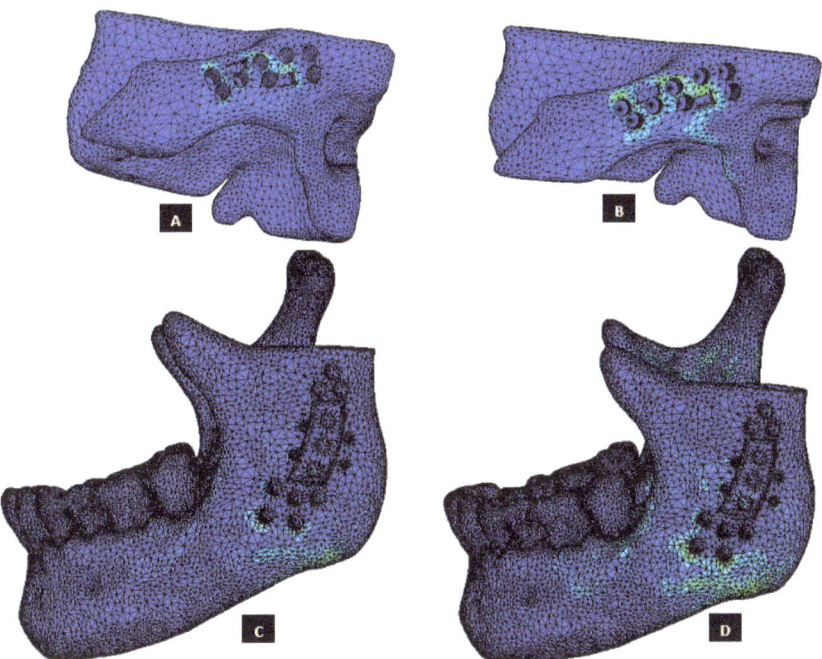

Figure 31. Stress distribution in the host bone components during FE simulations of the total TMJ replacement with custom designed TMJ prostheses with medial notches. (**A,B**) Show von Mises stress in the fossa bone under normal and worst-case/over-load configurations, respectively. (**C,D**) Show von Mises stress in the mandibular bone under normal and worst-case/over-load configurations, respectively.

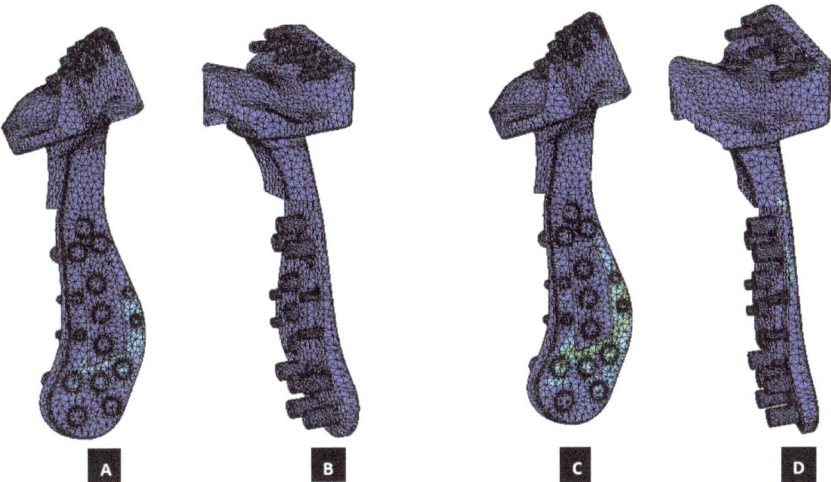

Figure 32. Peak von Mises stress in patient-specific 'notched' TMJ prosthesis (with medial notches) during FE simulations of two different loading scenarios. (**A,B**) Show von Mises total TMJ prosthesis under normal loading configuration. (**C,D**) Show von Mises stress profile in the prosthesis during FE simulation of worst-case/over-loading scenario.

3.1. Stress and Strain in Native Bone

Small difference in the stress and micro-strain occurred in the host bone adjacent to screws in condylar and fossa components of both total TMJ prosthesis systems. For both types of implant designs, von Mises stress in the bone surrounding fixation screws was in the range of 3.2–4.7 MPa and 7.1–13.6 MPa under normal loading and over–loading, respectively (see Table 5, Figures 29 and 31). These results are comparable to the findings reported by [31] who studied stress distribution in the screws of a condylar implant and host bone. von Mises strain in the bone surrounding prosthetic screws ranged from 1210 microstrain to 1983 microstrain during normal loading, and from 1564 microstrain to 3586 microstrain during over-loading condition. A strain higher than 4000 microstrain can cause hypotrophy of bone [32]. The highest micro-strain in host bone in this study is below the hypertrophy limit. Moreover, use of more screws at appropriate locations would further lower the chances of higher strains capable of bone formation around the screws.

3.2. Stress and Strain in TMJ Implants

FE simulations resulted in lower stress in anterior part of the condylar and fossa prostheses (see Figures 30 and 32) similar to what found in [20,28] although their FE studies included only the condylar TMJ implants with different loading conditions. The von Mises stress found in condylar and fossa components of both types of implants were lower than the yield strength of their materials Figure 33, Ti-6Al-4V and UHMWPE, respectively. The trends in the stress and strain profiles under normal and over-loading conditions were similar in both types of total TMJ prostheses. von Mises stresses of higher magnitude were developed in condylar neck, posterior part of condylar head, and inferior region of ramal component compared with rest of the condylar/ramus prosthesis. For fossa component, magnitude of von Mises stress and strain was higher in the posterior region on the articulating surface, Figures 33 and 34.

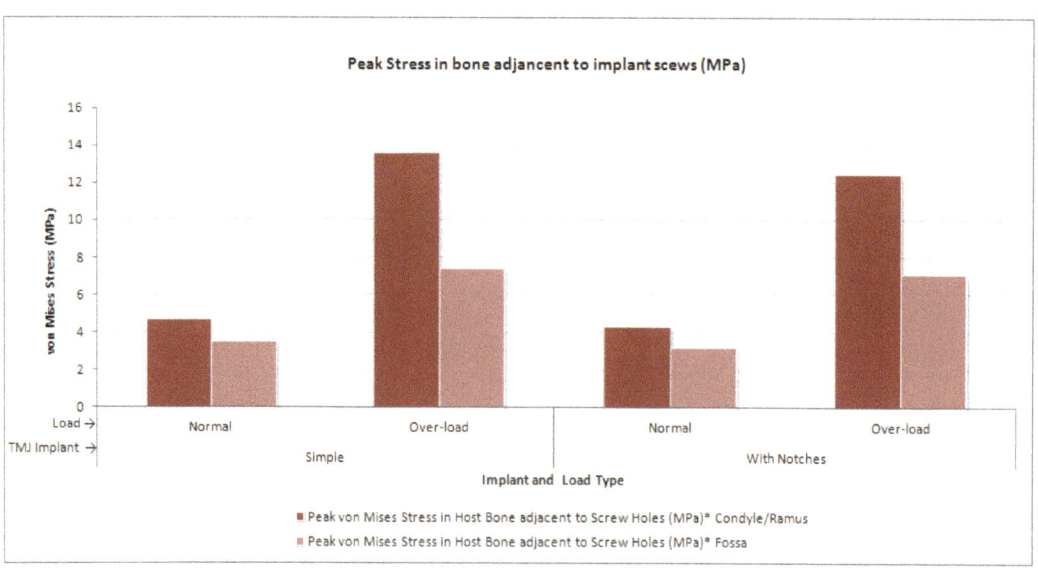

Figure 33. Peak von Mises stress in the mandibular and fossa bone adjacent to fixation screws of total TMJ prostheses during FE simulations under normal and worst-case/over-load configurations.

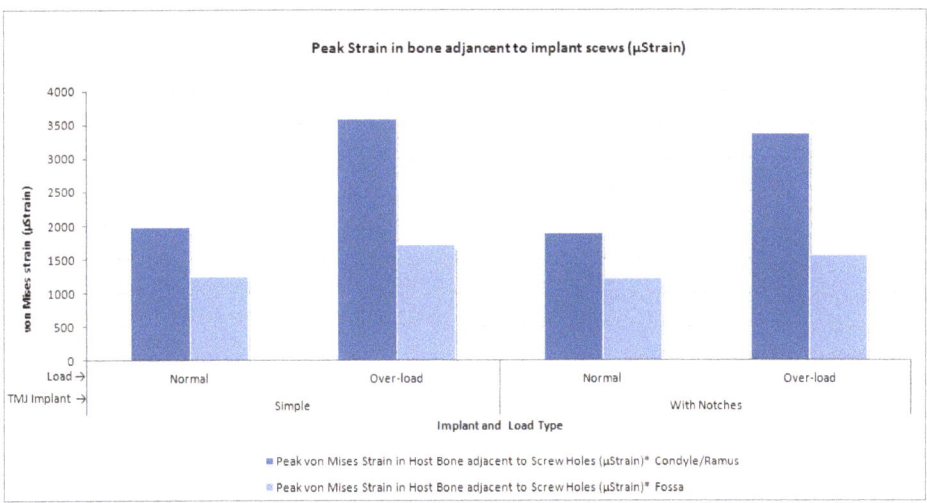

Figure 34. Peak micro-strain in the mandibular and fossa bone adjacent to fixation screws of total TMJ prostheses during FE simulations under normal and worst-case/over-load configurations.

Other than where the actual loads and constraints were applied, the stress concentration was highest around the inserted screw and the screw holes in the host bone. The medial notches in the condylar and fossa prostheses are designed to provide improved stability by promoting post-implantation bone growth into the perforated surfaces of the notches. Von Mises stress in the notch regions of the condylar and fossa implants were less than that in the screw regions, indicating that the notches may not act as stress risers in the device. Stress profile in the host bone portion where the medical notches of the implant are inserted show stresses lower than that at the screw holes, but higher than those in other parts of the host bone. This indicates that the stress developed in the notches under functional loading may augment bone growth into the perforated notches, thereby maximizing the opportunity for improved stability of the prostheses. Also, in all simulations, the peak von Mises stresses in the condylar component were higher than those in the fossa component of the total TMJ prostheses (see Figure 35). This may have resulted from the model constraints which allowed mobility of the condylar/ramal prosthesis along with natural mandible and kept the artificial fossa fixed in its position along with the host fossa bone.

3.3. Stress and Strain in Screws

Peak stress and strain the implant fixation screws are summarized in Table 4 and Figure 36. In fixation screws for condylar and fossa components, the highest magnitude of stress values occurred at the neck portion of screws. However, the highest stresses in all screws were found to be less than the ultimate stress as well as yield point of the screw material (Ti-6Al-4V). This trend of stress profile in screws is similar to that reported in [31] who studied stress distribution in a stock condylar prosthesis and screws using FE method. The highest von Mises stresses found in screws in the present study are much lower than those reported in [31]. This discrepancy suggests that screws used for fixation of the condylar component of patientspecific total TMJ prostheses undergo lower load and resultant stress while transferring the functional loads between implant and host bone. This further suggests that the custom-designed implants offer better adaptation to the host bone (compared to their stock counterparts) and partly transfer the load directly to the bone in contact (e.g., at the location where condylar collar of the implant sits superiorly on natural ramus after condylectomy), thereby reducing the exposure of screws to higher loads and stresses.

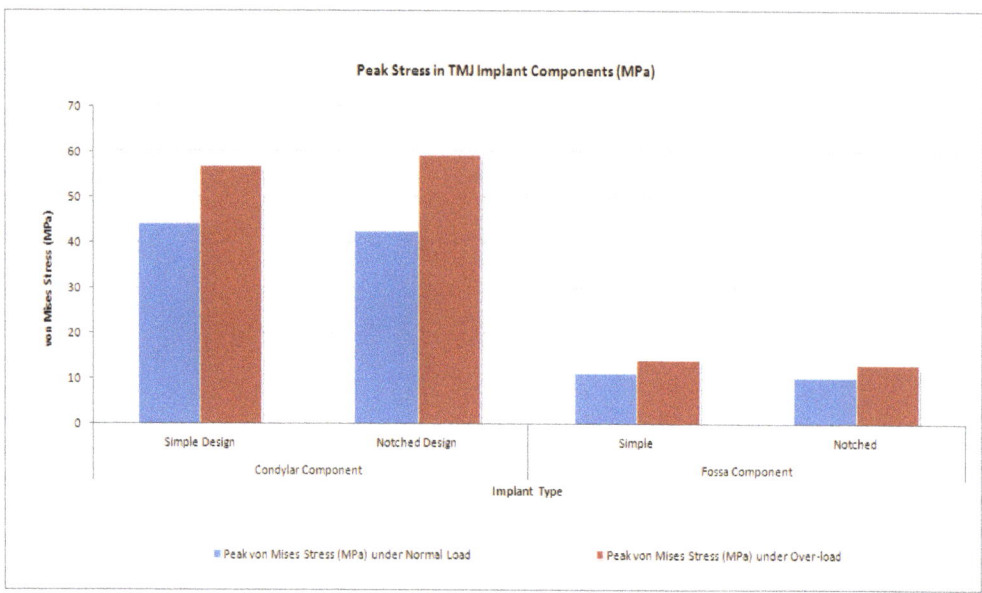

Figure 35. Peak von Mises stress in condylar/ramus and fossa components of patient-specific total TMJ prostheses during FE simulations under normal and worst-case/over-load configurations.

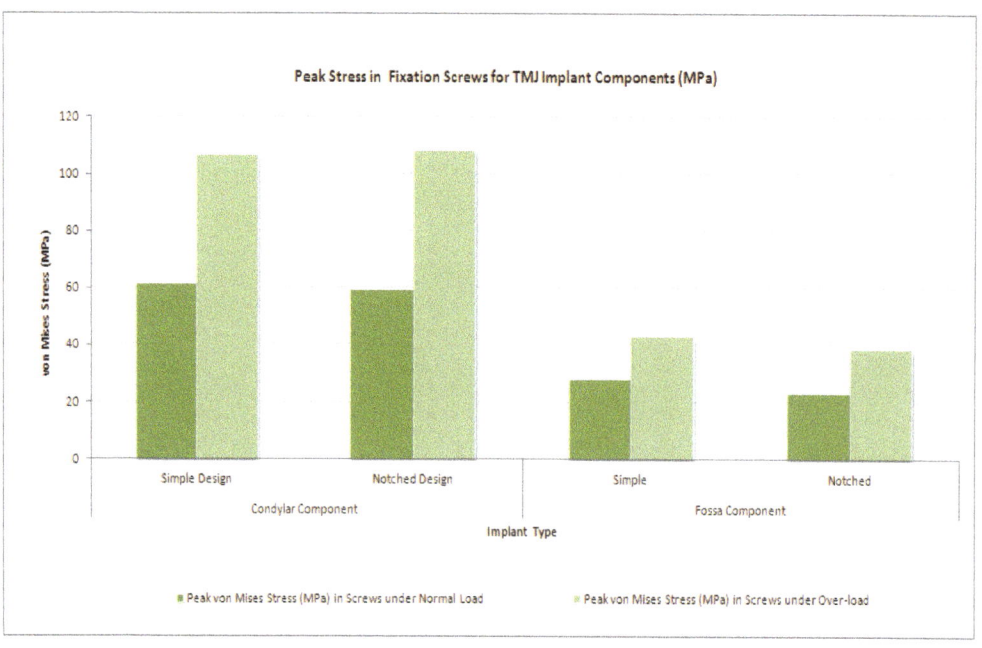

Figure 36. Peak von Mises stress in the fixation screws for condylar/ramus and fossa components of patient-specific total TMJ prostheses during FE simulations under normal and worst-case/over-load configurations.

Among the screws, the highest stresses occurred in the neck portion of two condylar screws—one placed most inferiorly, and another placed most posteriorly and at the curvature of the ramal part of the implant. This is contradictory to what found in [31,33] reporting highest stresses in the condylar screw placed most superiorly (near the neck of implant). However, both these studies included a stock condylar TMJ implant in which the implant collar did not contact or adapt to the host bone as it does in the present study. Also, these researchers applied a vertically downward force at the top of prosthetic condylar head whereas we applied load in mandibular angle region.

Screws used for both condylar and fossa components showed von Mises stresses of higher order at their interfaces with prostheses, especially in the region of screw neck and at the site of prosthesis-bone junction. This may have happened as the screws carried more load when they served as a medium of load-sharing between the prosthesis and host bone. As listed in Table 4, maximum von Mises stress generated in the screws was relatively higher than that in the corresponding prosthetic component these screws were used for fixation of.

4. Discussion and Conclusions

In view of scarcity of published literature about design methods and biomechanical analysis of total TMJ implants, the present study provides good reference work for patient-specific design and biomechanical evaluation of such designs through FE simulations. Few published studies have investigated biomechanics of the artificial TMJ implants. In our knowledge, no study has reported FE analysis of total TMJ prostheses. The present study can serve as a reference for the clinicians regarding advantageous features of the patient-specific total TMJ implants. Moreover, design methodology and FE findings of this study can provide industrial designers with reference data for improving their products, especially the custom-designed products intended for patients with complex and challenging anatomic situations.

Limitations of this study must be considered when reviewing implant designs and evaluating FE results. The main focus of this study was the patient-specific design and biomechanical analysis of the total TMJ prostheses. The skill of the surgical approach (preauricular incision or retromandibular incision) and patient-related issues such as long-term effects were not considered. Therefore, the surgeons should be careful when applying the findings from this study to clinical situation. The present study used only one set of material properties for patient's host bone. Future investigations should assess effect of altered bone quality on the performance of total TMJ replacement. Also, material properties of bone were assumed to be homogeneous and isotropic. Although this represents a major simplification, other studies have demonstrated that this is acceptable [20,26,34]. Only two loading conditions (with loads applied at the mandibular angle) were used in this study. Other muscle forces which are normally present would also affect the mandibular biomechanics. Although several studies have suggested that forces from other muscles do not exert major effects in the mandible [20,26,35], future work should consider using more sophisticated FE models. It will be beneficial to evaluate the effect of screw positions on biomechanical performance of total TMJ prostheses. Future work should also include more comprehensive non-linear and dynamic FE simulations using different implant materials.

In summary, this study demonstrates that our custom-design approach offers potential for stable and durable total TMJ reconstruction, and that the FE models can reproduce information useful in design and assessment of total TMJ prostheses. Findings of this study provide a good basis for future work focusing on developing a more refined and standardized method for custom design of total TMJ prostheses, and pre-clinical FE tests for design verification and validation.

Author Contributions: Conceptualization, T.G.; Data curation, S.M.I.; Formal analysis, S.M.I.; Investigation, S.M.I.; Methodology, S.M.I.; Project administration, T.G.; Resources, T.G.; Supervision, T.G. All authors have read and agreed to the published version of the manuscript.

Funding: This research received no external funding.

Institutional Review Board Statement: This research was conducted using the cadaveric mandibles from the Anatomical Gift Program of the Wright State University which were imaged at Miami Valley Hospital and biomechanically tested for a PhD dissertation project.

Informed Consent Statement: This research used cadaveric mandibles therefore no consent statement was applicable.

Conflicts of Interest: The authors declare no conflict of interest.

References

1. Mercuri, L. The use of alloplastic prostheses for temporomandibular joint reconstruction. *J. Oral Maxillofac. Surg.* **2000**, *58*, 70–75. [CrossRef]
2. Mercuri, L.G. Alloplastic temporomandibular joint replacement: Rationale for the use of custom devices. *Int. J. Oral Maxillofac. Surg.* **2012**, *41*, 1033–1040. [CrossRef] [PubMed]
3. Quinn, P.D. Lorenz Prosthesis. *Oral Maxillofac. Surg. Clin. N. Am.* **2000**, *12*, 93–104. [CrossRef]
4. Quinn, P.D. Alloplastic Reconstruction of the temporomandibular joint. *Sel. Read. Oral Maxillofac. Surg.* **1999**, *7*, S31–S40.
5. Driemel, O.; Braun, S.; Müller-Richter UD, A.; Behr, M.; Reichert, T.E.; Kunkel, M.; Reich, R. Historical development of alloplastic temporomandibular joint replacement after 1945 and state of the art. *Int. J. Oral Maxillofac. Surg.* **2009**, *38*, 909–920. [CrossRef] [PubMed]
6. Wolford, L.M.; Mehra, P. Custom-made total joint prostheses for temporomandibular joint reconstruction. *Bayl. Univ. Med. Cent. Proc.* **2000**, *13*, 135–138. [CrossRef]
7. Wolford, L.; Pitta, M.; Reiche-Fischel, O.; Franco, P. TMJ Concepts/Techmedica custom-made TMJ total joint prosthesis: 5-year follow-up study. *Int. J. Oral Maxillofac. Surg.* **2003**, *32*, 268–274. [CrossRef]
8. Wolford, L.M. Temporomandibular joint devices: Treatment factors and outcomes. *Oral Surg. Oral Med. Oral Pathol. Oral Radiol. Endod.* **1997**, *83*, 143–149. [CrossRef]
9. Driemel, O.; Braun, S.; Müller-Richter UD, A.; Behr, M.; Reichert, T.E.; Kunkel, M.; Reich, R. Historical development of alloplastic temporomandibular joint replacement before 1945. *Int. J. Oral Maxillofac. Surg.* **2009**, *38*, 301–307. [CrossRef]
10. Mercuri, L.G. Alloplastic temporomandibular joint reconstruction. *Oral Surg. Oral Med. Oral Pathol. Oral Radiol. Endod.* **1998**, *85*, 631–637. [CrossRef]
11. Mercuri, L.G.; Edibam, N.R.; Giobbie-Hurder, A. Fourteen-year follow-up of a patient-fitted total temporomandibular joint reconstruction system. *J. Oral Maxillofac. Surg.* **2007**, *65*, 1140–1148. [CrossRef] [PubMed]
12. Ingawalé, S.; Goswami, T. Temporomandibular joint: Disorders, treatments, and biomechanics. *Ann. Biomed. Eng.* **2009**, *37*, 976–996. [CrossRef] [PubMed]
13. Wolford, L.M.; Dingwerth, D.J.; Talwar, R.M.; Pitta, M.C. Comparison of 2 temporomandibular joint total joint prosthesis systems. *J. Oral Maxillofac. Surg.* **2003**, *61*, 685–690. [CrossRef] [PubMed]
14. van Loon, J.P.; de Bont, L.G.M.; Boering, G. Evaluation of temporomandibular joint prostheses Review of the literature from 1946 to 1994 and implications for future prosthesis designs. *J. Oral Maxillofac. Surg.* **1995**, *53*, 984–996. [CrossRef]
15. Van Loon, J.P.; De Bont LG, M.; Stegenga, B.; Spijkervet FK, L.; Verkerke, G.J. Groningen temporomandibular joint prosthesis. Development and first clinical application. *Int. J. Oral Maxillofac. Surg.* **2002**, *31*, 44–52. [CrossRef]
16. U.S. Food and Drug Administration. Available online: https://www.fda.gov/medical-devices/dental-devices/temporomandibular-disorders-tmd-devices (accessed on 16 May 2022).
17. Swanson, S.A.V.; Freeman, M.A.R. (Eds.) *The Scientific Basis of Joint Replacement*; Wiley and Sons: New York, NY, USA, 1977; 182p.
18. Ingawalé, S.M. Mandibular Bone Mechanics and Evaluation of Temporomandibular Joint Devices. Ph.D. Thesis, Wright State University, Dayton, OH, USA, 2012.
19. Komistek, R.D.; Dennis, D.A.; Mabe, J.A.; Anderson, D.T. In vivo kinematics and kinetics of the normal and implanted TMJ. *J. Biomech.* **1998**, *31*, 13. [CrossRef]
20. Hsu, J.T.; Huang, H.L.; Tsai, M.T.; Fuh, L.J.; Tu, M.G. Effect of Screw Fixation on Temporomandibular Joint Condylar Prosthesis. *J. Oral Maxillofac. Surg.* **2011**, *69*, 1320–1328. [CrossRef]
21. Morscher, E.W. Implant stiffness and its effects on bone and prosthesis fixation. In *Hip Surgery Materials & Developments*; Cabanela, M.E., Sedel, L., Eds.; Mosby: St. Louis, MO, USA, 1998; pp. 9–20.
22. Galante, J.O.; Lemons, J.; Spector, M.; Wilson, P.D., Jr.; Wright, T.M. The biologic effects of implant materials. *J. Orthop. Res.* **1991**, *9*, 760–775. [CrossRef]
23. Lemons, J.E. Metals and alloys. In *Total Joint Replacement*; Petty, W., Ed.; WB Saunders: Philadelphia, PA, USA, 1991; p. 811.
24. Sanford, W.M.; Maloney, W.J. Ultra-high molecular weight polyethylene. In *Hip Surgery Materials & Developments*; Cabanela, M.E., Sedel, L., Eds.; Mosby: St. Louis, MO, USA, 1998; pp. 45–56.
25. Westermark, A.; Leiggener, C.; Aagaard, E.; Lindskog, S. Histological findings in soft tissues around temporomandibular joint prostheses after up to eight years of function. *Int. J. Oral Maxillofac. Surg.* **2011**, *40*, 18–25. [CrossRef]

26. Hsu, J.; Huang, H.; Tu, M.; Fuh, L. Effect of bone quality on the artificial temporomandibular joint condylar prosthesis. *Oral Surg. Oral Med. Oral Pathol. Oral Radiol. Endod.* **2010**, *109*, e1–e5. [CrossRef]
27. Groning, F.; Liu, J.; Fagan, M.J.; O'Higgins, P. Validating a voxel-based finite element model of a human mandible using digital speckle pattern interferometry. *J. Biomech.* **2009**, *42*, 1224–1229. [CrossRef] [PubMed]
28. Ramos, A.; Completo, A.; Relvas, C.; Mesnard, M.; Simões, J. Straight, semi-anatomic and anatomic TMJ implants: The influence of condylar geometry and bone fixation screws. *J. Cranio-Maxillofac. Surg.* **2011**, *39*, 343–350. [CrossRef]
29. Ichim, I.; Swain, M.; Kieser, J.A. Mandibular biomechanics and development of the human chin. *J. Dent. Res.* **2006**, *85*, 638–642. [CrossRef] [PubMed]
30. Kurtz, S.M. *UHMWPE Biomaterials Handbook: Ultra High Molecular Weight Polyethylene in Total Joint Replacement and Medical Devices*; Academic Press: London, UK, 2009; p. 543.
31. Roychowdhury, A.; Pal, S.; Saha, S. Stress analysis of an artificial temporal mandibular joint. *Crit. Rev. Biomed. Eng.* **2000**, *28*, 411–420. [CrossRef]
32. Roberts, W.E.; Huja, S.; Roberts, J.A. Bone modeling: Biomechanics, molecular mechanisms, and clinical perspectives. *Semin. Orthod.* **2004**, *10*, 123–161. [CrossRef]
33. Kashi, A.; Chowdhury, A.R.; Saha, S. Finite Element Analysis of a TMJ Implant. *J. Dent. Res.* **2009**, *89*, 241–245. [CrossRef] [PubMed]
34. Mesnard, M.; Ramos, A.; Ballu, A.; Morlier, J.; Cid, M.; Simoes, J.A. Biomechanical Analysis Comparing Natural and Alloplastic Temporomandibular Joint Replacement Using a Finite Element Model. *J. Oral Maxillofac. Surg.* **2011**, *69*, 1008–1017. [CrossRef]
35. Ingawale, S.M.; Johnson, R.M.; Goswami, T. Biomechanical Evaluation of Cadaver Mandible under Cyclic Compressive Loads. *Ital. J. Maxillo-Facial Surg.* **2014**, *25*, 55–63.
36. Martins, D.; Couto, R.; Fonseca, E.; Carreiras, A. Numerical analysis of the mechanical stimuli transferred from a dental implant to the bone. *J. Comput. Appl. Res. Mech. Eng.* **2021**, *11*, 1–11. [CrossRef]

Article

The Effect of Surface Processing on the Shear Strength of Cobalt-Chromium Dental Alloy and Ceramics

Liaisan Saleeva [1], Ramil Kashapov [2,3], Farid Shakirzyanov [4], Eduard Kuznetsov [1], Lenar Kashapov [2,3], Viktoriya Smirnova [5], Nail Kashapov [2,3], Gulshat Saleeva [1], Oskar Sachenkov [5,*] and Rinat Saleev [1]

1. Kazan State Medical University, 420012 Kazan, Russia; saleeva.100mat@yandex.ru (L.S.); my@ekuznetsov.ru (E.K.); gulshat.saleeva@kazangmu.ru (G.S.); rinat.saleev@kazangmu.ru (R.S.)
2. Joint Institute for High Temperatures of the Russian Academy of Science, 125412 Moscow, Russia; kashramiln@gmail.com (R.K.); lenkashapov@kpfu.ru (L.K.); nail.kashapov@kpfu.ru (N.K.)
3. Institute of Engineering, Kazan Federal University, 420008 Kazan, Russia
4. Kazan State University of Architecture and Engineering, 420043 Kazan, Russia; faritbox@mail.ru
5. N.I. Lobachevsky Institute of Mathematics, Kazan Federal University, 420008 Kazan, Russia; yaikovavictoriya@mail.ru
* Correspondence: 4works@bk.ru

Abstract: Porcelain fused to metal is widespread dental prosthetic restoration. The survival rate of metal-ceramic restorations depends not only on the qualifications of dentists, dental technicians but also on the adhesive strength of ceramics to a metal frame. The goal of the research is to determine the optimal parameters of the surface machining of the metal frame to increase the adhesion of metal to ceramics. Adhesion of cobalt-chromium alloy and ceramics was investigated. A profilometer and a scanning electron microscope were used to analyze the morphology. To estimate the adhesion the shear strength was measured by the method based on ASTM D1002-10. A method of surface microrelief formation of metal samples by plasma-electrolyte treatment has been developed. Regimes for plasma-electrolyte surface treatment were investigated according to current-voltage characteristics and a surface roughness parameter. The samples were subjected to different surface machining techniques such as polishing, milling, sandblasting (so-called traditional methods), and plasma-electrolyte processing. Morphology of the surface for all samples was studied and the difference in microrelief was shown. The roughness and adhesive strength were measured for samples either. As a result, the mode for plasma- electrolytic surface treatment under which the adhesive strength was increased up to 183% (compared with the traditional methods) was found.

Keywords: dental prosthesis; porcelain fused to metal; metal-ceramic; adhesion; profilometry; plasma-electrolyte processing; shear strength

1. Introduction

The modern development of dental science and practice makes it possible to maintain the dental health of the population at a high level. Despite effective preventive measures, new technologies of treatment and dental prosthetics, dental morbidity remains at a high level worldwide [1,2]. In most cases, dental rehabilitation requires the use of dentures to restore the integrity of the dentition [3–6].

One of the most common and traditionally used dental prosthetic restorations is non-removable fixed dentures, in particular porcelain fused to metal (PFM) dentures. Many years of experience in the use of these prostheses shows their effectiveness, high survival over time and improves the aesthetics and quality of life of patients [7–10]. Despite this, in clinical practice there are often failures and complications associated with both errors in complex clinical planning and violations of laboratory stages during their manufacture.

There are enough publications in medical literature devoted to the issues of preserving the vitality of teeth selected as supports for metal-ceramic crowns, decementation of prostheses, inflammatory phenomena of the dentoalveolar papillae and others [11–13]. All of these complications have a human factor (dentist—dental technician) and can be caused by the individual characteristics of the patient. Also the manufacturing technologies of dentures are not quite perfect and require constant improvement, the development of new high-quality and biologically inert materials.

As materials used in the manufacture of prostheses is porcelain which is applied to the metal surface. Glass ceramics are obtained by melting glass using the method of directed crystallization. It consists of silicon oxide, commonly known as quartz (SiO_2) with negligible aluminum content. The basic principle of obtaining a solid ceramic material is its molding when the molten glass cools. At the same time, during its sequential heating, a controlled crystallization process occurs, as a result of which crystals appear and grow. The process of transformation from pure glass to partially crystalline is called "ceramization".

Noble metal alloys are widely used in dentistry [14]. However, base-metal alloy is an economical alternative to expensive gold alloy [15]. Nickel-chromium (Ni-Cr) and cobalt-chromium (Co-Cr) alloys are most widely used when cost and rigidity are taken into account [16,17]. However, the immediate advantage of the Co-Cr alloy is comparable performance to other base metal alloys, but without an allergenic nickel component [18], while titanium has low quality of metal-ceramic bond [19]. In prosthetic dentistry special grades of stainless steels are used, the so-called alloy steels: for stamping AISI 321 or AISI 321, for casting 20Cr18N9S2. The composition of stainless steels includes: 72% iron, 0.12% carbon, 18% chromium, 9–10% nickel, 1% titanium, 2% silicon. Alloy steels contain a minimum amount of carbon (its increase leads to an increase in hardness and a decrease in ductility of steel) and an increased content of specially introduced elements that ensure the production of alloys with the desired properties. Chromium gives resistance to oxidation. Nickel is added to the alloy to increase ductility and viscosity. Titanium reduces brittleness and prevents intercrystalline corrosion of steel. Silicon is present only in injection-molded steel and improves its fluidity. Stainless steel has good malleability and poor casting qualities.

The main problem in the manufacture of PFM restorations is the problem of poor adhesion of the ceramic coating to the metal frame [20–22]. Increasing the adhesion of materials will significantly increase the service lifetime of dentures, reduce costs, which in turn will improve the quality of life of patients.

One of the criteria for good adhesion is the absence of impurities on the two surfaces that will be compared. With regard to ceramic coating, the issue is at a high stage of solution and the dental technician only needs to observe the technology of applying the mass to the metal frame by excluding the ingress of impurities from the ambient air of the dental laboratory, then with regard to the preliminary preparation of the metal frame itself, a number of issues arise that require more in-depth study [23,24]. In particular, this concerns the use of sandblasting technology of the metal frame at the stages of cleaning from the molding mass after casting, preliminary mechanical grinding with milling cutters and finishing with an abrasive sand with a size of 50 μm. Unfortunately, the last stage does not exclude the possibility of introducing abrasive sand particles into the metal frame, which in the future may cause poor local adhesion and cause chipping of the ceramic veneering.

In addition, the shape, sharpness of the peaks and cavities of the microrelief, size, depth of surface defects and internal defects determine the strength of the material [25,26]. A review of the literature has shown that common clinical problems with ceramic materials are chips, marginal fractures and fractures of the main mass [27,28]. Volumetric fractures are still one of the main causes of failures in the use of PFM prostheses, but literary sources also suggest the long-term survival of various ceramic prostheses [29–31].

Due to the presence of surface roughness, developed cracks may not spread randomly, but occur at points with a higher stress concentration. The theory that crack nucleation begins at points of stress concentration caused by surface roughness was proposed by Mecholsky et al. [32], who loaded samples with grooves and furrows both perpendicular and parallel to the direction of loading.

Separately, a number of works devoted to the study of the processing of mating surfaces can be noted. Thus, Imbriglio et al. [33] investigated aluminum oxide and silicon carbide. A negative correlation was shown between the reduced powder size and adhesion. At the same time, the adhesion for silicon carbide did not statistically depend on the reduced powder size, and, as a consequence, roughness (the range of roughness 0.2–0.4 μm was studied).

Budhe et al. [34] investigated the adhesion to shear strength between aluminum and wooden samples during various surface processing. Thus, for aluminum samples, the maximum adhesion was achieved with a roughness of about 2 μm, but with an increase or decrease in roughness, the amount of adhesion decreased. The relationship between shear strength and surface roughness for wooden samples was nonlinear, the maximum adhesion was achieved at a roughness of about 1.5 μm, and then with a decrease in roughness, adhesion decreased. Similar results for shear strength of aluminum and steel were obtained by Ghumatkar et al. [35]. Murat et al. [36] investigated the pull-off adhesion testing for a medium-density fiberboard lined with polyvinyl chloride, and a nonlinear relationship between the amount of adhesion and roughness was also shown.

The purpose of the study is to increase the adhesion of cobalt-chromium alloy and ceramics by surface machining technology. Regimes for plasma-electrolyte treatment were investigated to find optimal surface microgeometry to increase the adhesion.

2. Materials and Methods

2.1. Study Protocol

The contact surfaces of the samples were subjected to different surface machining techniques such as polishing, milling, sandblasting (so-called traditional methods), and plasma-electrolyte processing. The traditional method of surface machining was carried out using a milling cutter NTI 060. 14.2 mm (NTI, Alfeld, Germany), a sandblasting machine (Bego Easyblast, Bremen, Germany) using sand (Renfert Cobra, Hilzingen, Germany) with particles size 125, 90, 50 μm at 0.25 MPa air-abrasion pressure approximately 10 mm from the sample surface during 60 s.

After surfacing, the samples were labeled. Then morphological and elemental analyses were carried out. After that samples with ceramic layers were manufactured. Shear strength was determined for joint samples.

2.2. Manufacturing of Samples and Surface Machining

The wax patterns were fabricated using basic wax "Stoma" (Kharkiv Oblast, Ukraine). The cobalt chromium metal alloy "I-bond NF" (Interdent d.o.o., Slovenia) (Co 63%, Cr 24%, W 8%, Mo 3%, So 1%) was used to cast a metal framework using Bego Fornax T (BEGO Bremer Goldschlägerei Wilh. Herbst GmbH & Co. KG, Germany). The geometry of samples is shown in Figure 1a.

Then the mating/contact surfaces were processed (the blue area in Figure 1a). The metal frame was veneered with IPS Inline ceramic mass (Ivoclar vivadent, Germany), firing was carried out in a Programat E5000 furnace (Ivoclar vivadent, Germany) by following instructions of the manufacturer [37].

At the first stage of manufacturing PFM restoration, an opaque layer was applied to two separate samples (see Figure 1), which was then sintered in an oven under firing mode I in Table 1. After firing, the samples were cooled and treated with an Emmevi (Hansgrohe-Axor, Germany) steam jet in a pressure mode of 0.6 MPa. The second layer was re-applied with an opaque layer and the same treatment was carried out. The next step was to apply a layer of porcelain to connect the two samples according to the scheme shown in Figure 1b, and baked in a connected form in the oven in mode II (Table 1). The thickness of the opaque layer and ceramics are shown in Figure 1c. Two connected samples were cooled after sintering and subjected to steam jet treatment. At the last stage, a layer of dentin was applied along the edge of the junction of the two samples and baked in mode III (Table 1). After cooling, the sample was cleaned using a steam jet machine.

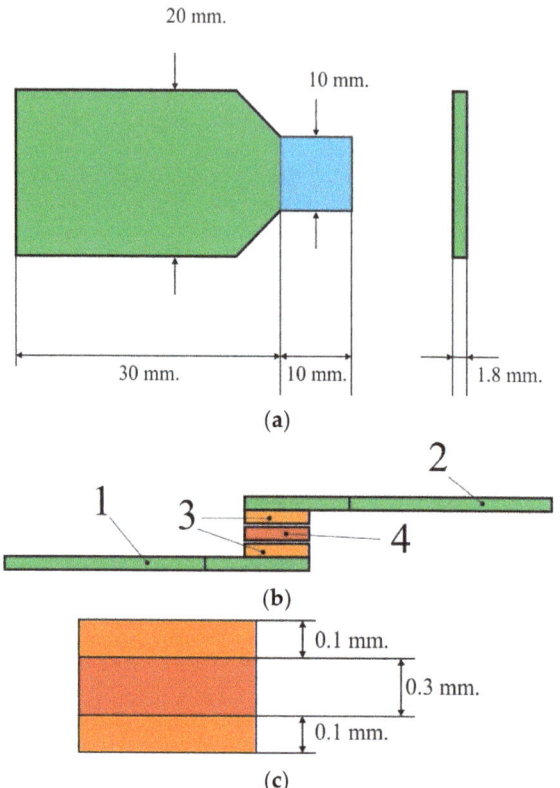

Figure 1. Sketch of the sample (**a**)—the sample, contacts surface is highlighted by blue; (**b**) the scheme of joint the samples: 1, 2—samples, 3—opaque layer, 4—ceramic layer; (**c**) the scheme of the joint layer: opaque and ceramic layers thickness.

Table 1. Firing parameters for ceramic layer.

Mode	T, °C	B, °C	S, min	t, °C/min	H, min	V$_1$, °C	V$_2$, °C
I	930	403	6	100	2	450	929
II	910	403	4	60	1	450	909
III	900	403	4	60	1	450	899

where T—firing temperature, B—stand-by temperature, S—closing time, t—heating rate, H—holding time, V$_1$—vacuum on temperature, V$_2$—vacuum off temperature.

2.3. The Technique of Plasma-Electrolyte Formation of the Microrelief of the Surface of Cobalt-Chromium Alloys

The method of surface microrelief formation of metal samples by plasma-electrolyte treatment consists of the use of gas discharges with liquid electrodes. The treatment process occurs as a result of the gas discharge combustion on the surface of a metal electrode dipped in an electrolyte solution. Depending on the polarity of the electrode and the type of material, various processes can occur: surface polishing, oxide ceramic coating formation, changes in surface roughness, application of metallic nanostructured coatings etc. To achieve our goal—control changes in roughness—it is necessary to use the cathode polarity of the active processed electrode.

The flat plate of the sample (9 in Figure 2) was fixed and immersed to a certain depth by the electrode system (3 in Figure 2). The work used a DC power supply (1 in Figure 2) with a continuously adjustable voltage, consisting of a diode bridge (diodes SD 246) and a laboratory autotransformer 1 M with a voltage range from 1 to 400 V (depending on the experimental conditions, a smoothing capacitor filter (C = 1560 µF) is connected to the power supply). Additional resistance (5 in Figure 2) also was added. The voltage and discharge current were measured using two digital universal measuring devices APPA 305 (6 in Figure 2) and APPA 109N (7 in Figure 2), the relative measurement error is 0.8%. An oscilloscope FLUKE scopemeter 190-062 (4 in Figure 2) was used to monitor the system. Electrolytic bath (2 in Figure 2) was filled with electrolyte—sodium chloride aqueous solution of concentration: 1%, 3% and 5% by weight. For each concentration of solution was carried out removal of the current-voltage characteristics of the plasma-electrolyte treatment. Temperature measurements were carried by thermocouple (8 in Figure 2).

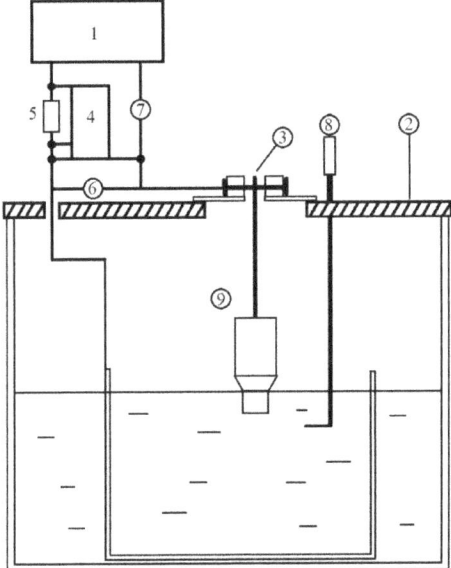

Figure 2. Scheme of the experimental installation of plasma-electrolyte formation of the microrelief of the surface: 1—electric power supply, 2—an electrolytic bath, 3—an electrode system, 4—an oscilloscope, 5—an additional resistance, 6—a voltmeter, 7—an ammeter, 8—a thermocouple and 9—a fixed sample.

The main parameters affecting the treatment process are the magnitude and shape of the applied voltage, the current strength of the discharge circuit and the temperature of the electrolyte. In the experiments, a smoothed voltage form obtained by using a capacitive filter was used, and by adjusting it, we changed the processing mode. The use of an active processed cathode polarity electrode leads to local melting of its surface under the action of randomly occurring single microdischarges. Depending on the discharge power and the temperature of the electrode itself, the formation of various microholes is observed, which in turn collectively form the overall surface roughness. The formation of microholes occurs as a result of melting of the surface and partial release of the electrode material into the electrolyte. Different treatment regimen forms different surface roughness. So a large number of samples for the case of plasma-electrolyte treatment is explained by the variety of choice of combustion regimes for discharges with liquid electrodes.

2.4. Morphological and Elemental Analysis

The study of the surface morphology was carried out with a scanning electron microscope (SEM) "EVO 50 XVP" (Carl Zeiss, Jena, Germany) with a probe microanalysis system "INCA Energy—350" (Oxford Instruments, Abingdon, UK). Roughness parameters were determined using a TR-200 profilometer (TIME GROUP Inc., Beijing, China). According to the curves of the surface profilograms, the values of the roughness parameters were calculated: Ra is the mean deviation of the profile; Rq is the mean square deviation of the profile. As a result of the research, it was found that the total error is less than 5%.

2.5. Apparent Shear Strength Determining

Researchers divide methods for determining adhesion into two groups: the pull-off strength measurement of adhesion and the apparent shear strength of single-lap-joint adhesively bonded specimens. Nowadays there are a number of standardized methods for their determination: ISO 4624:2002, ASTM C633-01, ASTM D7234-05, ASTM D1002-10, ASTM D4541, ISO 9693-1:2012. Mainly these standards were developed for paints and varnishes, for thermal spray coating materials, for rigid plastics, for metal to metal. For metal-ceramic materials, standardized methods have not been found during the study of this issue (except ISO 9693-1:2012). ISO 9693-1:2012 is focused on normal strength, but in the research the focus was on shear strength. It leads from the assumption that surface morphology critically influences shear [32]. Therefore, in this work, an approach to determine the apparent strength is based on the developed standards for adhesively bonded metal specimens. In this regard, for determining the apparent shear strength of the adhesive ceramic layer taken as a basis ASTM D1002-10 (Standard test method for apparent shear strength of single-lap-joint adhesively bonded metal specimens by tension loading) [38]. Similar approach was used in other researches [34–36].

The samples were tested on a universal testing machine UTS 110M-100 (Ivanovo, Russia). The range of measured loads is 0.001–100 kN, load measurement error less than 1% from the statement, up to 1/100 from the value of the permissible load. The test method is to obtain the value of the tensile load of failure of two samples bonded together with ceramics. The forces tending to shift one half of the sample relative to the other showed in Figure 3, strain rate was equal to 1 mm/min. To decrease the bending stresses during the shear test, the thickness of the metal frames was equal to 1.8 mm. Shear strength can be calculated by the following equation:

$$\tau = \frac{F}{A} \quad (1)$$

where F is the failure load (N), A is the area (mm^2) of the contact surface (ceramic joint).

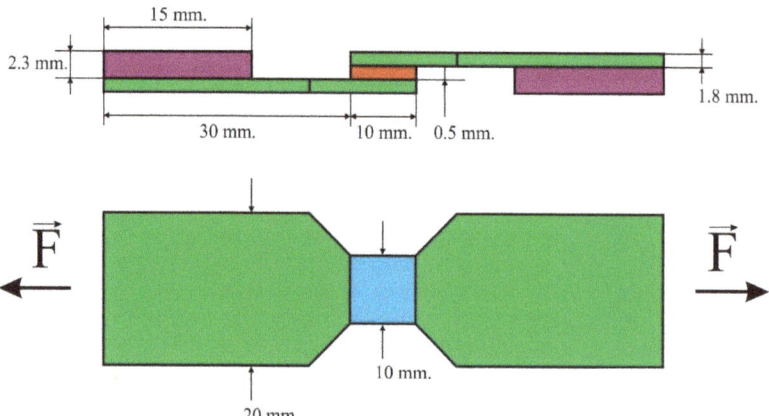

Figure 3. Sketch of the sample, dimensions and loading scheme: the ceramic layer is highlighted by red; the cobalt-chromium sample is highlighted by green; the centering plates highlighted by magenta; the contact surfaces highlighted by blue.

2.6. Statistical Processing of Experimental Data

Statistical analysis of the data was carried out using the MatLab software. The results are given in the following format: average ± half of the confidence interval ($p < 0.05$). Non-parametric measurements are given in the following format: median (Me) and interquartile range of 25–75 (Q1–Q2). A normal distribution check was carried out using the Jarque-Bera test. To compare groups two-sample t-test ($p < 0.05$) and Kolmogorov–Smirnov test were used. The interpolation of the data was carried out using the least-squares method.

3. Results

Samples were prepared according to the study protocol and divided by groups with the following marks: polishing ($n = 3$)—PL; milling ($n = 3$)—MC; abrasive blasting, with an abrasive size of 50 μm ($n = 12$), abrasive size 90 μm ($n = 3$), abrasive size 125 μm ($n = 3$)—AB50, AB90, AB125 respectively; plasma-electrolyte processing ($n = 16$)—PZ.

3.1. The Surface Microrelief Formation by Plasma-Electrolyte Treatment

During plasma-electrolyte treatment, the surface relief formation occurs due to the combustion of individual microdischarges that melt the surface. For local melting, it is necessary that at the ignition point of the gas discharge, the temperature of the cobalt-chromium alloy S&S Scheftner (Mainz, Germany) becomes higher than the solidus temperature of 1170 °C, preferably above the liquidus temperature of 1390 °C. The melting process is beginning when the solidus temperature is exceeded. A completely liquid metal can be obtained only at temperatures above 1390 °C. It is known that the casting temperature recommended by the manufacturer is 1490–1540 °C. At these temperatures the alloy has good fluidity and there is no burnout of alloying elements. Insufficient invested discharge power will not allow to achieve the required heating of the sample and lead to local melting with the formation of microholes. And a higher power can lead to overheating of the entire sample and its melting with loss of geometry, which is unacceptable. It should also be noted that the treatment proceeds more intensively on sharp and protruding surfaces, this is explained by the greatest intensity of the electric field in these places.

Thus, when processing samples, it is necessary to take into account their geometry, mass and properties of the alloy. With a sample thickness of a metal plate of 1.8 mm, surface melting occurred at a voltage of $U = 220$ V and a current strength of $I = 12$ A, and at lower voltage values, samples with a roughness parameter Ra from 0.77 to 2.51 μm were obtained.

At the initial moment of the treatment, the electrolyte is in contact with the surface of the metal sample. Depending on the magnitude of the applied voltage and the temperature of the electrolytic anode, one of the following processes will take place: only electrolysis; electrolysis eventually turning into the combustion of discharges on the surface of the sample; initial combustion of discharges without the stage of electrolysis. In the case of electrolysis, a linear dependence of the current strength on the voltage will be observed. According to the Lenz-Joule law, when an electric current passes through a metal electrode (cathode), the amount of heat released is directly proportional to the square of the current, the resistance of the electrode and the time during which the electric current flowed. An increase in voltage leads to an increase in current. This leads to an increase in the degree of cathode heating. On the cathode surface the process of boiling the electrolyte begins, in addition to the electrochemical release of gas bubbles. Combustion of discharges occurs in gas bubbles. The flow of the above processes is determined by the current-voltage characteristic (CVC) of the treatment, the behavior of which changes when the type of electrolyte, its concentration and temperature change. By changing the dependence of the current strength on the applied voltage, it is possible to determine the intensity of electrochemical reactions, vaporization near the surface of the metal electrode and combustion of single discharges.

Let us consider the CVC (Figure 4) of the plasma-electrolyte treatment of a cobalt-chromium alloy obtained using 1, 3 and 5% sodium chloride solutions. The CVC consists of two branches: the first is a gradual increase in voltage from 0 to 300 V, the second is a decrease in the applied voltage from 300 to 0 V.

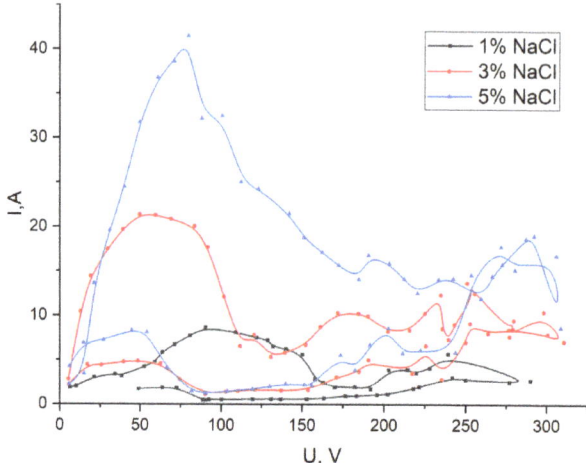

Figure 4. CVC of plasma-electrolyte treatment for NaCl solution: By grey line is 1%, red line is 3%, blue line is 5%.

For a 1% aqueous solution of sodium chloride, the combustion of single discharges with a pink-purple glow color occurs at a voltage of 120 V (grey line in Figure 4). When the applied voltage exceeds 100 V, local air-gas jets are observed departing from the cathode surface by 3–4 mm from the surface. On the CVC curve in this section, the beginning of a decrease in current strength is observed. A gradual increase in voltage leads to an increase in the number of single discharges. At a voltage of 177–181 V, local discharges of a different glow (yellow-orange) are additionally observed. At a voltage of 193 V, only yellow-orange glow discharges remain. The process of intensive heating of the electrode begins, it heats up to a temperature of 850–900 °C. At a given temperature of the electrode, a film boiling of the electrolyte will occur. Film boiling presupposes the presence of a stable vapor-air shell that covers the surface of the electrode. However combustion of discharges leads to a violation of the film boiling of the electrolyte and its splashing. The glow at the tip of the electrode turns white and corresponds to approximately a temperature of 1300 °C, in other areas of the surface it turns yellow and corresponds to a temperature of 1000 °C. Combustion of "yellow" discharges stops to prevail on the second branch of the CVC with a decrease in voltage to 216 V, mainly the combustion of electrical "pink-violet" discharges. Combustion of individual "yellow" discharges of low intensity occurs with a decrease in voltage to 163 V. When the voltage drops below 90 V, combustion stops and electrochemical reactions occur with boiling of the electrolyte near the electrode surface.

The use of a 3% aqueous solution of sodium chloride as an electrolyte changes the CVC curve while maintaining its outlines (red line in Figure 4). Namely, at 45 V—sound vibrations appear, at 60 V—the beginning of combustion of individual discharges on the cathode surface is observed. As the voltage increases, the number of discharges increases. The glow has a pink-purple color. At 100 V, larger yellow discharges appear. At 113 V, only yellow discharges burn. With increasing voltage, the intensity of radiation and acoustic vibrations increases. There are strong current fluctuations associated with intense splashing of the electrolyte near the cathode electrode. With a decrease in the applied voltage, the reverse transition occurs at 83 V. Local melting of the electrode surface is observed.

An increase in the concentration of sodium chloride in the electrolyte to 5% contributes to the displacement of the CVC to the region of lower voltages and an increase in the current strength (blue line in Figure 4). A voltage to 65 V increase leads to acoustic vibrations in the form of a crackle. At 67 V the combustion of individual discharges on the electrode surface begins. The color of the discharge radiation is yellow. At 88 V acoustic vibrations of a different frequency occur. With a decrease in voltage, the reverse transition from the combustion of discharges to the flow of only electrochemical reactions is observed at 63 V.

For all three CVCs with different concentrations of electrolyte, the presence of hysteresis is characteristic, with a reverse decrease in voltage, the current strength becomes significantly less. This is due to the heating of the electrolyte and the processed metal electrode. However, to a greater extent from the heating of the electrode, since it creates a film boiling that separates the electrolyte from the metal surface, and the absence of contact no longer allows electric current to be conducted due to electrochemical reactions. The current flow occurs only due to combustion of electrical discharges.

The obtained results of the CVC curves of plasma-electrolyte treatment indicate that a 3% solution of sodium chloride is the most optimal for use. With this sodium chloride content, there is no flow of high current densities for both 5% of the solution concentration and the risk of sample melting is reduced, and the use of 1% solution requires the use of higher voltage values to achieve the required energy supply. Based on the CVC curve for a 3% solution, the samples were processed to create roughness in the range from 120 to 200 V.

Varying plasma-electrolyte regimes generate different values of roughness. So, data of voltage, the temperature of the electrode, and the received roughness was interpolated by Delaunay triangulation (the natural neighbor interpolation method in Matlab). The obtained results are shown in Figure 5. The received interpolation allows selecting the plasma-electrolyte treatment mode to obtain a given roughness.

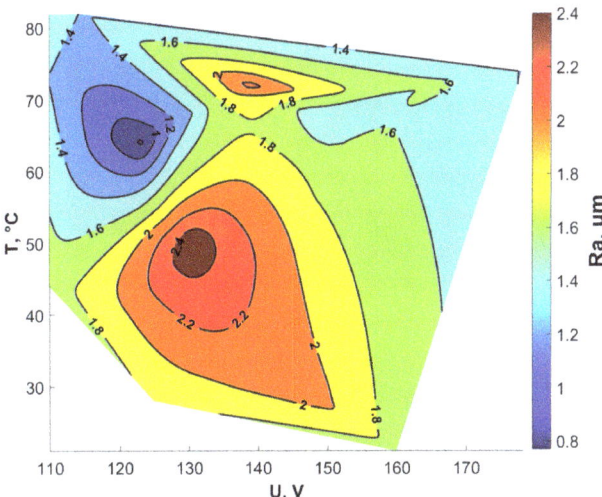

Figure 5. Roughness distribution depending on voltage and temperature of the electrode.

3.2. Analysis of the Microrelief of the Surface after Treatment

It should also be noted the gradient nature of the formation of the microrelief of the surface along the immersion of the sample in the electrolyte. The maximum size of the holes at the end of the sample, however, as they gradually move away from the edge, their size decreases. Using the example of changing the surface roughness parameters of the samples, it can be said that Ra in the longitudinal direction changed as follows: 2.734→2.064→1.728→1.375 μm.

Figure 6 shows pictures of the characteristic treated surface of the sample obtained by scanning electron microscopy. The images show that, depending on the depth of immersion of the electrode, the morphology of the surface changed. Conventionally, the surface can be divided into several sections, differing in morphology and size of microholes. The first region covers the end of the sample with a length of 2.5 mm, its morphology of the surface of the 1st region is very developed (Figure 6a). The diameter of microholes varies from ~3 μm to ~10 μm. Figure 6b shows the 2nd research area with 1000× magnification, which is 2.5 mm away from the edge of the sample, its surface morphology is developed. The diameter of microholes varies from ~1.5 μm to ~5 μm.

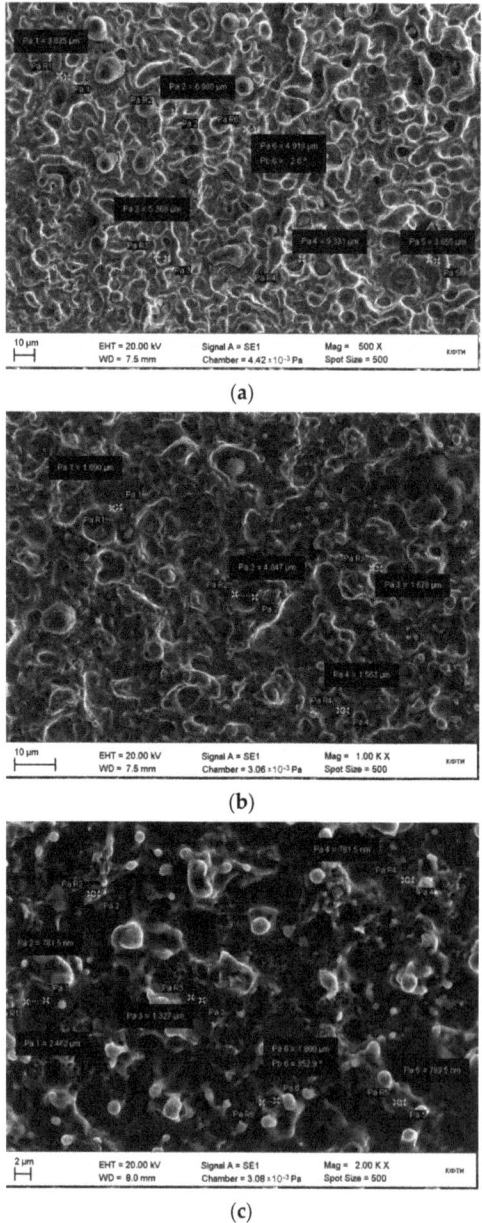

Figure 6. SEM images of the sample surface after treatment: (**a**)—the 1st research area with 500× magnification, (**b**)—the 2nd research area with 1000× magnification, (**c**)—the 3rd area of study of the sample with 2000× magnification.

Figure 6c shows the 3rd area of study of the sample with 2000× magnification, which is removed from the edge of the sample by 5 mm, its morphology is very developed. The diameter of microholes varies from ~500 nm to ~2 µm.

It is established that the magnitude of the applied voltage strongly affects the nature of the plasma processes and heat generation. At insufficient voltages, the processing effect is not observed, and at higher voltages, melting of the electrode may occur. The most optimal voltage range is 160–190 V. Also an interesting fact is the strong heating of the electrode and its glow to red when using a new electrolyte at room temperature at the initial moment of gas discharge combustion, but after a certain time the effect of incandescence of the cathode passes, and it acquires a natural color. At the end of the process, if the electrode is removed from the electrolyte in a heated state, without contact with water and its natural cooling in air, then it is covered with a dark oxide film. If the electrode is cooled in the electrolyte, there is a transition from film boiling to contact boiling, with local near-electrode boiling of the electrolyte, and the formation of a black oxide film does not occur.

Sandblasting and new plasma-electrolyte treatments were compared. The surface roughness parameter Ra of cobalt-chromium samples after various processing methods presented in Figure 7. Grouping in ascending order of Ra value: PL, MC, AB50, AB90, AB125. The PZ group was divided into three groups according to Ra values: PZ1, PZ2, and PZ3. The values of Ra in groups are described in details in Section 3.3.

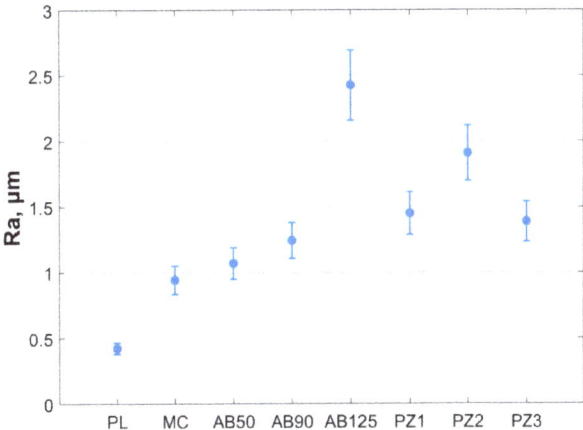

Figure 7. Surface roughness parameter distribution of samples, where bars represent mean (circle) and standard deviation (bars).

A comparison of the roughness parameters of standard surface treatment methods with plasma-electrolyte microrelief formation shows that this method is superior in its capabilities to sandblasting, milling and surface polishing. Thus, by adjusting the parameters of the processing process, it will be possible to obtain the required surface roughness.

A complex relationship exists between surface roughness and adhesion. As expected, when a larger particle size abrasive is used, the Ra value is seen to increase. However, the average roughness increases as the larger particles create higher peaks and deeper valleys in the surface profile [39]. Also in [39] it was shown that as the Ra value increases, the Rz and Rq values increase proportionally.

A comparative assessment of the surface morphology for various processing methods was carried out. SEM images of the surface for various processing methods presented in Figure 8. The surface after polishing is presented in Figure 8a–c. Irregularities in the form of cavities formed by the abrasive of the polishing tool are detected at 2000× and 3000× magnification. The surface after smearing the surface layer with a processing tool presented in Figure 8d–f. Surface after sandblasting using sand with a dispersion of 50 µm presented in Figure 8g–i. In these images can be observed the remains of sand particles pressed into the metal base. The morphology of the surface has many sharp protruding irregularities left as a result of the sliding of sand particles on the metal surface.

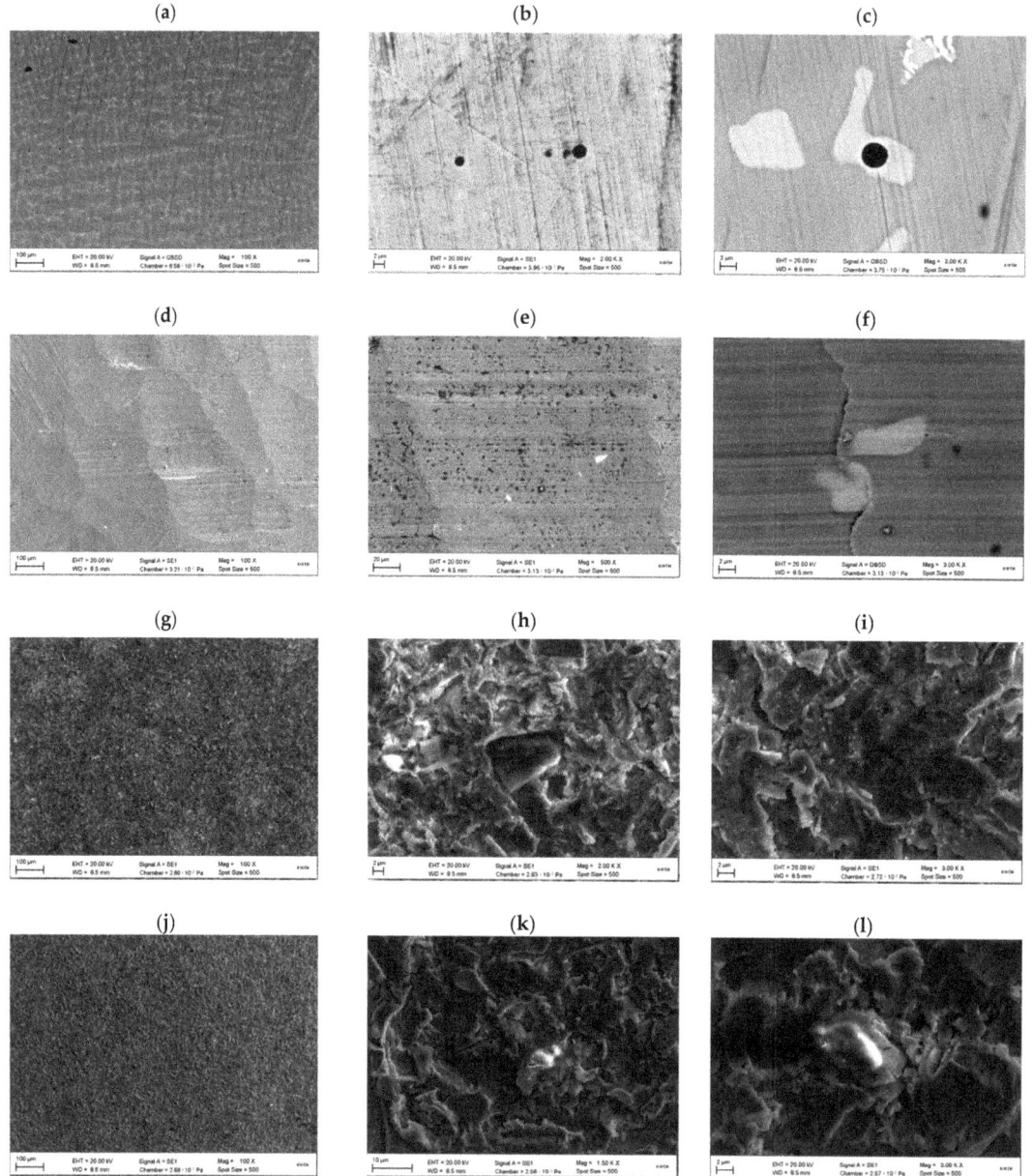

Figure 8. *Cont.*

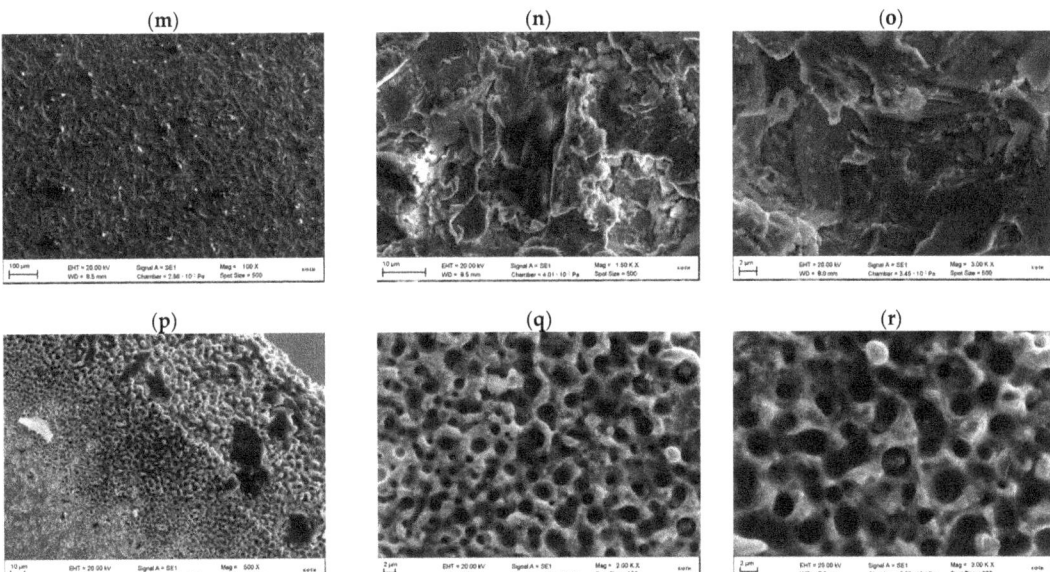

Figure 8. SEM images of the surface for various processing methods: (**a**–**c**)—surface after polishing (PL group); (**d**–**f**)—surface after milling (MC group); (**g**–**i**)—surface after sandblasting with an abrasive size of 50 μm (AB50 group); (**j**–**l**)—surface after sandblasting with an abrasive size of 90 microns (AB90 group); (**m**–**o**)—surface after sandblasting using with an abrasive size of 125 microns (AB125 group); (**p**–**r**)—surface after plasma-electrolyte treatment (PZ groups).

Figure 8j–l show images of the surface after sandblasting using sand with a dispersion of 90 μm. Figure 8m–o show images of the surface after sandblasting using sand with a dispersion of 125 μm. Sand particles are found on all the studied surfaces. The main difference is the size of the formed irregularities.

Figure 8p–r show images of the surface after plasma-electrolyte formation. A porous structure is observed, the protrusions have a spherical shape. The surface obtained by this method is fundamentally different in morphology from standard processing methods in the absence of sharp ledges, which later, when applying ceramics, can act as stress concentrators and lead to chipping of ceramics under cyclic loads.

3.3. Apparent Shear Strength

The distribution of shear strength by groups is presented in Figure 9. There was a significant decrease in adhesion strength for the PL group ($p < 0.05$). Shear strength in group AB90 was significantly higher than in group AB50 ($p < 0.05$) and insignificant higher than in group AB125. Groups AB50, AB125, and MC did not differ significantly. In plasma-electrolyte groups the shear strength of PZ2 was significantly higher than in PZ1 and PZ3. Ra and shear strength results in groups are presented in Table 2.

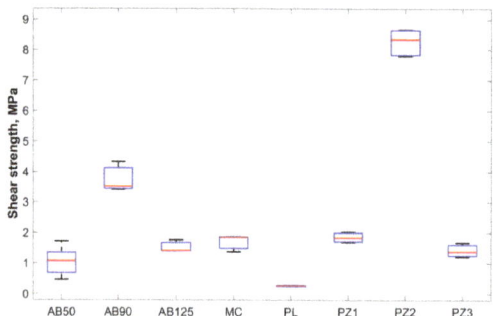

Figure 9. Shear strength distribution by groups, where are shown: the median (red line), the lower and upper quartiles (blue lines), and the minimum and maximum values (black lines) that are not outliers.

Table 2. Ra and shear strength results in groups.

Group	Ra, μm	τ, MPa
PL	0.422 ± 0.04	0.26 ± 0.05
MC	0.944 ± 0.11	1.69 ± 0.68
AB50	1.069 ± 0.12	1.05 ± 0.27
AB90	1.243 ± 0.13	3.75 ± 1.24
AB125	2.425 ± 0.26	1.52 ± 0.51
PZ1	1.136 ± 0.15	1.93 ± 0.10
PZ2	1.45 ± 0.16	8.35 ± 0.21
PZ3	1.91 ± 0.21	1.38 ± 0.07

Thus, when considering the relationship between the roughness parameter Ra and the shear strength for the PZ group a nonlinear relationship was noted. Due to the nonlinear dependence of shear strength on Ra for the PZ group, their averaging over the full data set is incorrect. So the data from the PZ group were divided according to the Ra values into three groups (see Figure 7). For the PZ1 group from PZ roughness Ra value was equal to 1.136 ± 0.15 μm and the shear strength was 1.93 ± 0.10 MPa, for the PZ2 group Ra was 1.45 ± 0.16 μm and the shear strength was 8.35 ± 0.21 MPa, for the PZ3 group Ra was 1.91 ± 0.21 μm and shear strength was 1.38 ± 0.07 MPa. Figure 10 shows the average values of shear strength and Ra value for all groups.

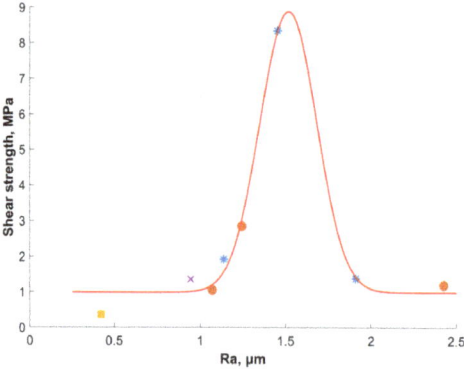

Figure 10. Distribution of average shear strength depending on Ra value, rectangle—PL, cross—MC, circle—AB (from left to right 50, 90, 125), stars—PZ (from left to right PZ1, PZ2, PZ3) and interpolation result—red line.

Assuming the homogeneity of all data, thereby excluding the influence of the chemical component and microrelief, it is possible to construct an exponential interpolation (red line in Figure 10) of the form:

$$\tau(Ra) = a + b \cdot \frac{1}{\sigma\sqrt{\pi}} Exp\left(\frac{(Ra - \mu)^2}{2\sigma^2}\right) \quad (2)$$

The parameters of interpolation (2) were as follows: $a = 0.984$ MPa, $b = 3.271$ MPa·μm, $\mu = 1.514$ μm, $\sigma = 0.165$ μm, error of interpolation was calculated as a squared norm of the residual (r^2) and was equal to 0.7564.

Due to the form of interpolation, the curve reaches a plateau in intervals of Ra values up to 1 μm and over 2 μm. In a strict sense, this fact illustrates a bad interpolation, but values of shear strength in these intervals are too small and these intervals are of no practical interest. Interpolation describes the results obtained quite qualitatively in the range of Ra values from 1 μm and up to 2 μm. Summarizing, the received interpolation and previous interpolation of roughness distribution depending on voltage and temperature of the electrode (see Figure 5) it is possible to select the appropriate mode of plasma-electrolyte treatment to obtain the required shear strength.

4. Discussion

The developed method of microrelief formation by plasma-electrolyte treatment allows to obtain the required surface roughness of metal frame by changing the parameters of gas discharge combustion. The microrelief arises due to combustion of individual microdischarges that melt the surface and create roughness. Size of microholes can be changed by controlling microdischarge parameters. This method of processing allows to obtain a clean surface, devoid of foreign impurities that may occur during abrasive processing. This method gives an opportunity to achieve the same results on the same parameters of plasma-electrolyte treatment. Also plasma-electrolyte treatment leads to the formation of a microrelief with spherical shape hollows and protrusions that can reduce the local stress concentration [40–42].

The main application of the plasma-electrolyte treatment is realized for surface polishing [43], heat treatment of products [44] and the formation of functional coatings [45]. In the presented work, the combustion of gas discharges with liquid electrodes was used to form a microrelief or controlled change in surface roughness. This application of the plasma-electrolyte treatment is achieved due to the possibility of local melting of the surface by single microdischarges. It should be mentioned that the study of the influence of the plasma-electrolyte treatment on the surface of the CoCr alloy was carried out on flat samples. Actually a dental crown has a complex geometry with different wall thicknesses. In the case of uniform burning of individual discharges on the surface of the dental crown, uneven heating of different parts of the crown is expected. But since the process takes place in an aqueous electrolyte solution, overheating of the CoCr alloy doesn't occur. However, there may be a change in the power of individual microdischarges depending on the geometry of the crown, which can lead to different roughness parameters, such as Ra. So, changes in roughness parameters for different geometries and types of metal frames of dental crowns should be investigated additionally.

For the proposed treatment method, as well as for classical methods, shear strength was measured. It was shown a nonlinear relationship between roughness parameters Ra and the shear strength value, such results correlate with other authors [34–36]. It was proposed to describe the obtained dependence by exponential interpolation. Such interpolation was received excluding the influence of the chemical component and microrelief over shear strength. Unlike plasma-electrolyte treatment, traditional methods (polishing, milling, and sandblasting) provide sharp geometry of hollows and protrusions.

As the result shear strength nonlinearly depends on the value of the roughness parameters. Similar trend of shear strength with respect to surface roughness was found by

Budhe et al. [34], Ghumatkar et al. [35] and Sekercioglu et al. [46]. Thus, the maximum shear strength appears near 1.5 µm Ra value for aluminum AA6061 [34] and steel S235JR—EN 10,025 [46]. Ghumatkar showed that the maximum shear strength appears near 2 µm Ra value for aluminum AA6063 and steel AISI1045 [35]. Comparing the dependence between roughness and shear strength for AA6061 and AA6063 it can be noted that maximum value remains about 5 MPa. Meanwhile the roughness (in terms of Ra) was about 1.75 µm for AA6061 and 2.25 µm for AA6063. Similar results obtained for AISI1045 and S235JR—EN 10,025—maximum shear strength localized at Ra values equal to 1.8 µm and 1.5 µm respectively. The influence of chemical adhesion with the frame microrelief is observed indirectly. It reflects on the absolute value of adhesion during the shear strength. So, the fluctuation of Ra value for maximum shear strength can be explained by chemical adhesion and the number of roughness points in the experiments.

A shear adhesion study showed that with standard metal surface treatments, the dispersion of test results was not big and it was sufficient to carry out tests on 3 samples. In contrast, plasma treated samples gave a large dispersion; so 16 samples were carried out. Subsequently, it turned out that the dispersion of results is greatly affected by temperature and current intensity.

So, the certain shear strength can be reached according to the received interpolation between mode and roughness and between roughness and adhesion. This increasing the value of the shear strength by 183% compared to the traditional method. The obtained results allow us to estimate the localization of the maximum shear strength depending on the roughness. Despite it, the determination of shear strength magnitude according to roughness and surface machining is still open, since it is influenced by other characteristics of the microrelief, chemical adhesion either.

The proposed method obtained samples with a value of Ra equal to 1.45 ± 0.16 µm and shear strength equal to 8.35 ± 0.21 MPa. The shear strength exceeds the same indicator for sandblasting by almost two times.

5. Conclusions

The high prevalence of dental diseases and untimely dental treatment leads to the loss of natural teeth. Missing or damaged teeth could be replaced by dental prosthesis, including fixed metal-ceramic restorations. The functional duration of dental prosthesis depends on many factors and one of them is the preservation of the integrity of the dental prosthesis by itself. Despite using PFM restorations various complications are possible such as decementation of crowns, the development of gingiva inflammation and chipping of the porcelaine layer. The reasons for the complications are patients' individual physiology, incorrect assessment of the clinical case, violation of clinical and laboratory stages, etc. These reasons include also technological imperfection of manufacturing of dental prosthesis and as a result the adhesion quality of metal frame veneering. Sandblasting is a widespread technology in dental laboratories, but it leads to impurities on the contact surfaces. And the impurities can negatively influence the adhesion. The formation of metal surface morphology and roughness affects adhesion, but nowadays there is no certain way to provide it.

A feature of the plasma-electrolyte treatment is the uniform microdischarges combustion over the entire surface of the crown. With intense combustion melting of the entire structure can occur. To avoid this, regimes were used at lower voltages and currents, which made it possible to implement microlocal melting of the surface without overheating the entire structure. In this regard, the purpose of the study was to determine the optimal parameters of the surface treatment of the metal frame to increase the adhesion of metal and ceramics and to develop a method for forming a microrelief of the surface.

To achieve the result, experimental samples were made of cobalt-chromium alloy. Metal frames were processed by 4 different methods: polishing, milling, sandblasting and plasma-electrolyte treatment. Ceramic layer was applied to the treated surface according to the manufacturer's recommendations. Morphological and profilometric studies of the samples allowed estimating the dependence between shear strength and the surface roughness. The developed method of microrelief formation by plasma-electrolyte treatment allows obtaining the required surface roughness of metal frame by changing the parameters of gas discharge combustion. At the same time, the geometry of the hollows and protrusions has a spherical shape that can reduce the local stress concentration. It was found that during sandblasting, abrasive sand particles remain on the metal surface and the geometry of the cavities and loads are sharp. According to the results of determining the CVC, a mode (159–178 V, 70–74 °C) was selected to obtain a given value of the roughness parameter Ra 1.45 ± 0.16 μm, which allowed increasing the shear strength of ceramics to the metal samples surfaces up to 8.35 ± 0.21 MPa.

Further research requires studying the effect of plasma-electrolyte microrelief formation on the resulting surface structure in terms of phase composition, microhardness, and Poisson's ratio. It is very important, since during the combustion of single microdischarges, the point of melting is rapidly cooled in the electrolyte. This cooling leads to an increase in microhardness. And most likely, in the study of microhardness from the surface into the depth of the sample, its gradient decrease will be observed. This will make it possible to create a buffer transition layer from ductile metal to brittle ceramic, thereby reducing the likelihood of ceramic chipping. Therefore, the logical continuation of this work is the study of changes in roughness parameters for different geometries and types of metal frames of dental crowns.

In the same time, the received results make it possible to state an optimistic forecast usage of the plasma-electrolyte processing for improving the adhesion. Of course pull-off strength should be measured for the same roughness values. The evaluation of dependence between normal strength and roughness is planned for future research. Additionally, cyclic tests are mapped out. It is also planned to carry out modeling of a dental prosthesis in the process of chewing by the finite element method.

The proposed method of surfacing allows increasing the shear strength between cobalt-chromium alloy and ceramics. Consequently such an approach increases the vitality of mounted dental construction. The usage of low voltages and currents allows the implementation of the developed method in dental labs and clinics. The plasma-electrolyte processing duration (about 1 min), allows to produce the surfacing during dental's appointment.

Author Contributions: Conceptualization, R.S. and G.S.; methodology, R.S., O.S. and R.K.; investigation, L.S., R.K., F.S., E.K., L.K. and N.K.; data curation, V.S. and O.S.; writing—original draft preparation, L.S., F.S., R.K. and O.S.; writing—review and editing, L.S., F.S., L.K., R.K. and O.S.; supervision, G.S. and R.S. All authors have read and agreed to the published version of the manuscript.

Funding: The results in Sections 2.1, 3.1 and 3.2 are obtained under support from the Russian Science Foundation (project No. 21-79-30062). The results in Sections 2.4, 2.6 and 3.2 are obtained under support from the Kazan Federal University Strategic Academic Leadership Program ("PRIORITY-2030").

Institutional Review Board Statement: Not applicable.

Informed Consent Statement: Not applicable.

Data Availability Statement: Not applicable.

Conflicts of Interest: The authors declare no conflict of interest to this work.

References

1. Kassebaum, N.J.; Smith, A.G.C.; Bernabé, E.; Fleming, T.D.; Reynolds, A.E.; Vos, T.; Murray, C.J.L.; Marcenes, W.; GBD 2015 Oral Health Collaborators; Abyu, G.Y.; et al. Global, Regional, and National Prevalence, Incidence, and Disability-Adjusted Life Years for Oral Conditions for 195 Countries, 1990–2015: A Systematic Analysis for the Global Burden of Diseases, Injuries, and Risk Factors. *J. Dent. Res.* **2017**, *96*, 380–387. [CrossRef] [PubMed]
2. Peres, M.A.; Macpherson, L.M.D.; Weyant, R.J.; Daly, B.; Venturelli, R.; Mathur, M.R.; Listl, S.; Celeste, R.K.; Guarnizo-Herreño, C.C.; Kearns, C.; et al. Oral Diseases: A Global Public Health Challenge. *Lancet* **2019**, *394*, 249–260. [CrossRef]
3. Eroshenko, R.E.; Stafeev, A.A. Analysis of the Prevalence of Dental Diseases Requiring Prosthodontic Treatment among the Rural Population of the Omsk Region. *Stomatologiya* **2018**, *97*, 9. [CrossRef] [PubMed]

4. Chisini, L.A.; Sarmento, H.R.; Collares, K.; Horta, B.L.; Demarco, F.F.; Correa, M.B. Determinants of Dental Prosthetic Treatment Need: A Birth Cohort Study. *Community Dent. Oral Epidemiol.* **2021**, *49*, 394–400. [CrossRef]
5. Nitschke, I.; Hahnel, S. Zahnmedizinische Versorgung älterer Menschen: Chancen und Herausforderungen. *Bundesgesundheitsblatt-Gesundh.-Gesundh.* **2021**, *64*, 802–811. [CrossRef]
6. Moeller, J.F.; Chen, H.; Manski, R.J. Diversity in the Use of Specialized Dental Services by Older Adults in the United States. *J. Public Health Dent.* **2019**, *79*, 160–174. [CrossRef]
7. Olley, R.C.; Andiappan, M.; Frost, P.M. An up to 50-Year Follow-up of Crown and Veneer Survival in a Dental Practice. *J. Prosthet. Dent.* **2018**, *119*, 935–941. [CrossRef]
8. Nejatidanesh, F.; Abbasi, M.; Savabi, G.; Bonakdarchian, M.; Atash, R.; Savabi, O. Five Year Clinical Outcomes of Metal Ceramic and Zirconia-Based Implant-Supported Dental Prostheses: A Retrospective Study. *J. Dent.* **2020**, *100*, 103420. [CrossRef]
9. Rammelsberg, P.; Lorenzo Bermejo, J.; Kappel, S.; Meyer, A.; Zenthöfer, A. Long-Term Performance of Implant-Supported Metal-Ceramic and All-Ceramic Single Crowns. *J. Prosthodont. Res.* **2020**, *64*, 332–339. [CrossRef]
10. Saker, S.; Ghazy, M.; Abo-Madina, M.; El-Falal, A.; Al-Zordk, W. Ten-Year Clinical Survival of Anterior Cantilever Resin-Bonded Fixed Dental Prostheses: A Retrospective Study. *Int. J. Prosthodont.* **2020**, *33*, 292–296. [CrossRef]
11. Galiatsatos, A.A.; Galiatsatos, P.A. Clinical Evaluation of Fractured Metal-Ceramic Fixed Dental Prostheses Repaired with Indirect Technique. Quintessence. *Int. Berl. Ger.* **2015**, *46*, 229–236. [CrossRef]
12. Pjetursson, B.E.; Valente, N.A.; Strasding, M.; Zwahlen, M.; Liu, S.; Sailer, I. A Systematic Review of the Survival and Complication Rates of Zirconia-Ceramic and Metal-Ceramic Single Crowns. *Clin. Oral Implant. Res.* **2018**, *29* (Suppl. 16), 199–214. [CrossRef]
13. Abakarov, S.I. Justification for Maintaining Vitality of Supporting Teeth When Using Ceramic and Metal-Ceramic Fixed Dentures. *Stomatologiya* **2021**, *100*, 52. [CrossRef]
14. Roberts, H.W.; Berzins, D.W.; Moore, B.K.; Charlton, D.G. Metal-ceramic alloys in dentistry: A review. *J. Prosthet. Dent.* **2009**, *18*, 188–194. [CrossRef]
15. Lopes, S.C.; Pagnano, V.O.; De Almeria Rollo, J.M.D.; Leal, M.B.; Bezzon, O.L. Correlation between metal-ceramic bond strength and coefficient of linear thermal expansion difference. *J. Appl. Oral Sci.* **2009**, *17*, 122–128. [CrossRef]
16. Moulin, P.; Degrange, M.; Picard, B. Influence of surface treatment on adherence energy of alloys used in bonded prosthetics. *J. Oral Rehabil.* **1999**, *26*, 413–421. [CrossRef]
17. Daftary, F.; Donovan, T. Effect of four pretreatment techniques on porcelain-to-metal bond strength. *J. Prosthet. Dent.* **1986**, *56*, 535–539. [CrossRef]
18. Park, W.U.; Park, H.G.; Hwang, K.H.; Zhao, J.; Lee, J.K. Interfacial Property of Dental Cobalt–Chromium Alloys and Their Bonding Strength with Porcelains. *J. Nanosci. Nanotechnol.* **2017**, *17*, 2585–2588. [CrossRef]
19. Czepułkowska, W.; Wołowiec-Korecka, E.; Klimek, L. The role of mechanical, chemical and physical bonds in metal-ceramic bond strength. *Arch. Mater. Sci. Eng.* **2018**, *1*, 5–14. [CrossRef]
20. Li, J.; Chen, C.; Liao, J.; Liu, L.; Ye, X.; Lin, S.; Ye, J. Bond Strengths of Porcelain to Cobalt-Chromium Alloys Made by Casting, Milling, and Selective Laser Melting. *J. Prosthet. Dent.* **2017**, *118*, 69–75. [CrossRef]
21. Saleeva, L.R.; Kashapov, R.N.; Kashapov, L.N.; Kashapov, N.F. Changes in the CoCr Alloys Surface Relief during Plasma Electrolytic Treatment. *IOP Conf. Ser. Mater. Sci. Eng.* **2019**, *570*, 012087. [CrossRef]
22. Vaska, K.R.; Nakka, C.; Reddy, K.M.; Chintalapudi, S.K. Comparative Evaluation of Shear Bond Strength between Titanium-Ceramic and Cobalt-Chromium-Ceramic: An in Vitro Study. *J. Indian Prosthodont. Soc.* **2021**, *21*, 276–280. [CrossRef]
23. Han, X.; Sawada, T.; Schille, C.; Schweizer, E.; Scheideler, L.; Geis-Gerstorfer, J.; Rupp, F.; Spintzyk, S. Comparative Analysis of Mechanical Properties and Metal-Ceramic Bond Strength of Co-Cr Dental Alloy Fabricated by Different Manufacturing Processes. *Materials* **2018**, *11*, 1801. [CrossRef]
24. Booth, J.A.; Hensel, R. Perspective on Statistical Effects in the Adhesion of Micropatterned Surfaces. *Appl. Phys. Lett.* **2021**, *119*, 230502. [CrossRef]
25. Menčík, J. *Strength and Fracture of Glass and Ceramics*; Elsevier: Amsterdam, The Netherlands; New York, NY, USA, 1992; ISBN 978-0-444-98685-6.
26. Fischer, H.; Schäfer, M.; Marx, R. Effect of Surface Roughness on Flexural Strength of Veneer Ceramics. *J. Dent. Res.* **2003**, *82*, 972–975. [CrossRef]
27. Molin, M.K.; Karlsson, S.L. A Randomized 5-Year Clinical Evaluation of 3 Ceramic Inlay Systems. *Int. J. Prosthodont.* **2000**, *13*, 194–200.
28. Krämer, N.; Frankenberger, R. Clinical Performance of Bonded Leucite-Reinforced Glass Ceramic Inlays and Onlays after Eight Years. *Dent. Mater.* **2005**, *21*, 262–271. [CrossRef]
29. Pallesen, U.; Van Dijken, J.W.V. An 8-Year Evaluation of Sintered Ceramic and Glass Ceramic Inlays Processed by the Cerec CAD/CAM System: Computer Processed Inlays. *Eur. J. Oral Sci.* **2000**, *108*, 239–246. [CrossRef]
30. Hayashi, M.; Wilson, N.H.F.; Yeung, C.A.; Worthington, H.V. Systematic Review of Ceramic Inlays. *Clin. Oral Investig.* **2003**, *7*, 8–19. [CrossRef]
31. Reiss, B.; Walther, W. Clinical Long-Term Results and 10-Year Kaplan-Meier Analysis of Cerec Restorations. *Int. J. Comput. Dent.* **2000**, *3*, 9–23.
32. Mecholsky, J.J. *Strength and Fracture of Glass and Ceramics, Glass Science and Technology*; Mencik, J., Ed.; Elsevier: Amsterdam, The Netherlands, 1992; Volume 12, 356p, ISBN 0-444-98685-5. *Adv. Mater.* **1993**, *5*, 224–225. [CrossRef]

33. Imbriglio, S.I.; Brodusch, N.; Aghasibeig, M.; Gauvin, R.; Chromik, R.R. Influence of Substrate Characteristics on Single Ti Splat Bonding to Ceramic Substrates by Cold Spray. *J. Therm. Spray Technol.* **2018**, *27*, 1011–1024. [CrossRef]
34. Budhe, S.; Ghumatkar, A.; Birajdar, N.; Banea, M.D. Effect of Surface Roughness Using Different Adherend Materials on the Adhesive Bond Strength. *Appl. Adhes. Sci.* **2015**, *3*, 20. [CrossRef]
35. Ghumatkar, A.; Budhe, S.; Sekhar, R.; Banea, M.D.; de Barros, S. Influence of Adherend Surface Roughness on the Adhesive Bond Strength. *Lat. Am. J. Solids Struct.* **2016**, *13*, 2356–2370. [CrossRef]
36. Kılıç, M.; Burdurlu, E.; Aslan, S.; Altun, S.; Tümerdem, Ö. The Effect of Surface Roughness on Tensile Strength of the Medium Density Fiberboard (MDF) Overlaid with Polyvinyl Chloride (PVC). *Mater. Des.* **2009**, *30*, 4580–4583. [CrossRef]
37. *IPS InLine Instructions for Use*; Ivoclar Vivadent AG: Schaan/Liechtenstein, Liechtenstein, 2014.
38. *ASTM D1002-10*; Standard Test Method for Apparent Shear Strength of Single-Lap-Joint Adhesively Bonded Metal Specimens. Tension Loading, Document Center Inc.: Silicon Valley, CA, USA, 2019.
39. Flanagan, J.; Schütze, P.; Dunne, C.; Twomey, B.; Stanton, K. Use of a blast coating process to promote adhesion between aluminium surfaces for the automotive industry. *J. Adhes.* **2018**, *96*, 580–601. [CrossRef]
40. Goryacheva, I.; Makhovskaya, Y. Combined Effect of Surface Microgeometry and Adhesion in Normal and Sliding Contacts of Elastic Bodies. *Friction* **2017**, *5*, 339–350. [CrossRef]
41. Losi, P.; Lombardi, S.; Briganti, E.; Soldani, G. Luminal Surface Microgeometry Affects Platelet Adhesion in Small-Diameter Synthetic Grafts. *Biomaterials* **2004**, *25*, 4447–4455. [CrossRef]
42. Makhovskaya, Y.Y. Adhesive Interaction of Elastic Bodies with Regular Surface Relief. *Mech. Solids* **2020**, *55*, 1105–1114. [CrossRef]
43. Zeidler, H.; Boettger-Hiller, F.; Edelmann, J.; Schubert, A. Surface Finish Machining of Medical Parts Using Plasma Electrolytic Polishing. *Procedia CIRP Open Access* **2016**, *49*, 83–87. [CrossRef]
44. Belkin, P.N.; Kusmanov, S.A.; Parfenov, E.V. Mechanism and technological opportunity of plasma electrolytic polishing of metals and alloys surfaces. *Appl. Surf. Sci. Adv.* **2020**, *1*, 100016. [CrossRef]
45. Qian, K.B.; Li, W.; Lu, X.; Han, X.; Jin, Y.; Zhang, T.; Wang, F. Effect of phosphate-based sealing treatment on the corrosion performance of a PEO coated AZ91D mg alloy. *J. Magnes. Alloy.* **2020**, *8*, 1328–1340. [CrossRef]
46. Sekercioglu, T.; Rende, H.; Gulsoz, A.; Meran, C. The effect of surface roughness on strength of adhesively bonded cylindrical components. *J. Mater. Process. Technol.* **2003**, *142*, 82–86. [CrossRef]

Article

The Effects of the Mechanical Properties of Vascular Grafts and an Anisotropic Hyperelastic Aortic Model on Local Hemodynamics during Modified Blalock–Taussig Shunt Operation, Assessed Using FSI Simulation

Alex G. Kuchumov [1,*], Aleksandr Khairulin [1], Marina Shmurak [1], Artem Porodikov [2] and Andrey Merzlyakov [3]

1. Department of Computational Mathematics, Mechanics, and Biomechanics, Faculty of Applied Mathematics and Mechanics, Perm National Research Polytechnic University, 614990 Perm, Russia; s.xayrulin@mail.ru (A.K.); shmurak2007@yandex.ru (M.S.)
2. Federal Center of Cardiovascular Surgery, 614990 Perm, Russia; porodikov.a@yandex.ru
3. Department of Continuum Mechanics and Computing Technologies, Faculty of Mechanics and Mathematics, Perm State National Research University, 614990 Perm, Russia; merzlyakov@psu.ru
* Correspondence: kychymov@inbox.ru; Tel.: +7-342-2-39-17-02

Abstract: Cardiovascular surgery requires the use of state-of-the-art artificial materials. For example, microporous polytetrafluoroethylene grafts manufactured by Gore-Tex® are used for the treatment of cyanotic heart defects (i.e., modified Blalock–Taussig shunt). Significant mortality during this palliative operation has led surgeons to adopt mathematical models to eliminate complications by performing fluid–solid interaction (FSI) simulations. To proceed with FSI modeling, it is necessary to know either the mechanical properties of the aorta and graft or the rheological properties of blood. The properties of the aorta and blood can be found in the literature, but there are no data about the mechanical properties of Gore-Tex® grafts. Experimental studies were carried out on the mechanical properties vascular grafts adopted for modified pediatric Blalock–Taussig shunts. Parameters of two models (the five-parameter Mooney–Rivlin model and the three-parameter Yeoh model) were determined by uniaxial experimental curve fitting. The obtained data were used for patient-specific FSI modeling of local blood flow in the "aorta-modified Blalock–Taussig shunt–pulmonary artery" system in three different shunt locations: central, right, and left. The anisotropic model of the aortic material showed higher stress values at the peak moment of systole, which may be a key factor determining the strength characteristics of the aorta and pulmonary artery. Additionally, this mechanical parameter is important when installing a central shunt, since it is in the area of the central anastomosis that an increase in stress on the aortic wall is observed. According to computations, the anisotropic model shows smaller values for the displacements of both the aorta and the shunt, which in turn may affect the success of preoperative predictions. Thus, it can be concluded that the anisotropic properties of the aorta play an important role in preoperative modeling.

Keywords: hemodynamics; modified Blalock–Taussig shunt; hyperelasticity; anisotropy; fluid–structure interaction

1. Introduction

The mortality rate of cardiovascular diseases in children is 5–16% globally [1]. Obstructive lesions of different segments of the right ventricular outflow tract (RVOT) are congenital heart defects, the treatment of which is possible only by surgical intervention. A modified Blalock–Taussig shunt (MBTS) is adopted as the first step in surgical treatment of RVOT. A shunt made of polytetrafluoroethylene (PTFE, Gore-Tex) is used to provide blood flow from the systemic circulation to the pulmonary circulation. In particular, the following pathologies are considered in the present study: pulmonary artery stenosis, pulmonary atresia, and Tetralogy of Fallot. However, MBTS operations cause a number of complications, such as excessive volumetric load, acute thrombosis, and low diastolic blood pressure leading to coronary insufficiency. Modern modeling techniques can be applied to predict and evaluate complications in the postsurgical period.

Computational modeling is used to study the local hemodynamics of MBTS surgery. The first such studies using numerical simulation were conducted in [2,3] using the computational fluid dynamics (CFD) method. The shunt size, its location, and its angles of proximal anastomoses for an idealized geometry were considered in these works. A similar study was [4]. CFD is a method determining the parameters of flow in a channel with solid walls. It allows for a detailed description of the local flow in a complex geometry. The hemodynamic effects of nonclosure of patent ductus arteriosus [5] on pulmonary blood flow were analyzed using the CFD method. The degree of shunt occlusion after MBTS surgery using CFD was determined by Arthurs et al. [6]. MBTS and right ventricle–pulmonary artery shunts were compared in [7,8]. Liu et al. conducted a series of studies [9–12] of MBTS shunts using real geometry to analyze the hemodynamic parameters of local blood flow, including the degree of shunt occlusion. In [10,13], the effect of MBTS location on blood flow distribution was investigated. Using the CFD method, the hemodynamics of the central shunt and modified Blalock–Taussig shunt were compared [14].

Vascular walls are flexible, which can affect the hemodynamics. The fluid–solid interaction (FSI) method is another method of computer modeling. This method, as opposed to the CFD method, takes into account the wall elasticity of the computational flow area. This numerical method is used to study the parameters of blood flow for different problems. Luo et al. [15] studied fluid flow in aortic and iliac arteries bifurcations with the use of FSI simulation. Stergiou et al. [16] investigated the occurrence and development of abdominal aortic aneurysm using the FSI method. Sousa et al. [17] analyzed the hemodynamics of the stenosis carotid bifurcation by FSI simulation. The deformability/elasticity of the vessel wall is an important factor in the study of blood flow. FSI simulation has not yet been used to study MBTS surgery. Therefore, this method was used in our study.

In this study, the shunt walls were modeled as an isotropic hyperelastic material, while the aortic walls were modeled as both anisotropic and isotropic elastic material. An experimental study of Gore-Tex shunts was performed to determine the parameters of the hyperelastic material of the shunt. FSI modeling of MBTS bypass surgery is an actual research method and has not been used before. The local distribution of velocities, pressures, wall shear stresses, and displacements of the aortic wall and shunt were analyzed in this study. The aim of the study was to show the effect of anisotropy on the main characteristics of local hemodynamics for the MBTS surgery. Results for the three methods for different shunt locations (central, right, and left) were obtained in this study.

2. Materials and Methods

Complex research, including experimental and computational studies, was carried out. The experimental part included tensile tests of artificial vascular grafts to determine mechanical properties and constitutive relation constants. FSI simulations of blood flow in the aorta after modified Blalock–Taussig shunt surgery were also performed. The results were analyzed for different shunt locations. The influence of the aortic mechanical properties on the local hemodynamics was studied.

2.1. Experimental Study on Mechanical Properties of Grafts

The study investigated the mechanical properties of Gore-Tex vascular grafts. These are most frequently used in surgical practice for shunt placement, including the modified Blalock–Taussig shunt operation. The Gore-Tex vascular graft is a tube made of a special material. It is a waterproof and vaporizable membrane. Constants of the strain energy density function for the hyperelastic material were obtained as a result of the experiments. The influence of the wall mechanical properties on the blood flow in an artificial vessel was evaluated considering the interaction between the vessel wall and the blood flow. The study employed the methods of computational fluid dynamics and mechanics. Calculation results were obtained for elastic and hyperelastic walls. The research considered local hemodynamics in the aorta with regard to anisotropy and hyperelastic properties; FSI modeling methods were applied. The results were analyzed to deduce the main factors influencing the local blood flow.

Gore-Tex shunts (W. L. Gore & Associates, Inc., Flagstaff, AZ, USA) of different sizes and thickness were used for tests (Figure 1). Currently, PTFE shunts are actively used in surgery. The test temperature was 37 °C. Experiments were performed under different loading to assess the effect of load rate on deformation.

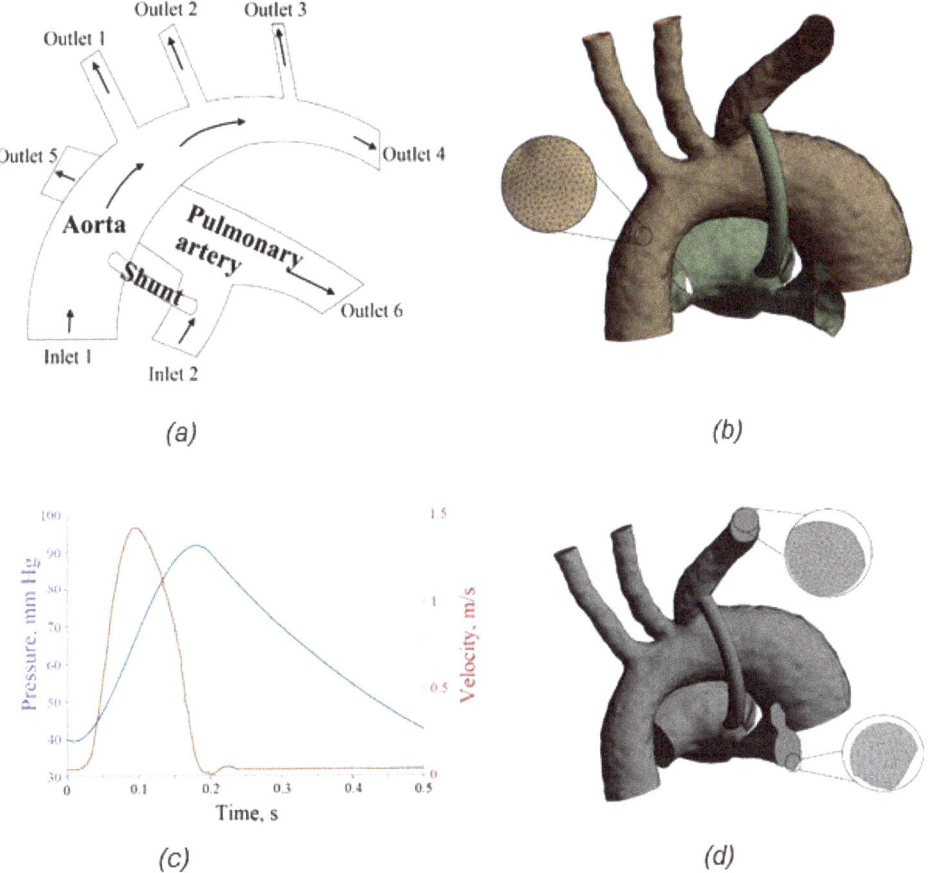

Figure 1. Meshes and boundary conditions of the aorta–shunt–pulmonary artery system: (a) boundary conditions, (b) solid mesh, (c) velocity and pressure profiles, (d) aorta fluid mesh model.

Most human tissues behave nonlinearly under a load. Such materials are called hyperelastic. A strain energy density function is used to describe the behavior of hyperelastic materials under a load.

The materials are incompressible. The results of the experimental study can be described by the five-parameter Mooney–Rivlin model:

$$W = c_{10}(\overline{I_1}-3) + c_{01}(\overline{I_2}-3) + c_{20}(\overline{I_1}-3)^2 + c_{11}(\overline{I_1}-3)(\overline{I_2}-3) + c_{02}(\overline{I_2}-3)^2 + \frac{1}{D_1}(J-1)^2 \tag{1}$$

and the three-parameter Yeoh model:

$$W = \sum_{i=1}^{3} c_{i0}(\bar{I}_1 - 3)^i, \tag{2}$$

where c_{ij} are material constants; $\bar{I}_1, \bar{I}_2$ are the first and the second invariant of the deviatoric strain tensors:

$$\bar{I}_1 = \lambda_1^2 + \lambda_1^2 + \lambda_1^2, \tag{3}$$

$$\bar{I}_2 = \lambda_1\lambda_2 + \lambda_2\lambda_3 + \lambda_1\lambda_3, \tag{4}$$

$$\bar{I}_3 = \lambda_1^2\lambda_1^2\lambda_1^2, \tag{5}$$

where λ_i are principal stretches in Equations (3)–(5). The materials are incompressible; $\bar{I}_3 = 1$.

The constants in the strain energy density function (Equations (1)–(2)) were determined in the ANSYS Workbench software (Ansys Workbench 18, Ansys Inc., Canonsburg, PA, USA) using a curve fitting procedure based on the experimental tensile diagrams obtained for the sample. The five-parameter Mooney–Rivlin model (Equation (1), Equations (3)–(5)) and the three-parameter Yeoh model (Equations (2)–(5)) are used to describe the behavior of mechanical properties of grafts in this study.

2.2. Mechanical Properties of Aorta

2.2.1. Ogden Model for Description of Isotropic Hyperelastic Behavior of Aorta

In the Ogden material model (Equation (6)), the strain energy density is expressed in terms of the principal stretches as:

$$W(\lambda_1, \lambda_2, \lambda_3) = \sum_{p=1}^{N} \frac{\mu_p}{\alpha_p}\left(\lambda_1^{\alpha_p} + \lambda_2^{\alpha_p} + \lambda_3^{\alpha_p} - 3\right), \tag{6}$$

where N, μ_p, and α_p are material constants. Under the assumption of incompressibility, one can rewrite this as

$$W(\lambda_1, \lambda_2) = \sum_{p=1}^{N} \frac{\mu_p}{\alpha_p}\left(\lambda_1^{\alpha_p} + \lambda_2^{\alpha_p} + \lambda_1^{-\alpha_p}\lambda_2^{-\alpha_p} - 3\right). \tag{7}$$

In general, the shear modulus is calculated as follows:

$$2\mu = \sum_{p=1}^{N} \mu_p \alpha_p, \tag{8}$$

where $N = 3$; by fitting the material parameters, the material behavior of rubbers can be described very accurately. For particular values of material constants, the Ogden (Equation (7)) model will reduce to either the neo-Hookean solid ($N = 1$, $\alpha = 2$) or the Mooney–Rivlin material ($N = 2$, $\alpha_1 = 2$, $\alpha_2 = -2$ with the constraint condition $\lambda_1\lambda_2\lambda_3 = 1$).

2.2.2. Holzapfel–Gasser–Ogden Model for Description of Anisotropic Hyperelastic Behavior of Aorta

The simplified form of the strain energy potential is based on that proposed by Holzapfel, Gasser, and Ogden [18] for modeling arterial layers with distributed collagen fiber orientations:

$$W = C_{10}(\bar{I}_1 - 3) + \frac{k_1}{2k_2}\sum_{\alpha=1}^{N}\{\exp[k_2\bar{E}_\alpha^2] - 1\}, \tag{9}$$

with

$$\bar{E}_\alpha = \kappa(\bar{I}_1 - 3) + (1 + 3\kappa)(\bar{I}_{4(\alpha\alpha)} - 1), \tag{10}$$

where W is the strain energy per unit of reference volume; C_{10}, κ, k_1, and k_2 are temperature-dependent material parameters; N is the number of families of fibers (N ≤ 3); $\bar{I}_1$ is the first deviatoric strain invariant; and $\bar{I}_{4(\alpha\alpha)}$ are pseudo-invariants of $\bar{C}$ and $\bar{E}_\alpha$.

The model (Equations (9)–(10)) assumes that the directions of the collagen fibers within each family are dispersed (with rotational symmetry) about a mean preferred direction. The parameter κ ($0 \leq \kappa \leq 1/3$) describes the level of dispersion in the fiber directions. If $\rho(\Theta)$ is the orientation density function that characterizes the distribution (it represents the normalized number of fibers with orientations in the range $[\Theta, \Theta + d\Theta]$ with respect to the mean direction), the parameter κ is defined as

$$\kappa = \frac{1}{4} \int_0^\pi \rho(\Theta) \sin^3 \Theta d\Theta. \tag{11}$$

It is also assumed that all families of fibers have the same mechanical properties and the same dispersion (Equation (11)). When $\kappa = 0$, the fibers are perfectly aligned (no dispersion). When $\kappa = 1/3$, the fibers are randomly distributed and the material becomes isotropic; this corresponds to a spherical orientation density function.

In this study, isotropic (Equations (7)–(8)) and anisotropic (Equations (9)–(11)) Holzapfel–Gasser–Ogden models are applied to describe the hyperelastic properties of the aorta.

2.3. FSI Simulations of Blood Flow in the Aorta–Pulmonary Artery–Shunt System

2.3.1. Problem Formulation

A concomitant pathology of congenital heart disease in children is impaired pulmonary circulation. This leads to abnormal lung growth and insufficient oxygenation. The use of a modified Blalock–Taussig shunt (MBTS) in such cases is one of the most common methods for eliminating pathology. Biomechanical modeling methods are used to objectify the choice of shunt parameters. An individualized three-dimensional model of the aorta–pulmonary artery–shunt system based on CT images (computed tomography) with contrast was built to analyze the local hemodynamics.

The study's protocol was approved by the Ethics Committee of the Perm Federal Center of Cardiovascular Surgery (Protocol No. 12 on 25 October 2021). Informed consent was obtained from parents of patients involved in the study.

Three-dimensional (3D) anatomical data were obtained via a 64-channel, dual-source multidetector-row CT scanner (Siemens Somatom Definition AS, Forchheim, Germany) with a 0.6-mm slice thickness and 0.6-mm slice interval, a 0.5 s rotation time, and a pitch of 0.25 (Gantry opening is 70 cm; the number of reconstructed slices is 67). The tube current was adjusted according to the body weight. Before CT examination, all patients were sedated. Intravenous propofol was given by anesthesiologist at a dose of 1–2 mg/kg body weight for induction. In some cases, the dose was increased to maintain sedation. Anatomical coverage extended from above the thoracic inlet to below the level of the L2 vertebra, including the origin of the celiac trunk. For vascular opacification, a non-ionic low-osmolar contrast agent containing 350 mg/mL was injected through the peripheral vein (the right ulnar vein if it was accessible). Contrast was administered with a mechanical injector at a dose of 2 mL/kg body weight. A flow rate of 1–2 mL/s was used, depending on the size and location of the venous access, as well as the size of the cannula used. Postprocessing was carried out on a dedicated workstation Singo via (Siemens Healthcare GmbH, Erlangen, Germany). Image reconstruction was performed in 3D volume rendering (VRT) maximum intensity projection (MIP) of multiplanar reconstruction (MPR) in coronary, sagittal oblique views.

Three variants of a modified Blalock–Taussig shunt were considered: the central, connecting the aorta with the pulmonary artery trunk; the right, connecting the left subclavian artery and the right pulmonary artery; and the left, connecting the brachiocephalic trunk and the left pulmonary artery.

The results obtained for two modifications of the model were compared: the first model the so-called "simplified" model accounting for the aortic wall's isotropy and the shunt elastic properties; the second model accounting the aortic wall's anisotropy and the shunt hyperelasticity. The calculations were performed in Ansys Workbench software (Ansys Workbench 18, Ansys Inc., Canonsburg, PA, USA). The two-way fluid–solid interaction problem of blood flow in the aorta–pulmonary artery system in children was solved.

2.3.2. Mathematical Problem Statement

The mass and momentum conservation equations (Equations (12)–(13)) for an incompressible fluid (Equation (14)) can be expressed as

$$\rho_f \left(\frac{\partial u}{\partial t} + (u \cdot \nabla) u \right) = \nabla \cdot \sigma \tag{12}$$

$$\nabla \cdot u = 0, \tag{13}$$

$$\sigma = -pI + \tau \tag{14}$$

$$\tau = \eta(\dot{\gamma}) D, \tag{15}$$

where ρ_f is the fluid density, p is the pressure, u is the fluid velocity vector, and u_g is the moving coordinate velocity. In the arbitrary Lagrangian–Eulerian (ALE) formulation, $(u - u_g)$ is the relative velocity of the fluid with respect to the moving coordinate velocity. Here, τ is the deviatoric shear stress tensor (Equation (15)). This tensor is related to the velocity through the strain rate tensor; in Cartesian coordinates it can be represented as follows:

$$D = \frac{1}{2}(+\nabla u^T). \tag{16}$$

The momentum conservation equation for the solid body can be written as follows:

$$\nabla \cdot \sigma_s = \rho_s \ddot{u}_s, \tag{17}$$

where ρ_s, σ_s, and $\ddot{u}_s$ are the density, stress tensor, and local acceleration of the solid, respectively.

It is known that blood vessels can be described as hyperelastic materials [18–20]. Because of a similar anatomical composition, the bile ducts can also be considered hyperelastic materials. For hyperelastic materials, the stress–strain relationship is written as follows:

$$\sigma_s = \frac{\partial W}{\partial \varepsilon}, \tag{18}$$

where ε is the strain tensor and W is the strain energy density function. The Mooney–Rivlin hyperelastic potential is shown in Equation (1).

The mathematical statement of blood flow in the aorta–shunt–pulmonary artery system is governed by Equations (12)–(18).

The FSI interface should satisfy the following conditions:

$$x_g = x_s \tag{19}$$

$$u_g = u_s \tag{20}$$

$$\sigma_g \hat{n}_g = \sigma_s \hat{n}_s. \tag{21}$$

The displacements of the fluid and solid domain should be compatible, as in Equation (19). The tractions at this boundary must be at equilibrium (Equation (21)). The no-slip condition for the fluid should satisfy Equation (20). In the above conditions, Equations (19)–(21) give the displacement, stress tensor, and boundary normal, respectively. The subscripts f and s indicate fluid and solid parts, respectively. Blood is assumed to be a Newtonian fluid. The blood density is equal to $\rho = 1060$ kg/m^3; the dynamic viscosity is constant and equal to $\mu = 0.0035$ Pa·s. The velocity profile during the systolic and diastolic phases of the left ventricle was applied at the aortic root inlet (Figure 1). The left ventricular systole period is t = 0.22 s. The period of ventricular diastole is t = 0.28 s. The total cardiac cycle is t = 0.5 s. The peak velocity is 1.4 m/s. A time-dependent pressure profile was used as the boundary conditions at the aortic outlets. Constant pressure of P = 20 mm Hg was applied at the pulmonary artery outlets.

2.3.3. Mesh and Convergence

The computational mesh of the fluid domain was generated using the Body Sizing and Inflation tools, respectively. Body Sizing allows one to set the mesh item type and size. The Inflation tool allows one to thicken the mesh in the near-wall regions to further reveal the near-wall effects (Figure 1a). The computational mesh for the solid domain was selected based on the study of the mesh convergence of the results.

Five different element sizes were selected to analyze the sensitivity to the grid density (Table 1). The element types used in all grids were hexahedral and tetrahedral. An analysis of the sensitivity to the mesh density was carried out based on the achievement of a relative difference ε_P^{min} = 0.21%, ε_V^{min} = 0.84% of the variation of the maximum values of pressure and velocity in the aorta–shunt–pulmonary artery system. Figure 2 shows a convergence plot for von Mises stress. The results of the study showed that the values of the maximum stress values for a coarse mesh differ significantly from a thickened mesh. Thus, for subsequent calculations, a denser mesh with a side size of a triangular finite element h = 0.2 mm was used.

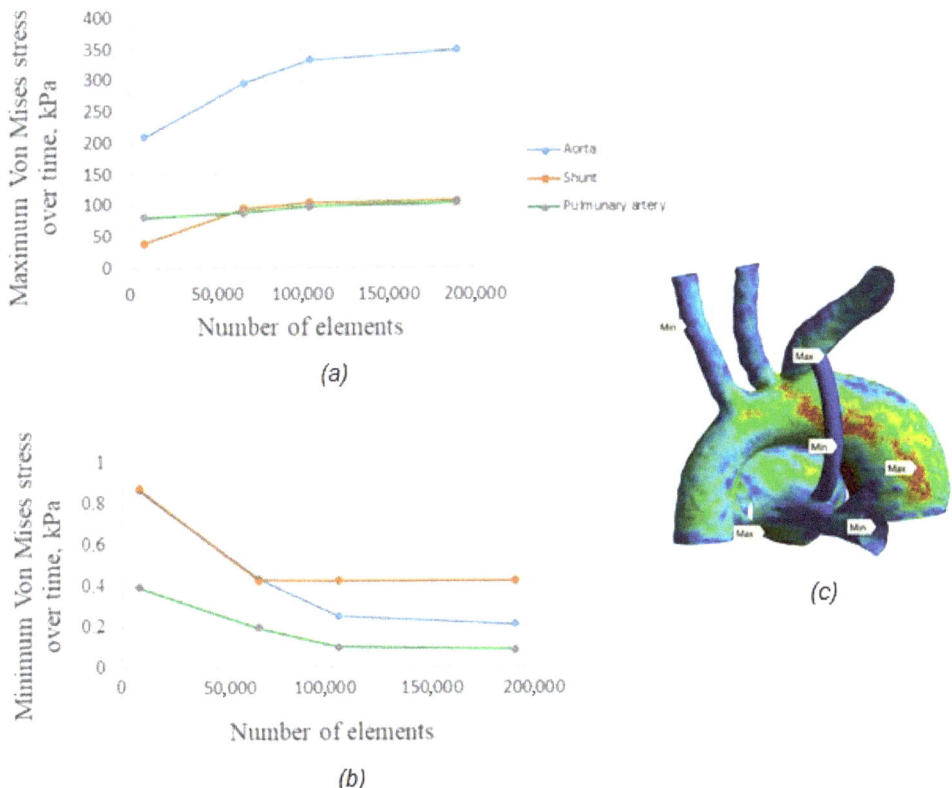

Figure 2. Maximum stress variations with respect to the number of mesh elements for solid domain: (**a**) mesh dependency tests for maximum von Mises stress, (**b**) mesh dependency tests for minimum von Mises stress, (**c**) maximum and minimum values for aorta, shunt, and pulmonary artery.

Table 1. Parameters of samples used in the study.

No.	Body Sizing, mm	Inflation			Number of Elements	Maximum Pressure, Pa	Maximum Velocity, m/s
		Transition Ratio	Maximum Layers	Growth Ratio			
1	0.95	0.5	3	1.2	80,353	17,634	3.85
2	0.8	0.4	5	1.4	159,379	17,952	4.38
3	0.63	0.3	7	1.3	349,926	17,892	4.47
4	0.5	0.35	8	1.6	709,578	18,470	4.71
5	0.38	0.32	10	1.3	1,544,745	18,509	4.75

3. Results

3.1. Results of the Experimental Study

Tensile and rupture tests were carried out. Various factors were analyzed, including the loading rate and geometric dimensions of the specimen. As a result of the experiment, the modulus of elasticity was determined for different specimens (Table 2). The elasticity modulus strongly depends on the diameter and thickness of the shunt: for a thickness over 0.5 mm, its value increases several times, and for a thickness less than 0.35 mm, it decreases strongly. The stress–strain curve for a shunt with a diameter of 4.5 mm, wall thickness of 0.35 mm, and length of 20 mm by a rupture test is shown in the Figure 3. The load rate was 30 mm/min and the preload was 0.5 MPa.

Table 2. Parameters of samples used in the study.

Sample Number	E (MPa)	Diameter, d (mm)	Wall Thickness (mm)
1	7.41	4.32	0.34
2	9.8	3.4	0.4
3	10.3	4.5	0.35
4	11.1	5.5	0.48
5	43.5	5	0.53

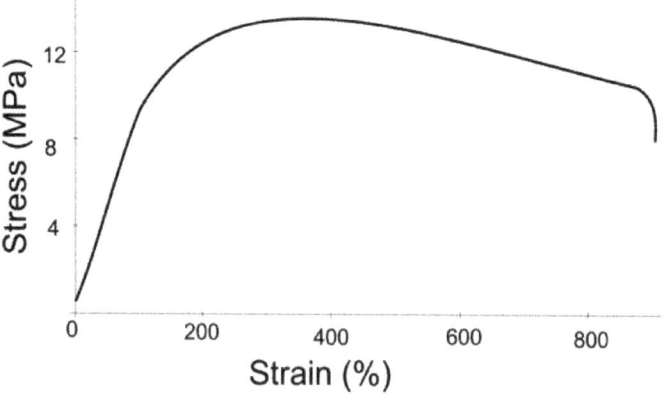

Figure 3. Stress–strain diagram by specimen rupture test.

The tensile ultimate strength σ_Y was also determined (Table 3); its value increases as the specimen diameter increases (Figure 4). The influence of loading rate on tensile strength σ_Y determination was analyzed. The tensile strength value remained practically unchanged when the load rate application changed from 50 to 250 mm/min. The shape of the tensile test curve for all specimens was the same. Stress–strain relationships were obtained as a result of tensile tests for two specimens (specimen no. 1, diameter of 5 mm, thickness of

0.5 mm, length of 20 mm; specimen no. 2, diameter of 3 mm, thickness of 0.35 mm, length of 20 mm). The constants for the strain density function were determined from the obtained dependences (Table 4) and stress–strain dependences were plotted (Figure 5).

Table 3. Mechanical properties of samples.

Sample Number	σ_Y (MPa)	Diameter, d (mm)	Wall Thickness, (mm)	Loading Rate, (mm/min)
1	11.6	3.4	0.4	30
2	13.6	4.5	0.35	30
3	14.5	5	0.53	30
4	17.0	6.2	0.85	50
5	16.9	6.2	0.85	250

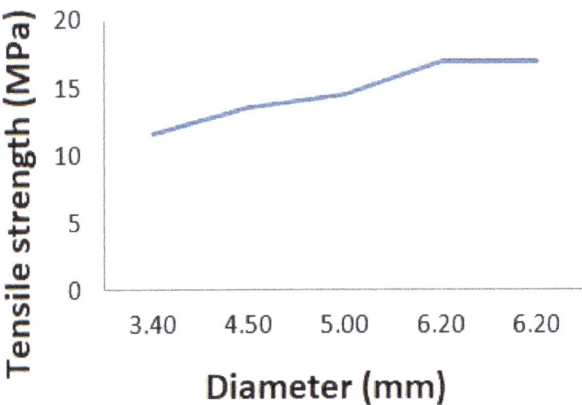

Figure 4. Plot of change in ultimate strength with specimen diameter.

Table 4. Values of hyperelastic models for two samples.

Strain Density Function	Constants, Specimen No. 1 (MPa)	Constants, Specimen No. 2 (MPa)
The five-parameter Mooney–Rivlin model	$C_{10} = -1.64$, $C_{01} = 2.59$, $C_{20} = 4.46 \times 10^{-7}$, $C_{11} = -2.39 \times 10^{-4}$, $C_{02} = 0.44$	$C_{10} = -2.2$, $C_{01} = 3.26$, $C_{20} = 3.86$, $C_{11} = -8.6 \times 10^{-4}$, $C_{02} = 0.62$
The three-parameter Yeoh model	$C_{10} = 0.11$, $C_{20} = -4.96 \times 10^{-6}$, $C_{30} = 1.67 \times 10^{-10}$	$C_{10} = 0.20$, $C_{20} = -6.73 \times 10^{-6}$, $C_{30} = 1.16 \times 10^{-10}$

3.2. Results of FSI Simulation of the Blood Flow

As a result of solving the problem, the distributions of hemodynamic parameters were obtained from three patients, including blood flow velocity, pressure, wall shear stress, time-averaged wall shear stress, and other parameters. The mechanical properties of the aorta–pulmonary artery–shunt system are shown in Table 5 in the considered computational domain. The most important results from the hemodynamic point of view were obtained at t = 0.09 s, corresponding to the maximum blood flow velocity. Similar results obtained for simple geometry (straight vessel) are presented in Supplementary Materials.

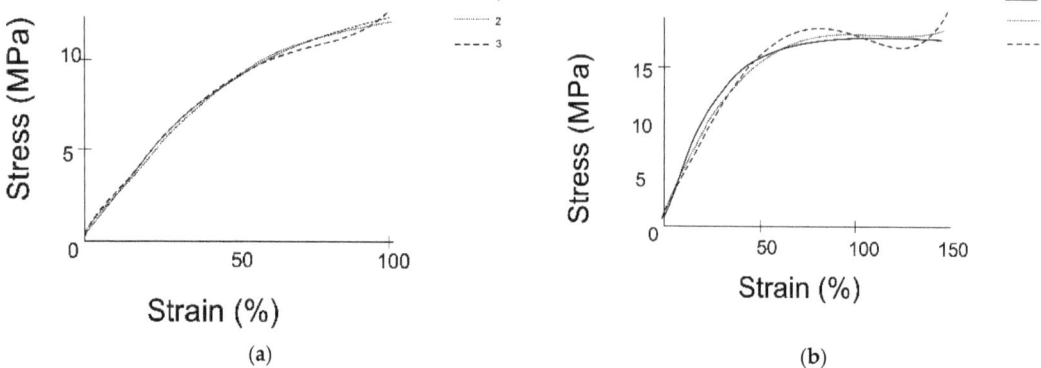

Figure 5. Stress–strain diagrams: (**a**) specimen 1, (**b**) specimen 2. The solid line (1) is the experimental curve, the dashed line (2) is the five-parameter Mooney–Rivlin strain energy density function, and the dotted line (3) is the three-parameter Yeoh strain energy density function.

Table 5. Mechanical parameters for aorta and shunt used in the study.

The Aorta		The Shunt	
Isotropic Hyperelastic Material	Anisotropic Hyperelastic Material)	Isotropic Elastic Material	Isotropic Hyperelastic Material
Ogden model: $\mu_1 = 1.274$ MPa $\mu_2 = -1.211$ MPa $\alpha_1 = 24.074$ $\alpha_2 = 24.073$	Holzapfel–Gasser–Ogden model: $\mu_1 = 2.363$ MPa $\mu_2 = 0.839$ MPa $\alpha_1 = 0.6$ $d = 0.001$ MPa^{-1}	$E = 10.3$ MPa $\mu = 0.49$	Experimental data (Table 3)

3.2.1. Velocity Distribution

Figure 6 shows the distribution of blood flow velocity characteristics. In the area of the aorta, the blood flow has a uniform distribution pattern. There is a local increase in the rate of blood flow in the region of the descending aorta and bifurcations. As one moves away from the descending part of the aorta, the blood flow velocity is equalized. The reverse situation is true in the pulmonary artery. In the pulmonary artery, there is mainly a vortex flow of blood in all patients. At the peak moment of systole, the maximum values of blood flow are observed in the area of the shunt.

3.2.2. Pressure distribution

Figure 7 shows the pressure distribution at the peak moment of systole. The distribution of pressure along the walls of the aorta and pulmonary artery is uneven. The highest values are concentrated on the walls of the ascending part of the aorta and its branches (left subclavian artery, left common artery, and brachiocephalic trunk), while the lowest values are observed on the walls of the pulmonary artery and shunt.

In the shunt zone, the maximum values are concentrated in the area of the junction with the aorta, then the pressures are distributed evenly up to the pulmonary artery.

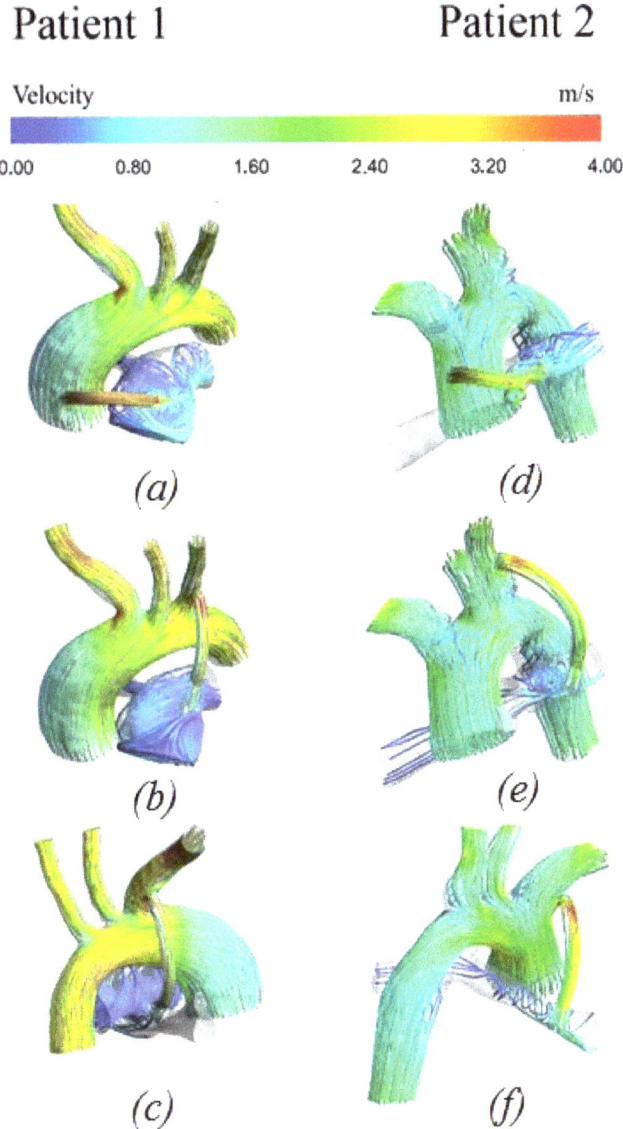

Figure 6. Velocity distribution with anisotropic properties of the aorta and hyperelastic properties of the shunt: (**a,d**) central shunt; (**b,e**) right shunt; (**c,f**) left shunt.

3.2.3. Wall Shear Stress

The distribution of shear stresses is important in the study of systemic blood flow. In the literature, particular importance is given to the distribution of the shear-wall shear stresses. Most authors associate hypoplasia of the intima of the vascular bed with high shear stress [21].

The wall shear stress indicates two problems: lipids remain on the vessel wall at low values, and they damage the vessel wall at high values, which also increases the ability of lipids to linger on the damaged intima.

Figure 8 shows the distribution of wall shear stress. The highest values are localized in the area of the shunt, which can lead to its thrombosis. Additionally, large values of parietal shear stresses are concentrated in the pulmonary artery in the vortex, with stagnant blood flow on the branches of the aorta (left subclavian artery, left common artery, and brachiocephalic trunk). The minimum values are observed in the areas of the descending part of the aorta and the beginning of the right and left pulmonary arteries.

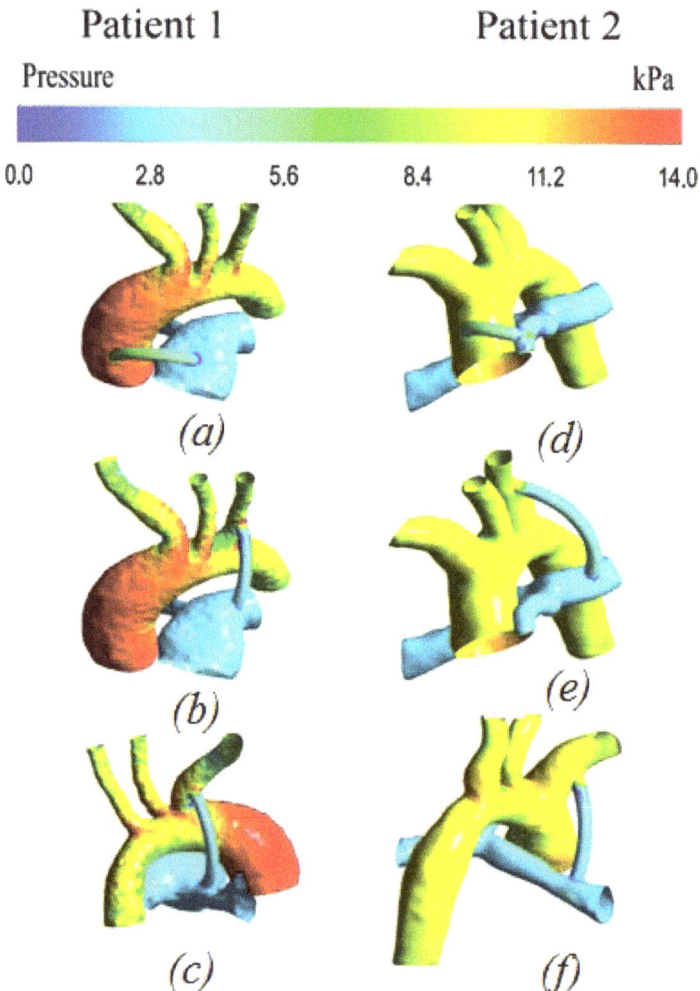

Figure 7. Pressure distribution with anisotropic properties of the aorta and hyperelastic properties of the shunt: (**a,d**) central shunt; (**b,e**) right shunt; (**c,f**) left shunt.

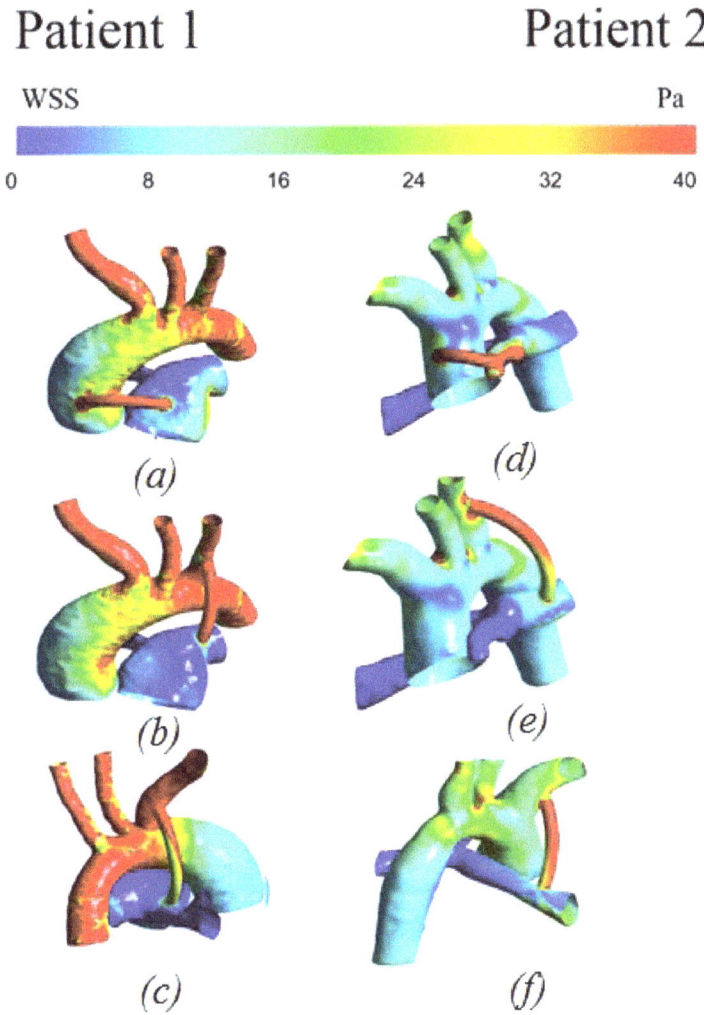

Figure 8. Distribution of wall shear stress with anisotropic properties of the aorta and hyperelastic properties of the shunt: (**a,d**) central shunt; (**b,e**) right shunt; (**c,f**) left shunt.

3.2.4. Distribution of Time-Averaged Shear Stress

Figure 9 shows the time-averaged shear stress. The values of the shear stress at the peak moment of systole are highest in the shunt area, causing shunt thrombosis. Additionally, large values are located in the area of vortex movement of blood in the underlying region of the pulmonary artery, as well as in local areas of the branches of the aorta, due to the special geometric characteristics of each geometry of patients.

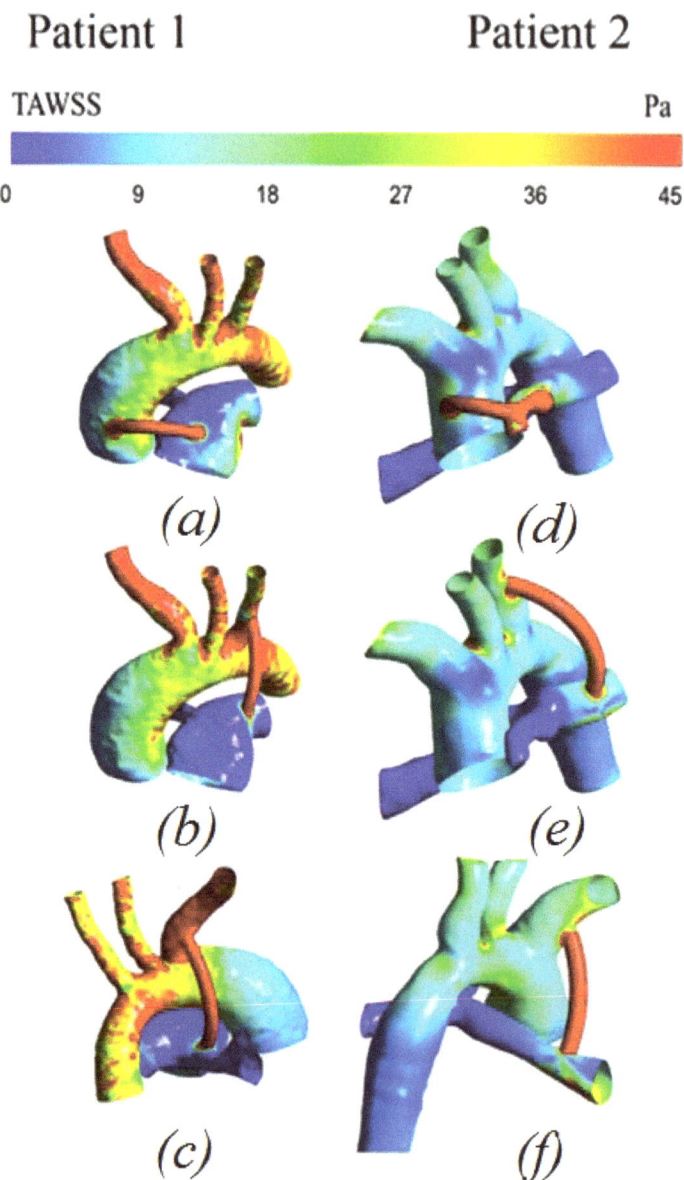

Figure 9. Distribution of time-averaged wall shear stress with anisotropic properties of the aorta and hyperelastic properties of the shunt: (**a,d**) central shunt; (**b,e**) right shunt; (**c,f**) left shunt.

3.2.5. Displacement Distribution

Figure 10 shows the distribution of displacements occurring in the system. The displacement values at the peak moment of systole are highest in the area of the shunt and the lateral part of the aorta with the central and right location of the shunt. With the left-sided shunt position, the maximum values are distributed only along the lateral part of the aorta.

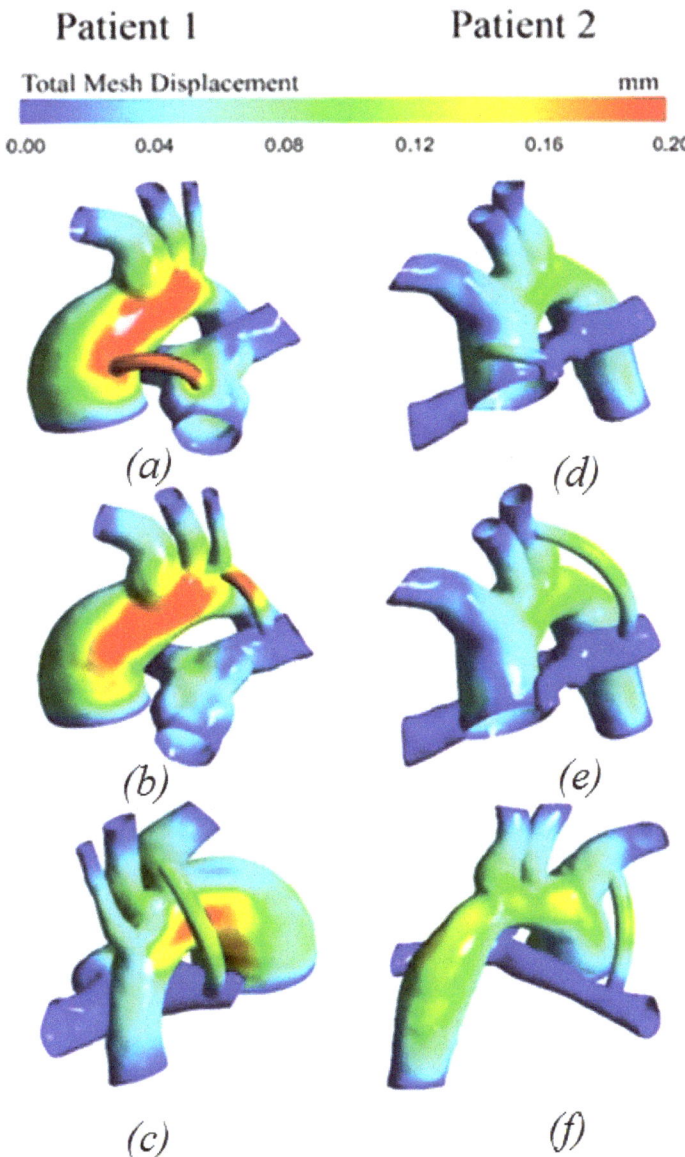

Figure 10. Distribution of displacements with anisotropic properties of the aorta and hyperelastic properties of the shunt: (**a,d**) central shunt; (**b,e**) right shunt; (**c,f**) left shunt.

3.2.6. Von Mises Stress Distribution

Figure 11 shows the distribution of stresses arising in the system. The stress values at the peak moment of systole are highest in the areas of blood flow separation and have a non-uniform distribution pattern. Additionally, high values are located in local areas of the aorta, characterized by the unevenness of the walls of the system.

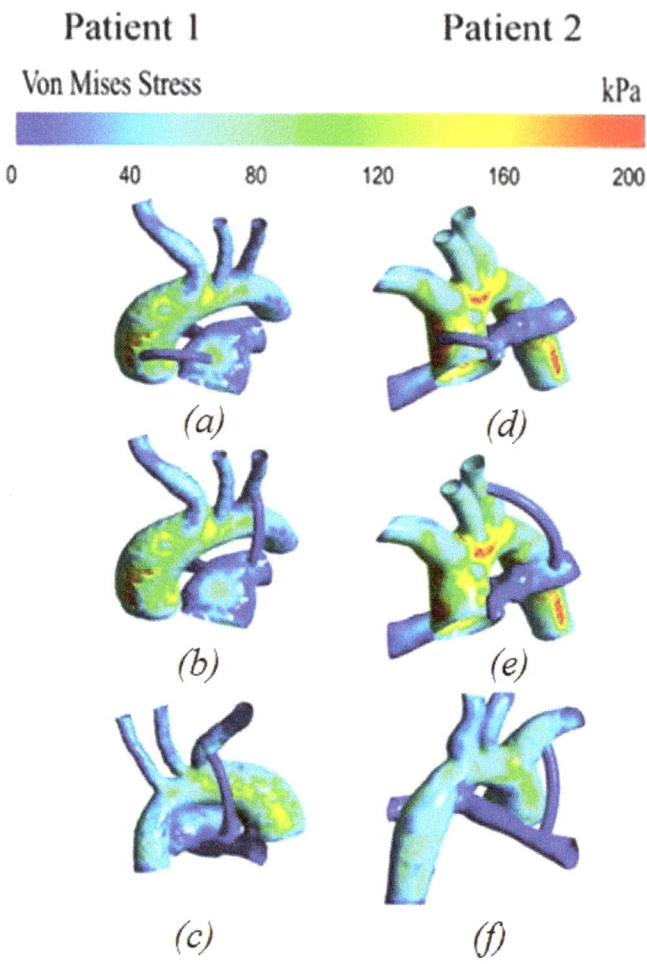

Figure 11. Distribution of stresses in the case of anisotropic properties of the aorta and hyperelastic properties of the shunt: (**a,d**) central shunt; (**b,e**) right shunt; (**c,f**) left shunt.

4. Discussion

4.1. Difference between Isotropic and Anisotropic Models

The distribution of hemodynamic parameters in the anisotropic and isotropic models of materials in patients has the same distribution pattern throughout the system. The dynamics of blood flow is identical in all patients. The numerical values are also the same (Figures 12 and 13). The differences between the anisotropic and isotropic properties of the aorta and the pulmonary artery are noticeable only when analyzing the stress–strain state, i.e., in displacements and stresses arising in the aorta–shunt–pulmonary artery system.

The opposite situation is seen with von Mises stress distribution (Figure 14). There is also a similar pattern of stress distribution throughout the system, with the exception of the central location of the shunt, where, according to the anisotropic model, increased stress values are mostly observed in the aorta. In addition, the anisotropic model of the material shows higher stress values than the isotropic model of the aortic material. The maximum stress on the wall of the anisotropic aorta is about 200 kPa; the maximum stress on the wall of the isotropic aorta is about 150 kPa.

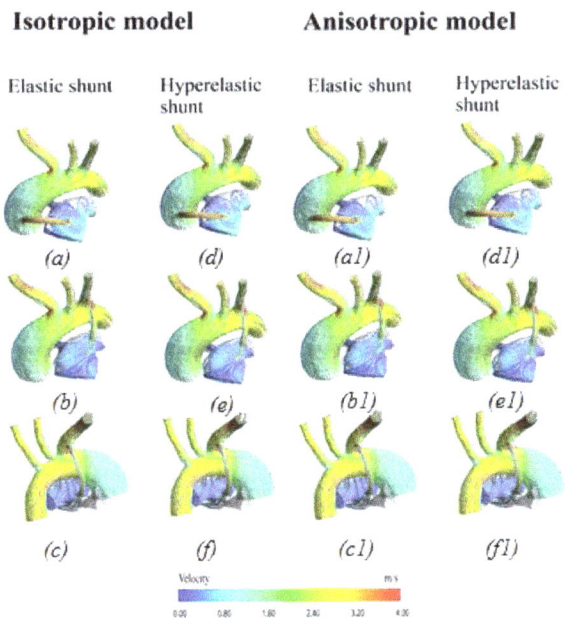

Figure 12. Velocity distribution: (**a,d,a1,d1**) central shunt; (**b,e,b1,e1**) right shunt; (**c,f,c1,f1**) left shunt.

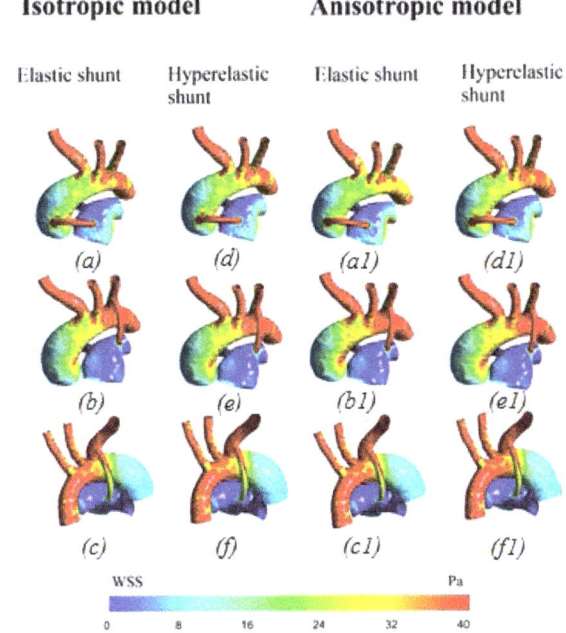

Figure 13. Wall shear stress distribution: (**a,d,a1,d1**) central shunt; (**b,e,b1,e1**) right shunt; (**c,f,c1,f1**) left shunt.

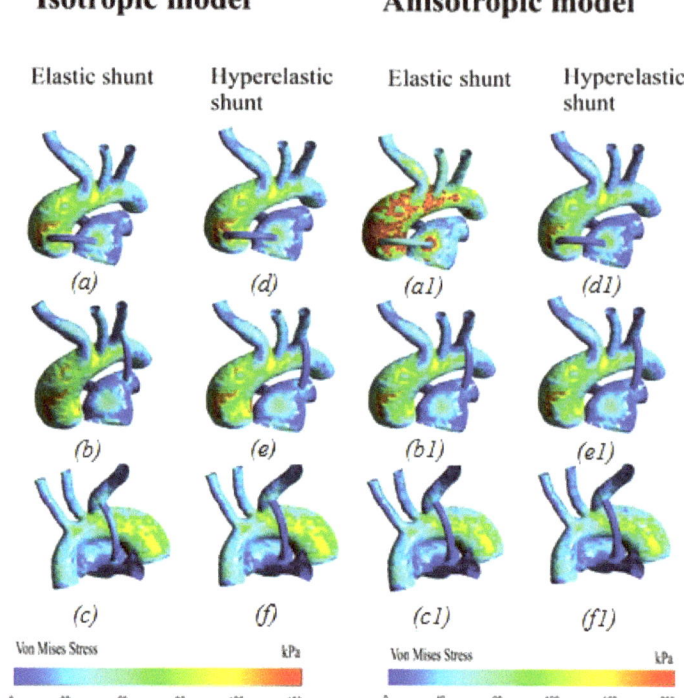

Figure 14. Von Mises stress distribution: (**a,d,a1,d1**) central shunt; (**b,e,b1,e1**) right shunt; (**c,f,c1,f1**) left shunt.

Along with the hemodynamic parameters, the parameters of the stress–strain state, such as displacements and von Mises stress, also affect the success of a surgical intervention [22].

It was shown that the anisotropic model of the aortic material shows higher stress values at the peak moment of systole, which in turn may be a key factor in determining the strength characteristics of the aorta and pulmonary artery, all other things being equal. Additionally, this mechanical parameter is important when installing a central shunt, since it is in the area of the central anastomosis that an increase in stresses on the aortic wall is observed.

Displacement distribution is also important. According to the computations, the anisotropic model shows smaller values of the displacements of both the aorta and the shunt, which in turn may affect the success of preoperative prediction. Thus, it can be concluded that the anisotropic properties of the aorta play an important role in preoperative modeling.

The time dependences of the volumetric flow rate of blood flow inside the shunt show that, for all locations of the shunt and taking into account the hyperelasticity of the shunt, the results are almost identical (Figure 15). However, in the case of the central position of the shunt, when the aortic walls were considered anisotropic material, the volumetric flow of blood within the shunt was different. The maximum deviation of this value was 12%, at 0.2 s of the cardiac cycle. For the elastic shunt, the volumetric flow rate throughout almost the entire time exceeded the analogous value for the hyperelastic shunt. This difference is essential for the further correct formation of pulmonary blood flow.

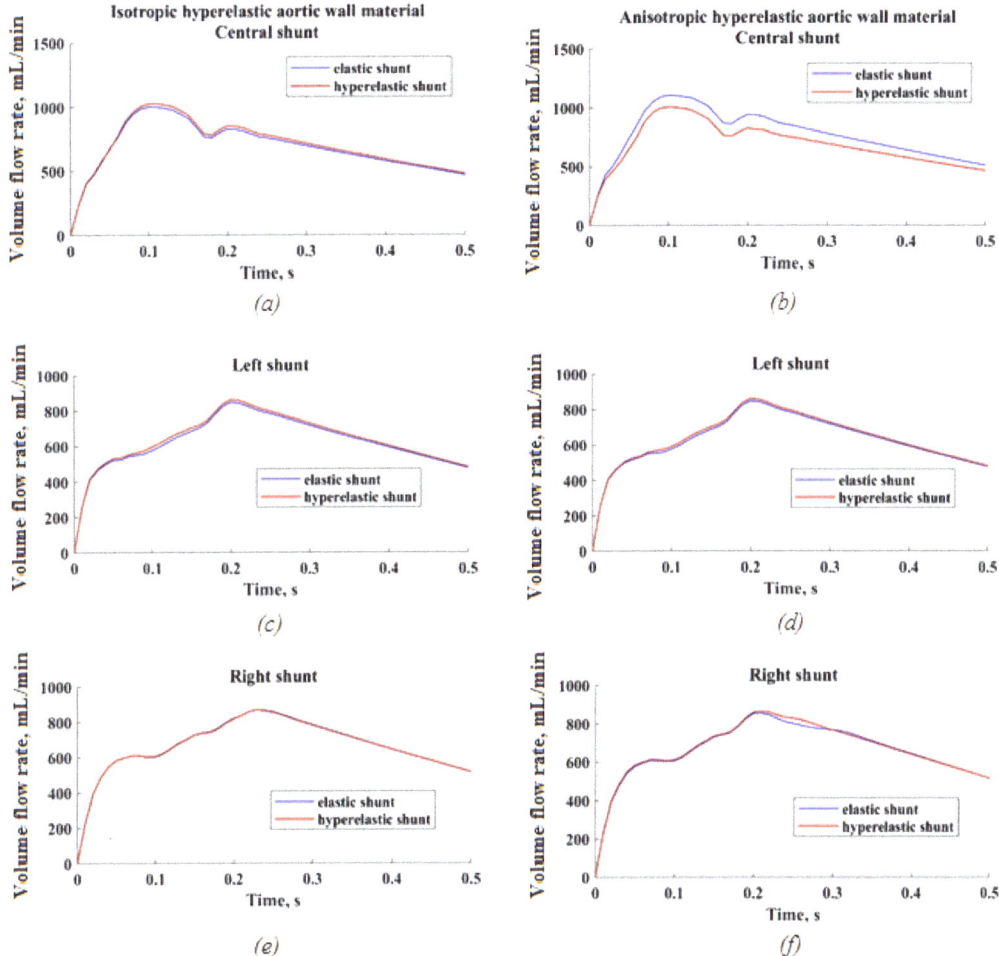

Figure 15. Flow rate through the shunt: (**a**) central shunt case (isotropic aortic wall), (**b**) central shunt case (anisotropic aortic wall), (**c**) left shunt case (isotropic aortic wall), (**d**) left shunt case (anisotropic aortic wall), (**e**) right shunt case (isotropic aortic wall), (**f**) right shunt case (anisotropic aortic wall).

Based on the same dependences of volumetric blood flow on time for the aortic wall, and taking into account its anisotropy, it can be concluded that this is reflected only in the case of a central shunt, when the shunt is an elastic material. Susequently, there is an excess of volumetric flow in 12% of cases with a hyperelastic shunt. Thus, taking into account the hyperelasticity of the shunts allows for obtaining more realistic results, reducing the possible negative consequences of operations.

Hemodynamic parameters are very important for the assessment of shunting [23]. To evaluate the effectiveness of shunting, it is worth considering indicators that can describe the probable risk of shunt thrombosis. Such indicators are wall shear stress and time-averaged wall shear stress.

As a result of the analysis of the distribution of indicators, high values of wall shear stress were revealed, and, consequently, time-averaged wall shear stress in the anastomotic region in all models. This, in turn, may indicate the risk of developing thrombosis [24–26]. It is also supported by the clinical and literature data [27–29].

4.2. Concluding Remarks

The obtained results of the distribution of von Mises stress show a doubling in this value for a central shunt, with an elastic material of the shunt in contrast to the hyperelastic material on the wall of the shunt itself. In addition, such a location is characterized by an increase in the stress values on the aortic wall and at the site of the shunt insertion into the pulmonary artery by 40–80 kPa. Taking into account the hyperelasticity of the shunt makes it possible to take into account some of its "damping" properties.

The values of shear stress on the wall for the central shunt also differ depending on the hyperelasticity of the shunt. Moreover, this is reflected in the descending aorta, where the difference is about 20%. It can be concluded that taking into account the hyperelasticity of the shunt material plays a particularly important role when the shunt is in the central position. In this case, the values of many hemodynamic characteristics change, which will further affect the development of shunt thrombosis and the distribution of pulmonary blood flow.

4.3. Limitations

We recognize several limitations of our study. Three variants of a modified Blalock–Taussig shunt were considered, central, right, and left, but these models are not always possible to realize. There are other organs nearby that can make it impossible to implement certain shunt locations.

The study does not present a retrospective clinical analysis of the MBTS operation for patients, which would allow us to objectify, to some extent, the calculation results obtained.

CT scans from only three patients were used for modeling in our study, which is a very small sample size. In future, with an increase in the number of subjects, these results can be used to guide clinical practice; this can be considered a pilot study towards this goal.

4.4. Possible Future Clinical Application

The development of non-invasive diagnostic and numerical methods in the contemporary surgery allows the estimation of the biomechanical processes in the human body. This circumstance increases the possibility of their use to improve existing and developing new personalized methods for diagnosing and predicting treatment. In particular, there is a growing need for the applications in the cardiovascular pediatric surgery.

Congenital heart disease is a general term for a range of birth defects that affect the normal way the heart works. The modified Blalock–Taussig shunt is commonly performed as early palliation in cyanotic congenital heart disease. One of the reasons is the use of subjective experience and the lack of individualized biomechanical models for the analysis of surgical interventions.

To predict and prevent postoperative complications, it is necessary to formulate and introduce new technological approaches, which, in particular, may consist in creating a software product (decision-making system in surgical interventions for gallstone disease and its complications). A proposed model of the blood flow in the system aorta–hunt–pulmonary artery makes it possible to assess hemodynamics in normal and pathological conditions, as well as to carry out a numerical assessment of modified Blalock–Taussig shunt to predict and prevent complications. The decision-making software based on such a biomechanical model will be able to evaluate the shunt position for the current patient, predict possible thrombosis risk, and evaluate mean flow rate after palliation surgery. Therefore, using the results of this paper, the surgeon can evaluate the circumstances of the operation for each patient before operation and evaluate the results of post-operative blood flow features.

5. Conclusions

There is no obviously ideal type and location for shunt insertion. Many studies have attempted to find a universal way to guide such a choice, in particular, using mathematical models. Our study offers a new step in the modeling of MBTS operations. In contrast to the well-known and more commonly used CFD method, we have proposed the FSI method, which takes into account not only the elasticity of the vessel wall but also its anisotropy. Changes in a number of the main hemodynamic parameters of local blood flow have been established depending on the accounting of elasticity properties. We consider this important for further research in modeling MBTS operations.

Modeling is an activity aimed at understanding and quantifying an event, object, or function. In state-of-the-art science, numerical modeling is often used to study a particular process. In our study, we also used already known tools (FSI) to simulate a specific event (MBTS). The use of the FSI method for MBTS operations has not been used before, so we called it a "new step".

We have carried out complicated FSI modeling for MBTS operations, taking into account the anisotropic properties of vessel materials. No one has done this before. We have established the importance of taking anisotropy into account when modeling central bypass surgery. Therefore, we consider this a "new step" of modeling in MBTS operations.

A comparison between the effect of isotropic and anisotropic aorta material properties was performed. It was shown that the anisotropic model of the aortic material showed higher stress values at the peak moment of systole, which may be a key factor in determining the strength characteristics of the aorta and pulmonary artery. Additionally, this mechanical parameter is important when installing a central shunt, since it is in the area of the central anastomosis that an increase in stresses on the aortic wall is observed. According to computations, the anisotropic model shows smaller values of the displacements of both the aorta and the shunt, which in turn may affect the success of preoperative prediction. Thus, it can be concluded that the anisotropic properties of the aorta play an important role in preoperative modeling.

Supplementary Materials: The following supporting information can be downloaded at: https://www.mdpi.com/article/10.3390/ma15082719/s1. Table S1. Parameters of computational mesh., Table S2. Mechanical parameters for aorta and shunt used in the study., Figure S1. Meshes and boundary conditions of the "straight vessel": upper–boundary conditions and fluid mesh, lower–solid mesh., Figure S2. Stress distribution at time t = 0.09 s: (a) isotropic properties of the vessel, (b) anisotropic properties of the vessel., Figure S3. Displacements distribution at time t = 0.09 s: (a) isotropic properties of the vessel, (b) anisotropic properties of the vessel., Figure S4. Comparison of parameters: (a) maximum values of displacements, (b) maximum Von Mises stress values, (c) maximum wall shear stress values.

Author Contributions: Conceptualization, A.G.K. and M.S.; methodology, A.G.K.; validation, A.K. and A.G.K.; investigation, A.G.K., A.K. and A.M.; resources, A.G.K.; data curation, A.G.K., A.M. and A.P.; writing—original draft preparation, A.G.K., A.K., A.P. and M.S.; writing—review and editing, A.G.K.; visualization, A.K. and A.M.; supervision, A.G.K.; funding acquisition, A.G.K. All authors have read and agreed to the published version of the manuscript.

Funding: A.G.K. acknowledges the financial support of the Ministry of Science and Higher Education of the Russian Federation in the framework of the program of activities of the Perm Scientific and Educational Center "Rational Subsoil Use". Artem Porodikov acknowledges financial support of RFBR and Perm Territory, project number 20-41-596005. Marina Shmurak acknowledges financial support of Perm National Research Polytechnic University in the framework of the Federal Academic Leadership Program «Priority-2030».

Institutional Review Board Statement: The study was conducted according to the guidelines of the Declaration of Helsinki, and approved by the Ethics Committee of S.G. Sukhanov Cardiovascular Center, Perm, Russia (protocol No. 12 on 25 October 2021).

Informed Consent Statement: Informed consent was obtained from parents of patients involved in the study.

Data Availability Statement: Not applicable.

Conflicts of Interest: The authors declare no conflict of interest.

References

1. Alsoufi, B.; Gillespie, S.; Kogon, B.; Schlosser, B.; Sachdeva, R.; Kim, D.; Clabby, M.; Kanter, K. Results of palliation with an initial modified blalock-taussig shunt in neonates with single ventricle anomalies associated with restrictive pulmonary blood flow. *Ann. Thorac. Surg.* **2015**, *99*, 1639–1647. [CrossRef] [PubMed]
2. Sant'Anna, J.R.M.; Pereira, D.C.; Kalil, R.A.K.; Prates, P.R.; Horowitz, E.; Sant'Anna, R.T.; Prates, P.R.L.; Nesralla, I.A. Computer dynamics to evaluate blood flow through the modified Blalock-Taussig shunt. *Rev. Bras. Cir. Cardiovasc.* **2003**, *18*, 253–260. [CrossRef]
3. Laganà, K.; Balossino, R.; Migliavacca, F.; Pennati, G.; Bove, E.L.; De Leval, M.R.; Dubini, G. Multiscale modeling of the cardiovascular system: Application to the study of pulmonary and coronary perfusions in the univentricular circulation. *J. Biomech.* **2005**, *38*, 1129–1141. [CrossRef] [PubMed]
4. Arnaz, A.; Pişkin, Ş.; Oğuz, G.N.; Yalçınbaş, Y.; Pekkan, K.; Sarıoğlu, T. Effect of modified Blalock-Taussig shunt anastomosis angle and pulmonary artery diameter on pulmonary flow. *Anatol. J. Cardiol.* **2018**, *20*, 2–8. [CrossRef]
5. Zhang, N.; Yuan, H.; Chen, X.; Liu, J.; Zhou, C.; Huang, M.; Jian, Q.; Zhuang, J. Hemodynamic of the patent ductus arteriosus in neonates with modified Blalock-Taussig shunts. *Comput. Methods Programs Biomed.* **2020**, *186*, 105223. [CrossRef]
6. Arthurs, C.J.; Agarwal, P.; John, A.V.; Dorfman, A.L.; Grifka, R.G.; Figueroa, C.A. Reproducing patient-specific hemodynamics in the Blalock-Taussig circulation using a flexible multi-domain simulation framework: Applications for optimal shunt design. *Front. Pediatr.* **2017**, *5*, 1–13. [CrossRef]
7. Bove, E.L.; Migliavacca, F.; de Leval, M.R.; Balossino, R.; Pennati, G.; Lloyd, T.R.; Khambadkone, S.; Hsia, T.Y.; Dubini, G. Use of mathematic modeling to compare and predict hemodynamic effects of the modified Blalock-Taussig and right ventricle-pulmonary artery shunts for hypoplastic left heart syndrome. *J. Thorac. Cardiovasc. Surg.* **2008**, *136*, 312–320.e2. [CrossRef]
8. Hsia, T.Y.; Cosentino, D.; Corsini, C.; Pennati, G.; Dubini, G.; Migliavacca, F. Use of mathematical modeling to compare and predict hemodynamic effects between hybrid and surgical norwood palliations for hypoplastic left heart syndrome. *Circulation* **2011**, *124*, 204–210. [CrossRef]
9. Liu, J.; Sun, Q.; Hong, H.; Sun, Y.; Liu, J.; Qian, Y.; Wang, Q.; Umezu, M. Medical image-based hemodynamic analysis for modified blalock-taussig shunt. *J. Mech. Med. Biol.* **2015**, *15*, 1–17. [CrossRef]
10. Zhao, X.; Liu, Y.; Ding, J.; Ren, X.; Bai, F.; Zhang, M.; Ma, L.; Wang, W.; Xie, J.; Qiao, A. Hemodynamic effects of the anastomoses in the modified blalock-taussig shunt: A numerical study using a 0D/3D coupling method. *J. Mech. Med. Biol.* **2015**, *15*, 1–19. [CrossRef]
11. Liiu, J.; Sun, Q.; Qian, Y.; Hong, H.; Liu, J. Numerical Simulation and Hemodynamic Analysis of the Modified Blalock-Taussig Shunt. In Proceedings of the Annual International Conference of the IEEE Engineering in Medicine and Biology Society, EMBS, Osaka, Japan, 3–7 July 2013; pp. 707–710.
12. Zhou, T.; Wang, Y.; Liu, J.; Wang, Y.; Wang, Y.; Chen, S.; Zhou, C.; Dong, N. Pulmonary artery growth after Modified Blalock-Taussig shunt: A single center experience. *Asian J. Surg.* **2020**, *43*, 428–437. [CrossRef] [PubMed]
13. Piskin, S.; Altin, H.F.; Yildiz, O.; Bakir, I.; Pekkan, K. Hemodynamics of patient-specific aorta-pulmonary shunt configurations. *J. Biomech.* **2017**, *50*, 166–171. [CrossRef] [PubMed]
14. Zhang, N.; Yuan, H.; Chen, X.; Liu, J.; Jian, Q.; Huang, M.; Zhang, K. Computational Fluid Dynamics Characterization of Two Patient-Specific Systemic-to-Pulmonary Shunts before and after Operation. *Comput. Math. Methods Med.* **2019**, *2019*, 1502318. [CrossRef] [PubMed]
15. Luo, K.; Jiang, W.; Yu, C.; Tian, X.; Zhou, Z.; Ding, Y. Fluid-Solid Interaction Analysis on Iliac Bifurcation Artery: A Numerical Study. *Int. J. Comput. Methods* **2019**, *16*, 1850112. [CrossRef]
16. Stergiou, Y.G.; Kanaris, A.G.; Mouza, A.A.; Paras, S.V. Fluid-structure interaction in abdominal aortic aneurysms: Effect of haematocrit. *Fluids* **2019**, *4*, 11. [CrossRef]
17. Sousa, L.C.; Castro, C.F.; António, C.C.; Azevedo, E. Fluid-Structure Interaction Modeling of Blood Flow in a Non-Stenosed Common Carotid Artery Bifurcation. In Proceedings of the 7th International Conference on Mechanics and Materials in Design, Albufeira, Portugal, 11–15 June 2017; pp. 1559–1564.
18. Holzapfel, G.A.; Gasser, T.C.; Ogden, R.W. A new constitutive framework for arterial wall mechanics and a comparative study of material models. *J. Elast.* **2000**, *61*, 1–48. [CrossRef]
19. Vassilevski, Y.V.; Salamatova, V.Y.; Simakov, S.S. On the elasticity of blood vessels in one-dimensional problems of hemodynamics. *Comput. Math. Math. Phys.* **2015**, *55*, 1567–1578. [CrossRef]
20. Amabili, M.; Balasubramanian, P.; Bozzo, I.; Breslavsky, I.D.; Ferrari, G.; Franchini, G.; Giovanniello, F.; Pogue, C. Nonlinear Dynamics of Human Aortas for Material Characterization. *Phys. Rev. X* **2020**, *10*, 011015. [CrossRef]
21. Malek, A.M.; Alper, S.L.; Izumo, S. Hemodynamic shear stress and its role in atherosclerosis. *JAMA* **2013**, *282*, 2035–2042. [CrossRef]
22. Lin, S.; Han, X.; Bi, Y.; Ju, S.; Gu, L. Fluid-structure interaction in abdominal aortic aneurysm: Effect of modeling techniques. *Biomed. Res. Int.* **2017**, *2017*, 7023078. [CrossRef]

23. Kuchumov, A.G.; Khairulin, A.R.; Biyanov, A.N.; Porodikov, A.A.; Arutyunyan, V.B.; Sinelnikov, Y.S. Effectiveness of blalock-taussig shunt performance in the congenital heart disease children. *Russ. J. Biomech.* **2020**, *24*, 65–83. [CrossRef]
24. Han, D.; Starikov, A.; Hartaigh, B.; Gransar, H.; Kolli, K.K.; Lee, J.H.; Rizvi, A.; Baskaran, L.; Schulman-Marcus, J.; Lin, F.Y.; et al. Relationship between endothelial wall shear stress and high-risk atherosclerotic plaque characteristics for identification of coronary lesions that cause ischemia: A direct comparison with fractional flow reserve. *J. Am. Heart Assoc.* **2016**, *5*, 1–10. [CrossRef] [PubMed]
25. Samady, H.; Eshtehardi, P.; McDaniel, M.C.; Suo, J.; Dhawan, S.S.; Maynard, C.; Timmins, L.H.; Quyyumi, A.A.; Giddens, D.P. Coronary artery wall shear stress is associated with progression and transformation of atherosclerotic plaque and arterial remodeling in patients with coronary artery disease. *Circulation* **2011**, *124*, 779–788. [CrossRef] [PubMed]
26. Campobasso, R.; Condemi, F.; Viallon, M.; Croisille, P.; Campisi, S.; Avril, S. Evaluation of Peak Wall Stress in an Ascending Thoracic Aortic Aneurysm Using FSI Simulations: Effects of Aortic Stiffness and Peripheral Resistance. *Cardiovasc. Eng. Technol.* **2018**, *9*, 707–722. [CrossRef]
27. Traub, O.; Berk, B.C. Laminar shear stress: Mechanisms by which endothelial cells transduce an atheroprotective force. *Arterioscler. Thromb. Vasc. Biol.* **1998**, *18*, 677–685. [CrossRef]
28. Badimon, L.; Vilahur, G. Thrombosis formation on atherosclerotic lesions and plaque rupture. *J. Intern. Med.* **2014**, *276*, 618–632. [CrossRef]
29. Küçük, M.; Özdemir, R.; Karaçelik, M.; Doksöz, Ö.; Karadeniz, C.; Yozgat, Y.; Meşe, T.; Sarıosmanoğlu, O.N. Risk factors for thrombosis, overshunting and death in infants after modified blalock-Taussig shunt. *Acta Cardiol. Sin.* **2016**, *32*, 337–342. [CrossRef]

Article

Innovative Design Methodology for Patient-Specific Short Femoral Stems

William Solórzano-Requejo [1,2,*], Carlos Ojeda [2] and Andrés Díaz Lantada [1,*]

1. Product Development Laboratory, Department of Mechanical Engineering, Universidad Politécnica de Madrid, C/José Gutiérrez Abascal 2, 28006 Madrid, Spain
2. Mechanical Technology Laboratory, Department of Mechanical and Electrical Engineering, Universidad de Piura, Piura 20009, Peru; carlos.ojeda@udep.edu.pe or grupo.biomecanica@udep.edu.pe
* Correspondence: wsrequejo@gmail.com or william.solorzano@alumnos.upm.es (W.S.-R.); andres.diaz@upm.es (A.D.L.)

Citation: Solórzano-Requejo, W.; Ojeda, C.; Díaz Lantada, A. Innovative Design Methodology for Patient-Specific Short Femoral Stems. *Materials* **2022**, *15*, 442. https://doi.org/10.3390/ma15020442

Academic Editor: Oskar Sachenkov

Received: 12 December 2021
Accepted: 4 January 2022
Published: 7 January 2022

Publisher's Note: MDPI stays neutral with regard to jurisdictional claims in published maps and institutional affiliations.

Copyright: © 2022 by the authors. Licensee MDPI, Basel, Switzerland. This article is an open access article distributed under the terms and conditions of the Creative Commons Attribution (CC BY) license (https://creativecommons.org/licenses/by/4.0/).

Abstract: The biomechanical performance of hip prostheses is often suboptimal, which leads to problems such as strain shielding, bone resorption and implant loosening, affecting the long-term viability of these implants for articular repair. Different studies have highlighted the interest of short stems for preserving bone stock and minimizing shielding, hence providing an alternative to conventional hip prostheses with long stems. Such short stems are especially valuable for younger patients, as they may require additional surgical interventions and replacements in the future, for which the preservation of bone stock is fundamental. Arguably, enhanced results may be achieved by combining the benefits of short stems with the possibilities of personalization, which are now empowered by a wise combination of medical images, computer-aided design and engineering resources and automated manufacturing tools. In this study, an innovative design methodology for custom-made short femoral stems is presented. The design process is enhanced through a novel app employing elliptical adjustment for the quasi-automated CAD modeling of personalized short femoral stems. The proposed methodology is validated by completely developing two personalized short femoral stems, which are evaluated by combining in silico studies (finite element method (FEM) simulations), for quantifying their biomechanical performance, and rapid prototyping, for evaluating implantability.

Keywords: biomechanics; hip replacement; short stems; custom-made medical devices; strain shielding; finite element analysis

1. Introduction

Hip arthroplasty is a surgical procedure in which the total or partial replacement of the hip joint is performed, using an artificial device, the hip/femoral prosthesis, which is a mechanical implant that replaces the joint primarily for two typical situations: to reduce pain and improve joint mobility due to progressive wear caused by osteoarthritis, or fracture of the femoral neck due to trauma or osteoporosis [1].

Koch's femoral model, which defines two different sets of stress lines with compressive loads along the medial side and tensile loads on the lateral one, has been used to design stems for total hip replacement (THR). Consequently, conventional stems use the medial side as the support, also called calcar, because bone tissue is more resistant to compression than to tension and its use reduces the likelihood of fracture. However, Koch's model does not accurately describe the biomechanics of the femur because it ignores muscle action; the forces generated by the iliotibial band and the vastus lateralis–gluteus medius complex create a tension band effect that converts the tensile stresses of the lateral femoral column into compressive ones [2,3].

Thus, it is proven that the cortical bone of the femur is subjected to compressive stresses in normal function, in accordance with its histological characteristics. This rethinking of the mode of load transfer on the entire proximal femur revolutionizes the design requirements of an anatomic cementless femoral implant [2].

The fixation of the cementless stem depends on the natural adherence between bone and stem; when properly adhered, the implant is stable. However, there is short- (primary) and long-term (secondary) stability. Primary stability depends on the tight insertion of the stem into the femoral canal; mechanically, it is quantified through the relative displacements that occur at the bone–stem interface [4]. The secondary stability is achieved through the bone ingrowth on its surface. This process is known as osseointegration; for this reason, the stem has a porous and textured coating. In addition, the material from which they are made should be biocompatible and not reactive to bone formation.

The ideal stem should restore the physiological load transfer of the femur; unfortunately, after its insertion, the load pattern is modified. Consequently, the natural response of the bone to the conventional (stiffer) stem is proximal bone resorption and distal bone formation, due to which the phenomenon of stress/strain shielding (SS) arises, which occurs when part of the loads is taken up by the stem and prevented from reaching the femur, resulting in decreased bone, reducing the implant support and increasing the risk of loosening and fracture. The effects of aseptic loosening and micro-displacement can cause difficulties for patients when performing daily activities. If this situation is prolonged, it can cause a lot of pain and revision surgery is likely to be performed; however, the bone surrounding the removed femoral component has less bone stock; therefore, the new implant must be longer and thicker to be stabilized. However, strain shielding may occur again; therefore, this phenomenon should be eliminated [5].

Short stems were designed as an alternative to conventional implants to preserve the proximal bone stock. Calcar loading with lateral flare stems [6–8], a type of short implant, is attributed to Santori et al. [2,9], whose idea was to eliminate the diaphyseal part of the conventional stem because it causes shielding, and Jasty et al. [10] reported that it became unusable once the implant was stabilized and bone ingrowth occurred.

Therefore, they deduced that, if this was true for a conventional prosthesis, it had to be true also for a stem that relies on a wide lateral flare for initial stability. It is recommended that the prosthesis be implanted initially in patients with good bone quality and normal anatomy. Contraindications for use are hip dysplasia, severe osteoporosis and previous hip osteotomies.

The objective of this stem is a physiological distribution with a proximal load transfer from the implant to the femur, restoring its biomechanics. In addition, by reducing the invasion of the femur, it may preserve good irrigation and nutrition, which would benefit the cellular action, and therefore the bone remodeling, and would consequently decrease the risk of avascular necrosis. Since the results of the implant are satisfactory both for stability and fixation, they would be even more so if it were personalized, eliminating the risks that are a consequence of errors in surgery due to poor selection and/or adaptation of the implant to the femoral cavity.

This article seeks to rethink the design methodology of customized hip prostheses, optimizing the stems for calcar loading by employing lateral flare stems, hereinafter referred to as short stem. The article describes how to obtain the virtual model of the proximal femur, and then explains the development of a novel elliptical fitting app that allows the morphological evaluation of the femur (geometric parameters and femoral cavity). Consequently, based on the morphology and the surgical procedure, the short stem is designed. Finally, the finite element method (FEM) is employed to verify the biomechanical advantages described and to validate the designed short stems.

2. Materials and Methods
2.1. Virtual Model

Virtual models of the proximal femur were used, obtained by downloading the CT scans of two male patients from the open-source virtual library *"The Cancer Imaging Archive"* with references TCGA-VP-A878 [11] and Pelvic-Ref-009 [12] corresponding to the geometric case 1 (GC1) and 2 (GC2). The medical images had a slice thickness of 2 and 3 mm in the axial plane, and 0.909 and 1 mm in the coronal and sagittal planes for GC1 and GC2, respectively, each image being 512 × 512 pixels. Both CT scans were imported into 3D Slicer® 4.10.2 (https://www.slicer.org, accessed on 3 January 2022) to segment the right femur and its cortical part (Figure 1A) using the threshold, level tracing, paint, erase and smooth tools. Then, the trabecular part was obtained through the logical operation of subtraction between the femur and its cortical bone (Figure 1B). The 3D Slicer® allowed us to export the segmentation of the femur, cortical and trabecular bone as meshes in STL format, and these files were imported into Meshmixer® 3.5 (Autodesk Inc., Mill Valley, CA, USA) for inspection, repair and smoothing. Finally, the virtual models were processed as solids in NX® 10 (Siemens PLM Software Solutions, Plano, TX, USA), matching their coordinate system with the femur's coordinate system so the axial, coronal and sagittal planes were the XY, XZ and YZ planes, respectively (Figure 1C).

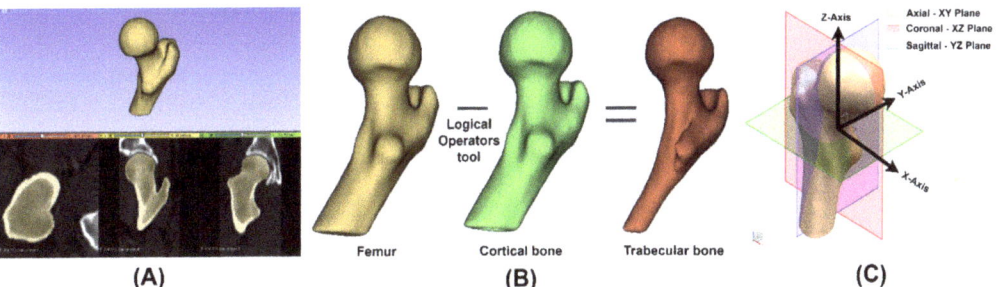

Figure 1. (**A**) Segmentation of the right femur using 3D Slicer®. (**B**) Process to obtain the trabecular bone. (**C**) Femoral coordinate system.

2.2. Elliptical Adjustment App

Trapezoidal, oval, elliptical and circular cross-sections have been used for the design of femoral stems. Previous studies [13,14] determined that the elliptical section produces a good stress distribution along the stem, allows its primary stability and improves its adaptability to different bone sections with changes in shape and size. Therefore, using Streamlit®, an open-source Python® library, an elliptical adjustment app was created to obtain the ellipse that best fits the bone section (https://github.com/solor5/elliptical_adjustment_app/blob/main/app.py, accessed on 3 January 2022). To utilize the application, the user must sample the bone section to be adjusted by employing the NX® point tool, and the coordinates of each point are exported in a DAT file (Figure 2), which is inserted into the app file uploader. The mathematical fundamentals that enable the elliptical fitting are explained below.

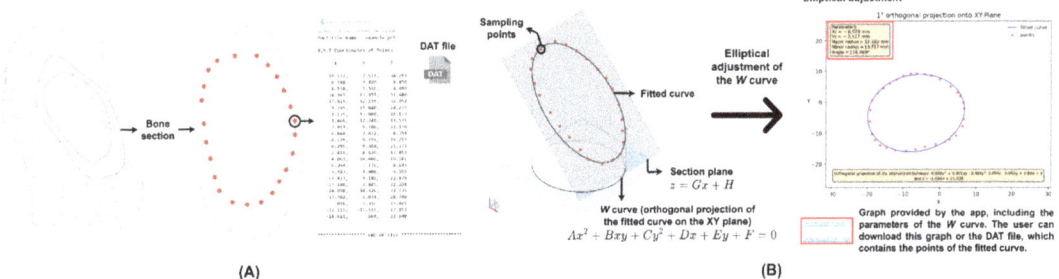

Figure 2. (**A**) Bone section sampling and DAT file. (**B**) Representation of the fitted curve and the elliptical adjustment of its orthogonal projection.

As a consequence of the position of the femur in NX® (Figure 1C), the bone section is located in an oblique plane perpendicular to the XZ; therefore, the X and Y coordinates of the points in the DAT file (p) allow the elliptical adjustment of the orthogonal projection of this section (Figure 2A). This conic (W) is represented by an implicit second-order polynomial (Q), defined by a vector of coefficients ($v = [A\ B\ C\ D\ E\ F]^T$):

$$W(v) = \left\{ p \in \mathbb{R}^2 \mid Q(p,v) = 0 \right\} \tag{1}$$

$$Q(p,v) = \begin{bmatrix} x^2 & xy & y^2 & x & y & 1 \end{bmatrix} \cdot \begin{bmatrix} A \\ B \\ C \\ D \\ E \\ F \end{bmatrix} = Ax^2 + Bxy + Cy^2 + Dx + Ey + F = 0 \tag{2}$$

If $P = \{p_1, p_2, \ldots, p_n\}$ is the set of points obtained from sampling the orthogonal projection of the bone section, it only includes the X and Y coordinates of the DAT file loaded in the app. The vector of coefficients must be adjusted to P; for this purpose, the algebraic distance (D_A) was used. This distance is widely employed because it simplifies the calculations and needs less computational resources [15]. Mathematically, it is obtained by replacing the coordinates of a point $p_i = (x_i, y_i)$ in the Q polynomial; hence, if p_i belongs to the ellipse, its distance will be 0.

$$D_A(p_i, W(v)) = Q(p_i, v) = Ax_i^2 + Bx_i y_i + Cy_i^2 + Dx_i + Ey_i + F \tag{3}$$

The least squares technique optimizes the fit by minimizing the square of the algebraic distance between the P points and the W curve, and it can be expressed as the squared norm of the product between the design matrix D_P, which contains information of P, and the v vector.

$$D_p = \begin{bmatrix} x_1^2 & x_1 & y_1 & y_1^2 & x_1 & y_1 & 1 \\ x_2^2 & x_2 & y_2 & y_2^2 & x_2 & y_2 & 1 \\ \vdots & \vdots & \vdots & \vdots & \vdots & \vdots \\ x_n^2 & x_n & y_n & y_n^2 & x_n & y_n & 1 \end{bmatrix} \tag{4}$$

$$min \sum_{i=1}^{n} D_A(P, W(v))^2 = min \sum_{i=1}^{n} Q(P,v)^2 = min \, \|D_P v\|^2 \tag{5}$$

To avoid the trivial solution of $v = \bar{0}_6$, the vector of coefficients is bounded [15]. Paton [16] analyzed the chromosome shape using a conic fit with a constraint of $\|v\|^2 = 1$, avoiding all coefficients being zero. Therefore, using this constraint, the solution can be an ellipse, hyperbola or parabola; however, because the bone sections, especially in the diaphyseal part, have an elliptical shape, the conic provided by the app will be an ellipse. The Lagrange multipliers allowed us to minimize the distance considering the constraint. Consequently, L is the Lagrange function to be optimized.

$$L = \|D_P v\|^2 - \lambda \left(\|v\|^2 - 1 \right) = v^T D_P^T D_P v - \lambda \left(v^T v - 1 \right) \tag{6}$$

Equating the gradient of L with respect to v to 0 for minimizing the function gives:

$$\nabla_v L = 0 \Leftrightarrow 2 D_P^T D_P v - 2\lambda v = 0 \tag{7}$$

$$D_P^T D_P v = \lambda v \tag{8}$$

The optimization leads to the eigenvector problem; then, λ and v must be an eigenvalue and an eigenvector of $D_P^T D_P$. If $D_P^T D_P v = \lambda v$, Equation (5) will be:

$$min\|D_P v\|^2 = min\ v^T D_P^T D_P v = min\ \lambda \|v\|^2 = min\ \lambda \tag{9}$$

As a result, the coefficient vector (v) that minimizes the algebraic distance will be the eigenvector of $D_P^T D_P$ corresponding to the smallest eigenvalue (λ). Once v is found, Q is defined; however, although CAD programs allow users to enter functions to draw a curve, this operation is often tedious. Therefore, the W ellipse can be defined with five parameters: the coordinates of its center (x_c, y_c), the largest (R) and smallest (r) radius and the angle (α) that rotates the curve counterclockwise. The coefficients of v are renamed:

$$v = [ABCDEF]^T = [a'\ 2b'\ c'\ 2d'\ 2e'\ f']^T \tag{10}$$

The five parameters can be calculated using the following equations [17]:

$$x_c = \frac{c'd' - b'e'}{b'^2 - a'c'} \tag{11}$$

$$y_c = \frac{a'e' - b'd'}{b'^2 - a'c'} \tag{12}$$

$$R = \sqrt{\frac{2\left(a'e'^2 + c'd'^2 + f'b'^2 - 2b'd'e' - a'c'f'\right)}{\left(b'^2 - a'c'\right)\left[\sqrt{(a'-c')^2 + 4b'^2} - (a'+c')\right]}} \tag{13}$$

$$r = \sqrt{\frac{2\left(a'e'^2 + c'd'^2 + f'b'^2 - 2b'd'e' - a'c'f'\right)}{\left(b'^2 - a'c'\right)\left[-\sqrt{(a'-c')^2 + 4b'^2} - (a'+c')\right]}} \tag{14}$$

$$\alpha = \begin{cases} 0; if\ b' = 0\ and\ a' < c' \\ \frac{\pi}{2}; if\ b' = 0\ and\ a' > c' \\ \frac{\arctan\left(\frac{2b'}{a'-c'}\right)}{2}; if\ b' \neq 0\ and\ a' < c' \\ \frac{\pi}{2} + \frac{\arctan\left(\frac{2b'}{a'-c'}\right)}{2}; if\ b' \neq 0\ and\ a' > c' \end{cases} \tag{15}$$

The fitted curve of the bone section is the intersection between an elliptical cylinder, with the W curve as its directrix and the Z-axis as its generatrix, and the section plane $z = Gx + H$ (Figure 2B), where the constants (G, H) are fitted using linear regression from the X and Z coordinates of each point of the DAT file. There are two ways to export the fitted curve from the app to NX®. The first one allows the user to obtain the points of the curve in DAT format by clicking on *Download DAT file* (Figure 2B), and then import these points to NX® and, with the spline tool, obtain the fitted curve. Likewise, the application provides a graph of the W curve containing the parameters that define it (x_c, y_c, R, r,α); these are introduced in the NX® ellipse tool and W is projected to the section plane to obtain the fitted curve (Figure 2B). The elliptical adjustment app allows the user to download the graph of the W curve (Figure 2B) and provides a 3D view of the fitted curves because the app can make several adjustments at the same time.

2.3. Morphological Study

Morphological study of the proximal femur is essential because it is the region that undergoes long-term bone resorption in most state-of-the-art implants [18]. During preoperative planning of THR, the surgeon chooses a suitable stem from among the prostheses manufactured in advance. For this purpose, he/she evaluates the patient's morphology using radiographs; however, the femur has specific and individual characteristics, and this technique does not provide detailed information about the femoral cavity, so the chosen stem may fill it poorly or exceed its dimensions, causing periprosthetic fractures.

In addition, the geometric parameters neck–shaft angle, anteversion and offset would be inadequate, which could result in a dislocated stem [19–24]. The three-dimensional femoral model obtained from the CT scan (Figure 1A) provides more accurate information that allows the morphological study of each patient because it is essential for the customized design of cementless stems, since precise dimensions of the femoral canal guarantee mechanical stability and avoid SS [25].

2.3.1. Neck–Shaft and Mechanical Angle

To measure the neck–shaft angle, modifications of the techniques described by Wang et al. [26] and Zhang et al. [27] were used. Previously, the femoral head was simulated as a sphere; thus, its centers are coincident. If this estimation is not possible due to the fracture of the femoral neck, the acetabulum can be used to define the sphere. Three reference planes are located: the first at the femoral neck isthmus (FNI), a plane parallel to the XY plane and rotated 45° clockwise with respect to the X-axis—45° is the supplement of the average neck–shaft angle according to the study by Gilligan et al. [28]; the second and third planes are located at the end of the lesser trochanter (LT) and 10 mm (LT-10) below and both are parallel to the XY plane (Figure 3A).

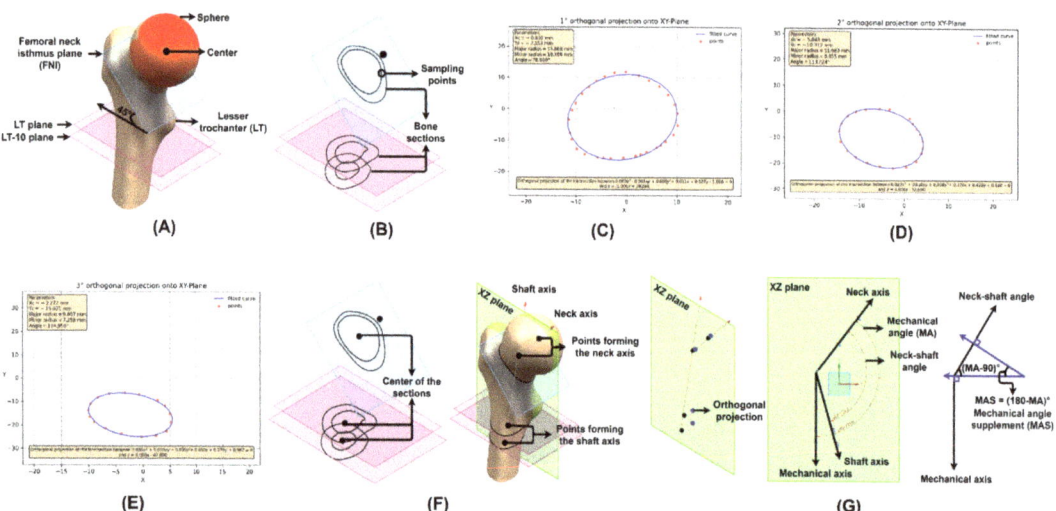

Figure 3. Steps to measure the neck–shaft and mechanical angle. (**A**) Estimation of the femoral head with a sphere and location of the FNI, LT and LT-10 planes. (**B**) Sampling of bone sections. The adjustment made by the app for sections (**C**) FNI, (**D**) LT and (**E**) LT-10. (**F**) Neck and shaft axis. (**G**) Neck–shaft and mechanical angle.

The proximal femur is cut through all three planes, generating bone sections. As described in the section *Elliptical adjustment app* (Section 2.2), the center of the bone section can be found by sampling it with NX® (Figure 3B). The fit performed by the app for the FNI, LT and LT-10 sections is shown in Figure 3C–E, respectively. The femoral neck axis passes through the centers of the sphere and the FNI section, and the shaft axis passes through the centers of the LT and LT-10 sections (Figure 3F). Both axes are orthogonally projected on the XZ plane and the angle between them is the neck–shaft angle. The mechanical axis is parallel to the Z-axis, and the angle between the neck and the mechanical axes is the mechanical angle (Figure 3G). Therefore, the neck–shaft and mechanical angle (MA) for GC1 are 126.4° and 141.9°, and for GC2 are 133.1° and 143°.

The neck–shaft angles of GC1 and GC2 are within the normal range of 90° to 135°; if the inclination is greater than 125°, it is called *coxa valga*, and if it is less than 120°, *coxa vara* [29]. If the stem selected by the orthopedist or designed by the engineer alters the patient's neck–shaft angle, valgus or varus position, a muscular imbalance is generated and, as a consequence, affects the load to which the joint is subjected after THR, favoring the loosening of the implant [30–32].

2.3.2. Anteversion

Yadav et al. [33] measure femoral anteversion three-dimensionally as the angle between the condylar plane, formed by the condylar and neck axes, and the femoral neck plane, composed of the neck and shaft axes. However, the virtual model of the proximal femur (Figure 1) does not include the condyles, and a new strategy to quantify anteversion is proposed, which consists of taking the XZ plane as a reference and redefining the femoral neck plane as the one formed by the neck and mechanical axes, since both planes are formed by an axis parallel to Z, favoring the measurement of anteversion: the angle between the new femoral neck plane and the XZ plane (Figure 4). The approximate anteversion for GC1 and GC2 is 13.5° and 3.6°, respectively.

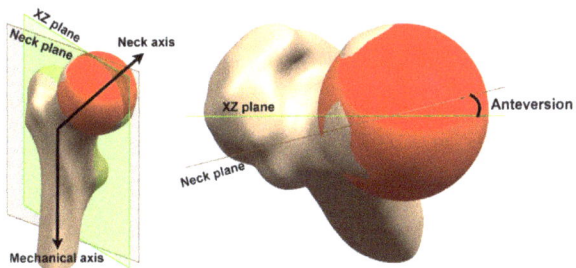

Figure 4. Anteversion.

Anteversion aims to restore the femoral center of rotation [34]. Its reduction leads to increased external rotation of the leg, increases torsional moments on the prosthesis [35–37] and may be associated with an increased risk of loosening [38]. Moreover, it has a strong influence on hip contact forces [39]; therefore, the correct anteversion angle allows an optimal range of motion with minimal risk of instability [40,41].

2.3.3. Offset

The offset is the perpendicular distance between the shaft axis and the center of the femoral head. Because the femoral head was simulated as a sphere to measure the neck–shaft angle (Figure 3A), it has implicit offset information, so it is not necessary to quantify the offset on the condition that the sphere is used in the custom design. This parameter improves physical function, increases hip stability, maintains postoperative pelvic balance and minimizes the risk of dislocations [31,42,43]. Several studies have shown that an increase in offset correlates with a reduced neck–shaft angle, increased range of motion, increased lever arm and abductor strength. If not restored, it increases the reactive force of the joint, consequently causing wear and leading to implant failure [42–45].

2.3.4. Femoral Cavity

The customized stem design determines the areas of contact with the cortical bone, which results in differences in biomechanics and fixation between implants. The goal is to achieve initial stability through fixation with adequate bone contact [46], hence indicating the importance of studying the femoral cavity, as it geometrically delimits the dimensions of the stem and prevents early loosening and periprosthetic fractures. In addition, unlike the geometric parameters of the femur, which correlate with each other, the femoral cavity has highly variable characteristics specific to each person, so it is not proportional to the external femoral geometry. The study of the femoral cavity for the design of conventional stems consists of orthogonal cuts that section the bone [1]. However, the cutting planes will host the sketches that will compose the stem, which is obtained from the interpolation of them.

As shown in Figure 5A, if the conventional analysis is performed, the result does not mimic the lateral side of the proximal femur, increasing the SS because the biomechanics are not restored, since this methodology is optimized to adapt the contact between implant and bone in the calcar and the femoral diaphysis. For this reason, another technique is needed to study the cavity and design the personalized short stem.

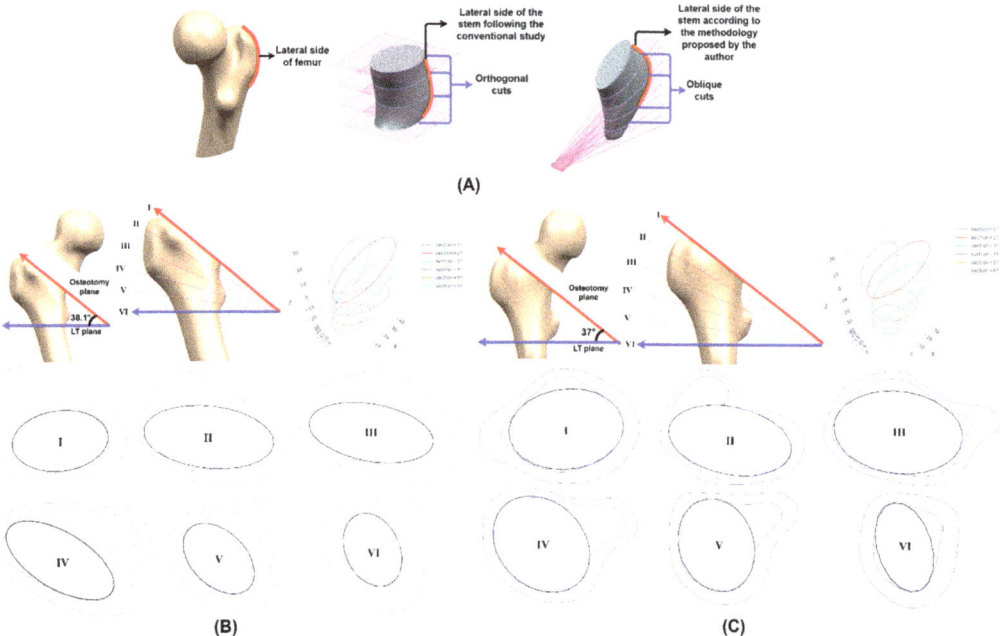

Figure 5. (**A**) Methodologies to study the femoral cavity. Femoral cavity analysis of (**B**) GC1 and (**C**) GC2.

To replicate the curvature of the lateral side, it was necessary to create an arch whose origin generates oblique planes that allow the study of the canal and the design of the stem that adapts to it. To generate the arch, the LT plane was used (Section VI, Figure 5), and then an oblique plane (Section I, Figure 5) was placed below the FNI plane because, according to the study by Solórzano et al. [18], this is the area with the highest risk of fracture of the proximal femur. Finally, the arc created from both planes, whose interior angle is the mechanical angle supplement (MAS) (Figures 3G and 5), was divided into five equal parts, producing the planes II, III, IV and V (Figure 5).

It is possible to obtain more study planes by dividing the arch into more parts; however, the design becomes more complex and the stem less organic.

From the oblique planes, the bone sections used to study the cavity were obtained, and each one was sampled following the procedure described in the *Elliptical adjustment app*; see Section 2.2. The app provided the three-dimensional scheme of the fitted curves and the individual fitting graph of each section, which contained the ellipse parameters and allowed the import of the fitted curve to NX® to check that it is properly adapted to the original bone section (Figure 5). These fitting curves constrained the stem geometry and allowed the study of its implantability.

2.4. Custom Design

Gómez-García et al. [47] mentioned that, in general, the short stem design has five basic defined characteristics: the anatomical region they occupy, geometric characteristics of the design, areas where stress transmission occurs, osteotomy and insertion. The short stem occupies and transmits stresses toward the metaphysis, due to which it is known as a metaphyseal stem. Therefore, throughout this section, geometry, osteotomy and insertion are integrated into the custom design of the short stem. As mentioned in the *Morphological study* (Section 2.3), the three femoral parameters play an important role in muscle action and range of motion; therefore, their preservation is crucial in order not to alter the femoral biomechanics.

2.4.1. Osteotomy

This is the procedure that repairs damaged joints by cutting and remodeling the bones. In THR, its role is to remove the femoral neck to place a stem inside its cavity and remodel the acetabulum to align with the implant and create an artificial joint that restores the patient's mobility. Hereafter, the term osteotomy is used to refer to the femoral neck removal. Dimitriou et al. [48] determined that the cutting plane, called the osteotomy plane, affects the implantation section, the bone section resulting from the removal of the neck through which the prosthesis enters in the cavity (I Section, Figure 5), and the postoperative position of the non-customized femoral stem altering the neck–shaft angle and anteversion due to the complex morphology of its proximal canal. Therefore, they suggest that the osteotomy be optimized considering the alignment of the stem that restores the femoral mechanical response, to avoid generating a muscular imbalance that accelerates loosening. However, in customized implants, this is achieved through individual analysis and design. Consequently, its role is the evaluation of the implantability, since the design of the personalized stem must guarantee the correct interaction between bone and implant (fit) and be able to enter through the implantation section (filling), to prevent fractures during surgery. Recalling the subsection *Femoral cavity* (Section 2.3.4), the I plane was located below the FNI and the angle that it formed with the LT plane or VI section was the MAS; this occurs because the I plane is the osteotomy plane and must consider a cutting zone below the fracture, which would occur in the FNI, and restore femoral parameters such as the neck–shaft angle through the mechanical one (Figure 5).

2.4.2. Insertion

The custom short stem design is characterized by mimicking the curvature of the lateral side of the proximal femur (Figure 5A). This lateral widening requires a new implantation method to achieve femoral reaming, which consists of gradually opening the cavity using calibrated elements similar to the stem until the appropriate size is achieved for insertion, while respecting the greater trochanter and the gluteal muscles. This technique has been called "round the corner" and is possible due to the absence of the distal part of the stem [2,9]. "Round the corner" requires that the reamers and final implant are first inserted in the varus position and then progressively tilted into the correct alignment while descending the femoral metaphysis (Figure 6).

This technique facilitates the use of minimally invasive approaches such as Micro-Hip [49], but precludes the use of intramedullary guides and may also result in a varus position when the tip of the stem touches the lateral side of the femur, contributing to a possible fracture, so the use of fluoroscopy during insertion is advisable [50].

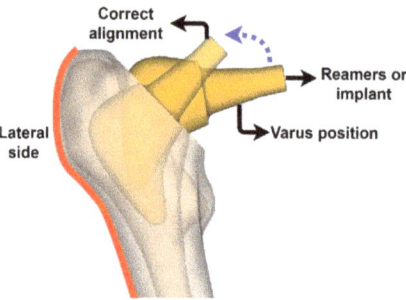

Figure 6. "Round the corner" technique.

2.4.3. Implantability

The designers, despite having the virtual model, often do not consider in the design process the osteotomy and the insertion method, key aspects that determine whether the customized prosthesis is implantable or not. Consequently, a methodology is proposed to study implantability by ensuring that the prosthesis adapts to the canal and its insertion is possible. From a geometric point of view, to use the "round the corner" technique, the limits of the implant sections must be projections of the implantation curve on the planes used in the cavity analysis (Figure 5), due to the rotation performed to place the stem in the correct alignment (Figure 6). The orthogonal projection of the implantation or osteotomy curve can be of two types: the first consists of projecting the I section on the oblique planes (S1); in the second, the I curve is projected on the II plane, and the result is projected on the III plane and so on until reaching the IV plane (S2). Interpolating the generated curves, two solids are formed; however, not only these bodies must be evaluated, but the intersection (S3) and union (S4) of both must also be included in the implantability analysis. These four solids represent the **constraint of the implantation section** (Figure 7A).

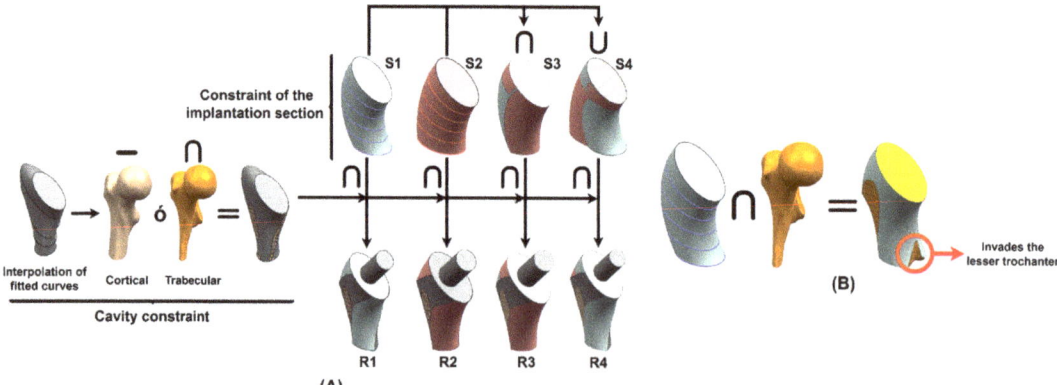

Figure 7. Assessment of implantability (**A**) without and (**B**) with the trabecular bone.

However, the **cavity constraint** should be included in the analysis. As in the previous case, the adjusted curves obtained from the study of the femoral canal are interpolated, forming a solid that approximates the patient's cavity; therefore, it is necessary to rectify areas of overestimation, which invade the cortical part of the femur, intercepting it with the trabecular bone or subtracting the cortical one, achieving, as a result, the maximum volume of the customized stem (Figure 7A). Both constraints must be considered to ensure the implantability of the prosthesis; therefore, the solids, which are a physical representation of the constraints generated by the patient's cavity and the implantation section, have to be intercepted. As a result, four regions are produced, R1, R2, R3 and R4, which are a consequence of the intersection of S1, S2, S3 and S4 with the cavity constraint (Figure 7A).

Regarding the cavity constraint, the question may arise as to why the trabecular bone, which contains exact information about the patient's cavity, is not used directly; this is because, when intercepting the trabecular bone with the constraint of the implantation section, the result invades the lesser trochanter, which, according to the study of Solórzano et al. [18], is a moderately critical area of the proximal femur (Figure 7B). Therefore, fitted curves that adapt to the femoral canal, and do not invade the lesser trochanter, allow proper implant design, avoiding periprosthetic fractures.

To test the implantability, the four regions and the cortical part already included in the osteotomy for GC1 (Figure 8B) and GC2 (Figure 8C) were fabricated to imitate the "round the corner" technique, certifying that the regions enter through the I section and fit the cavity properly.

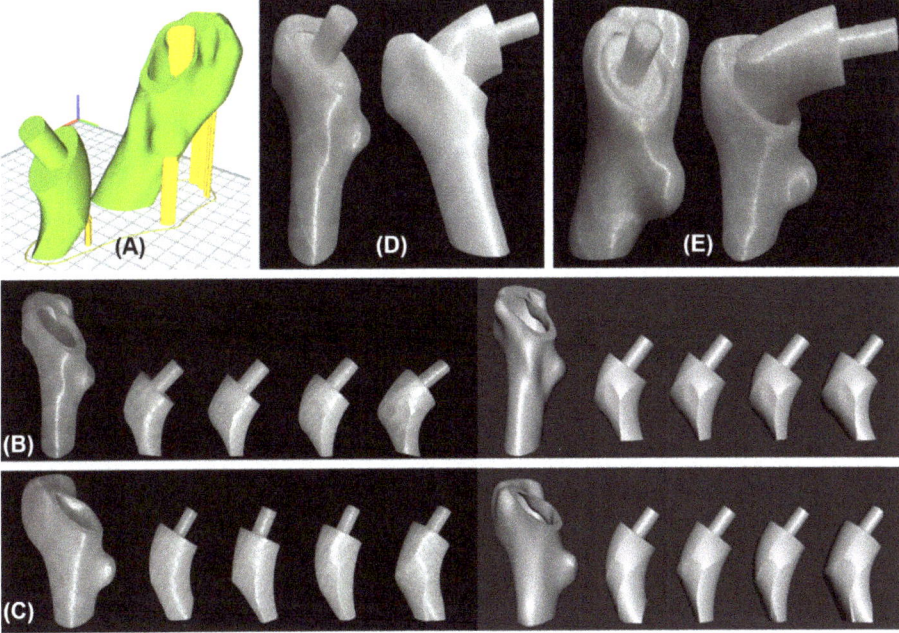

Figure 8. Prototype fabrication. (**A**) Planning using Ultimaker Cura®. PLA prototypes with their respective STLs for (**B**) GC1 and (**C**) GC2. Evidence of experiments performed for (**D**) GC1 and (**E**) GC2 demonstrating the insertion of regions 1, 2 and 3 (left), but not 4 (right).

Therefore, using fused material deposition printing, PLA prototypes were produced from the STL files of the solids, which were laminated in Ultimaker Cura 4.8.0® (Ultimaker, Geldermalsen, Netherlands; Figure 8A) and manufactured using the Ender 3 Pro® (Creality, Shenzhen, China) printer.

The results showed, for both geometric cases, that regions 1, 2 and 3 are implantable solids; therefore, the customized stem was designed from them; however, region 4 is not implantable because it did not enter through the cavity (Figure 8D,E). Now, the choice of two geometric cases with different femoral morphology makes sense since it gives reliability to the conclusions obtained from experimentation; however, to generalize this behavior, further testing with other patients is needed. Moreover, to emphasize that this study is possible thanks to the new methodology proposed to study the femoral cavity since, had the conventional technique been used, the restriction of the implantation section would be an elliptical cylinder. As a result, the implantable solid would not adapt to the lateral side of the femur, which would impair its biomechanics after surgery.

2.4.4. Stem

The implantable regions, because they adequately fit and fill the patient's cavity and enter through the implantation section, were the stem of the customized short implant. However, because of the Boolean operations performed, they were not uniform (Figure 8), hindering bone ingrowth; therefore, they were smoothed, preserving their shape using the Meshmixer® smooth tool with a smoothing scale of 50, and, to facilitate its insertion through the femoral canal, the edge of the VI section was rounded by 5 mm.

2.4.5. Neck and Receiving Taper

Mimicking European standards, the custom stem taper was 12/14 since, according to the study by Morlock et al. [51], this receiving taper is the most commonly used in that continent. The 12/14 model is defined by a proximal diameter of 12 mm, distal of 14 mm and a height of 20 mm, resulting in a taper angle of 5°43'30".

To model the taper, a plane perpendicular to the femoral neck plane was defined and, to preserve the neck–shaft angle in the design, it was rotated (90-MAS)° counterclockwise with respect to the Y-axis. In addition, following the recommendation of Wen-Ming et al. [24], it was placed at the middle of the sphere, which approximates the femoral head, obtaining an oblique plane where the sketch of the 12 mm circumference was drawn. To maintain the height of the cone, a plane 20 mm below the oblique plane was positioned and the sketch of the 14 mm circumference was drawn following the direction of the femoral neck axis. For the neck, the initial curve of the stem was needed, which was obtained by projecting the first section onto the osteotomy plane. This whole process is illustrated in Figure 9A.

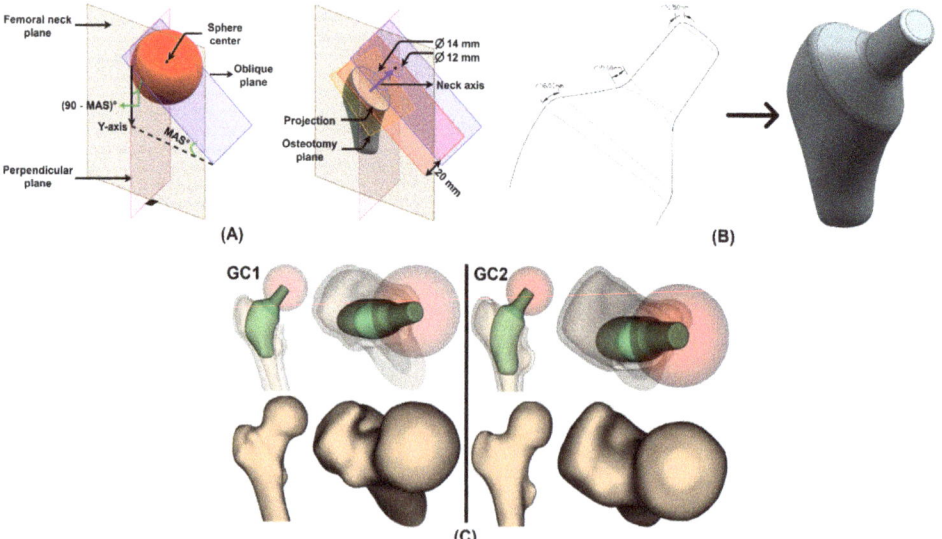

Figure 9. Neck and receiving taper (**A**) sketches and (**B**) solid. (**C**) Comparison between implanted and intact femur for GC1 and GC2.

Based on the drawn curves, which form the neck and the receiving taper, a solid was obtained and integrated into the stem through the Boolean union operation; to avoid stress concentration, the edges of the curves were rounded as shown in Figure 9B. The described process was repeated for each region (R1, R2 and R3) to obtain the final custom stem designs (V1, V2 and V3). Once the design of the short stem has been completed with the described methodology that considers the patient's features, such as his anatomy, and those that depend on the surgery—osteotomy and insertion—the adjustment, filling and implantability of the prosthesis are guaranteed. However, to carry out the design, elements of the femoral morphology study were used, such as the LT plane, the MAS, the femoral neck axis and plane and the sphere that fits the femoral head, thanks to which it was possible to restore the neck–shaft angle, anteversion and offset of each geometric case, as visualized in Figure 9C. Therefore, none of the proposed designs is expected to fail due to muscle imbalance, modification of the range of motion or impingement with the acetabulum. Another point in favor of this technique is its simplicity, since the designer does not need to manually define the implant–bone contact zones because the program created provides the curve that best fits the bone section guaranteeing primary stability, thus avoiding human error in the process. The next step was to perform the finite element analysis (FEA) to select which of the three options is the best stem for each geometric case.

2.5. Finite Element Model

2.5.1. Mesh

NX® was used for performing FEA, employing its Nastran solver. There are two finite element models: intact and implanted femur. In the study of the intact femur, only the cortical and trabecular bone are involved; on the contrary, in the implanted femur, the two osteotomized bones and the customized stem interact. The cortical and trabecular bone for GC1 was meshed with an element size of 1.87 mm; for GC2, the size was 1.3 mm, both for the intact and implanted femur. The stems for each geometric case were meshed with an element size of 0.9 mm. All bodies used CTETRA 10 as the element. The selection of these element sizes and type is a consequence of the convergence analysis performed using the p-method and h-method with an admissible error of 2%, which considers the quality of the results and the speed of calculation. Because most of the remodeling processes occur in full osseointegration [52], the meshes are joined through the "surface-to-surface bonding" tool.

2.5.2. Bone Properties

The biomechanical behavior of bone is extremely complex due to its anisotropic and viscoelastic nature. However, it exhibits elastic behavior under usual mechanical conditions. The femur being a long bone, the analysis was performed considering the transversely isotropic properties of cortical bone and, according to the literature [53], it has been assumed that the trabecular bone presents a large-scale isotropy. Bone properties were estimated using the apparent density (ρ_{app}), which was obtained employing the "Segment Statistics" tool of 3D Slicer® [18] and considering its relationship with the Hounsfield units (HU). Rho et al. [54] determined a linear relationship between HU and apparent density for the proximal femur:

$$\rho_{app} = 131/1000 + 1.067 HU/1000 \, [\text{g/cm}^3] \tag{16}$$

The Young's modulus of cortical bone in the longitudinal direction ($E_{z,cortical}$) and the stiffness of trabecular one ($E_{trabecular}$) were estimated using the equation described by Keyak et al. [55] and rectified by Schileo et al. [56]:

$$E_{z,cortical} = E_{trabecular} = 14{,}900 \left(0.6 \rho_{app}\right)^{1.86} [\text{MPa}] \tag{17}$$

In addition, the Young's modulus (E_x, E_y) and shear modulus (G_{yz}, G_{zx}) in the transverse direction, for cortical bone, were calculated using Pithioux's laws [57]:

$$E_x = E_y = 0.6 E_z \tag{18}$$

$$G_{yz} = G_{zx} = 0.25 E_z \tag{19}$$

Poisson's coefficients in the longitudinal (ν_{yz}, ν_{zx}) and transverse (ν_{xy}) directions of cortical bone were obtained from the literature, being 0.25 and 0.4, respectively [58]; the value of 0.3 for the Poisson's coefficient (ν) of trabecular bone was taken from experimental data [59]. The shear modulus in the longitudinal direction (G_{xy}) of cortical bone [60,61] and the shear modulus (G) of trabecular bone were obtained from the following equations:

$$G_{xy} = \frac{E_x}{2(1+v_{xy})} \quad (20)$$

$$G = \frac{E}{2(1+v)} \quad (21)$$

Table 1 summarizes the physical and mechanical properties of both bones for GC1 and GC2.

Table 1. Physical and mechanical properties of the cortical and trabecular bone of each geometric case.

Properties	Cortical Bone		Trabecular Bone	
	GC1	GC2	GC1	GC2
HU	1458	1197	779	745
ρ_{app} (g/cm^3)	1.69	1.41	0.96	0.93
E_x (MPa)	9140.76	6534.52		
E_y (MPa)	9140.76	6534.52	5363.09	4993.1
E_z (MPa)	15,234.61	10,890.87		
G_{xy} (MPa)	3264.56	2333.76		
G_{yz} (MPa)	3808.65	2722.72	2062.73	1920.42
G_{zx} (MPa)	3808.65	2722.72		
v_{xy}	0.4	0.4		
v_{yz}	0.25	0.25	0.3	0.3
v_{zx}	0.25	0.25		

2.5.3. Stem Properties

The material must be biocompatible to promote osseointegration, and the bone must grow close to the implant surface and fill the grooves or pores that have been deliberately introduced to firmly embed the stem and reduce the bone resorption, be immune and inert to corrosion by body fluids and tissues, be strong and ductile to withstand the mechanical demands of the patient's daily activity, have low density, be light so as not to affect gait and not have magnetic properties, to perform a clinical evaluation after surgery using medical imaging such as MRI or CT [62–65].

Among the materials employed in the manufacture of femoral prostheses, the most used is Ti6Al4V because its Young's modulus is close to that of bone and it has proven to be more biocompatible than stainless steel and cobalt–chromium–molybdenum [65]; it also meets the requirements mentioned above. Nevertheless, titanium implants are retained in bone by mechanical and chemical stabilization, as, through direct contact between calcium atoms and the titanium oxide surface, they create an inorganic interface, leading to osseointegration [46]; however, wear caused by friction between bone and implant liberates metal ions that react biologically with the body, including aluminum ions, which have been linked to the development of diseases such as Alzheimer's and cytotoxicity caused by excessive concentrations of vanadium [66,67].

A substitute for Ti6Al4V may be the Ti alloy Ti-15Mo-2.7Nb-3Al-0.2Si, also known as Ti21S, because it reduces the aluminum content, eliminates vanadium, improving its cytotoxicity, and presents an extremely low Young's modulus, good strength and ductility, excellent corrosion resistance and biocompatibility, which makes this material suitable for biomedical applications [68]. Additive manufacturing (AM) technologies allow the fabrication of specific and intricate patient geometries, reduce stiffness due to inherent porosity and roughness, have been shown to promote bone ingrowth and employ efficient material usage [69–72]. Therefore,

Ti6Al4V ELI (extra low interstitials) and Ti21S were defined as the stem material for the FEA (Table 2) since both are used in the AM of femoral stems.

Table 2. Mechanical properties of the stem material.

Properties	Ti6Al4V ELI [73]	Ti21S [68]
E (GPa)	114	52
G (GPa)	42.5	19.6
v	0.34	0.33
σ_y (MPa)	795	709

2.5.4. Boundary Conditions

Solórzano et al. [18] studied the mechanical behavior of the proximal femur against the nine loads proposed by Bergmann et al. [74], including the ISO force [75], widely used to test femoral stems. They concluded that the representative loads that increase the risk of fracture are ISO and jogging; consequently, both were used to evaluate the differences between the biomechanics of the intact and implanted femur. The jogging load for the intact femur is only composed by the contact forces (F_X, F_Y and F_Z); however, when the stem is implanted, the moments that stress the fixation in the acetabulum appear; hence, the implanted femur is subjected to contact forces and frictional moments (M_X, M_Y and M_Z). This load depends on the body weight of each patient; nevertheless, in this study, the same load state (shown in Table 3) was used for GC1 and GC2 of our recently explained procedure [18], since, being equal, the boundary conditions allowed us to evaluate and compare the influence of the femoral morphology in the stem design.

Table 3. Standardized loads for the intact and implanted femur.

	Jogging [74]	ISO [75]
F_X (N)	−884.8	-
F_Y (N)	−15	-
F_Z (N)	−3222	−2300
M_X (Nm)	−0.69	-
M_Y (Nm)	0.76	-
M_Z (Nm)	0.09	-

To apply the load on the intact femur, the body was first placed in a frontal position and rotated (90-MAS)° clockwise with respect to the Y-axis, and then it was rotated at an angle equal to the patient's anteversion clockwise with respect to the Z-axis, to finally place the load on the cortical femoral nodes that make up the acetabular region—the region located from the beginning to the middle of the femoral head. Since the prostheses were designed using the geometric parameters of the patient, in the implanted femur, the load was applied on the flat part of the receiving taper, which happened to be the same position as in the intact femur—since the cone was designed considering the middle of the femoral head (sphere), the mechanical angle and the anteversion. Therefore, no torque was produced by displacement of the forces, allowing a fair comparison between the intact and implanted femur. In both situations, the movement of the femur was limited through the fixed constraint in the flat part of the cortical and trabecular bone (Figure 10).

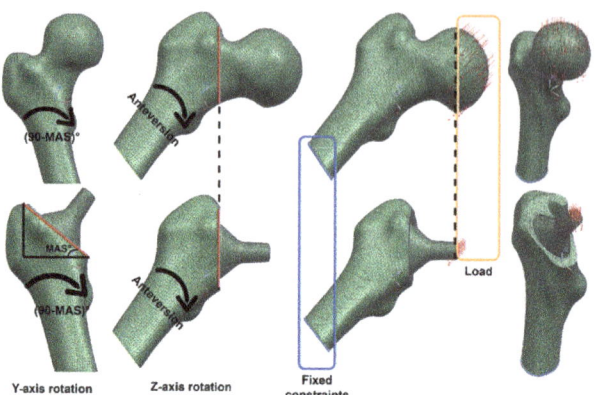

Figure 10. Boundary conditions for the intact and implanted femur.

Once the meshes were generated, the material and the boundary conditions for each of the bodies were defined, and the simulations were carried out in NX®—first for the intact femur of both geometric cases (GC1 and GC2) with the two defined load states (jogging and ISO load), and then for the implanted femur for both geometric cases with the load states and using the two materials (Ti6Al4V ELI and Ti21S) in each of the 3 stems (V1, V2 and V3). Once the results of each simulation were obtained, they were processed to extract the information useful in the evaluation of the SS.

2.5.5. Postprocessing

According to Wolff's law [76], the adaptation of the bone to the mechanical stimulus causes the remodeling process. However, according to the definition of the mechanostat [77], bone adapts towards a target strain; hence, osteocytes sense this stimulus and send biochemical signals that activate cellular action to remodel bone. In vitro or in vivo studies even use strain gauge rosettes to quantify the strain of the femur and study the relationship between in vivo loading and bone adaptation. Strain data recorded in extensometer studies are usually summarized in terms of principal strains. Therefore, it is necessary to represent the multiaxial strain state as an equivalent metric, i.e., to reduce the complicated and directionally specific strain state to a scalar quantity that is independent of direction. This metric is called "equivalent strain"; it was first introduced by Mikić and Carter [78] with the aim of incorporating strain gauge data in the context of bone adaptation models. Turner et al. [79], through clinical testing of patients with femoral prostheses, evaluated changes in their bone density and found that it adequately modeled bone remodeling. It is easy to interpret, direction-invariant and a positive scalar, because, mathematically, it is the norm of the strain tensor (ε_{ij}):

$$\varepsilon_{ij} = \begin{bmatrix} \varepsilon_x & \varepsilon_{xy} & \varepsilon_{xz} \\ \varepsilon_{xy} & \varepsilon_y & \varepsilon_{yz} \\ \varepsilon_{xz} & \varepsilon_{yz} & \varepsilon_z \end{bmatrix} = \begin{bmatrix} \varepsilon_1 & 0 & 0 \\ 0 & \varepsilon_2 & 0 \\ 0 & 0 & \varepsilon_3 \end{bmatrix} \qquad (22)$$

To perform postprocessing, the proximal femur was cut longitudinally using a plane coincident with the Y-coordinate of the elliptical adjustment performed for the implantation section (I) of each geometric case (Figure 11A). Mimicking the position of the strain gauges and the orthopedist's analysis of bone density using medical imaging, shielding and bone remodeling over the outer medial (M) and lateral (L) sides of the proximal femur were evaluated using the equivalent strain, in the region bounded by section I and VI (Figure 11B), since these regions undergo more bone resorption in the proximal femur.

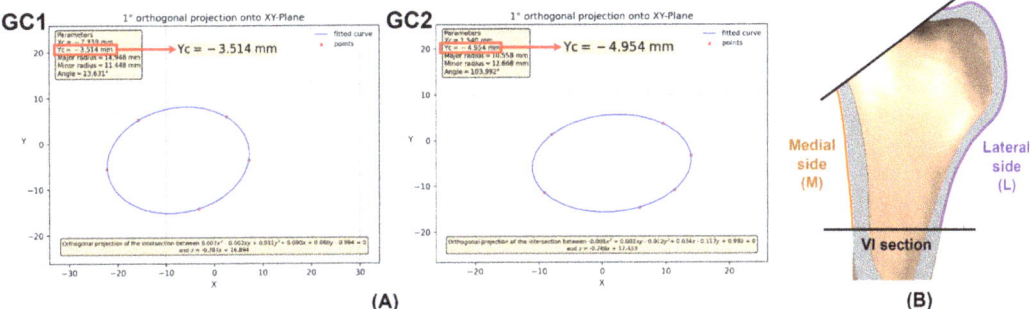

Figure 11. (**A**) Elliptical adjustment of the implantation section. (**B**) Medial and lateral side of the proximal femur.

3. Results and Discussion

3.1. Remodeling Curve and Regression Graph

The bone adapts towards a target strain, and, if this is greater than desired, the bone mass increases, and if it is less, it decreases. The dead zone is defined as the zone where bone resorption and bone formation are in equilibrium. All these characteristics are summarized graphically in the bone remodeling curve (Figure 12), which relates the variations in apparent density to the mechanical stimulus. Likewise, bone density is directly proportional to stiffness and strength and inversely proportional to its ductility; it is understood that an increase or decrease in density causes undesirable mechanical performance.

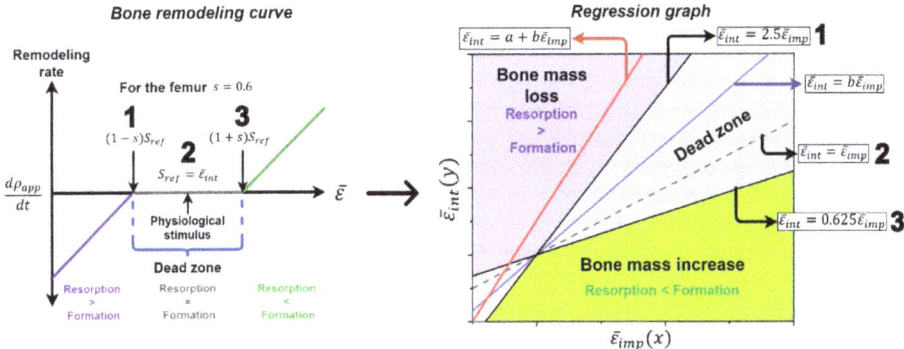

Figure 12. Remodeling curve and regression graph.

Iatrogenic remodeling is related to the bone changes caused by the implant; this type of remodeling should be avoided by the designer and the orthopedist since it contributes to implant loosening and periprosthetic fractures and complicates revision surgeries. Therefore, the ideal stem is one that does not change the femoral biomechanics, does not cause iatrogenic bone remodeling and integrates perfectly through bone ingrowth. However, each stem leads to a specific change in the mechanical response of the femur. Consequently, the designer wants the implant to keep the femur within the dead zone and not cause an excessive increase or decrease in its density. To analyze bone adaptation, the equivalent strain of the mesh element before ($\bar{\varepsilon}_{int}$) and after ($\bar{\varepsilon}_{imp}$) stem insertion was obtained; then, the bone remodeling curve was defined, where S_{ref} is $\bar{\varepsilon}_{int}$, and in order to establish the dead zone, the "s" value was necessary, which, according to the study by Turner et al. [79], is 0.6. Once the parameters were established, the $\bar{\varepsilon}_{imp}$ was located on the abscissae to determine whether it was inside or outside of the dead zone.

Despite defining whether or not the femoral region under study is in the dead zone, many designers analyze a region of the femur by averaging the mechanical stimulus of the mesh elements before ($\bar{\varepsilon}_{int,\ avg}$) and after ($\bar{\varepsilon}_{imp,\ avg}$) surgery, and calculate the respective strain shielding (SS_{avg}). However, the mean of the mechanical stimulus may not represent the loading pattern caused by the stem; consequently, the designer may reach erroneous conclusions using only the average parameters (Table 4).

Table 4. Errors caused by average equivalent strains.

Element	$\bar{\varepsilon}_{int}$ [%]	$\bar{\varepsilon}_{imp}$ [%]	SS
1	0.11	0.05	0.545
2	0.075	0.13	−0.733
3	0.5	0.3	0.4
4	0.08	0.1	−0.25
Average	7.875 ($\bar{\varepsilon}_{int,\ avg}$)	7.75 ($\bar{\varepsilon}_{imp,\ avg}$)	0.016 (SS_{avg})

Consequently, using concepts related to calculus and statistics, a method was found not only analytically but also graphically to evaluate strain shielding, bone remodeling and femoral biomechanics. This consists of transferring the information from the remodeling curve to a regression graph in an equivalent strain plane of the intact implanted femur, as shown in Figure 12. To obtain the regression graph, an assumption is made: the equivalent strain of the intact femur is dependent on the strain of the implanted femur ($\bar{\varepsilon}_{int} = f(\bar{\varepsilon}_{imp})$). This may seem contradictory; however, this assumption is very useful because, if a linear regression is performed between the values of the elemental strain before and after the insertion, the result is:

$$\bar{\varepsilon}_{int} = a + b\bar{\varepsilon}_{imp} \tag{23}$$

From this equation, it is possible to obtain the particular designs of femoral stems. For example, the ideal stem, defined as that which fully restores the femoral biomechanics, whose shielding is zero, describes its behavior through Equation (23) when $a = 0$ and $b = 1$. A stem that preserves the femoral biomechanics will be one whose fit results in the Equation (23) with $|a| \cong 0$; this means that the strain prior to THR is equal to that after but multiplied by a factor "b", and the trend in the mechanical response of the femur is maintained in a scaled manner. For this reason, the shielding of this type of implant is:

$$SS = \frac{\bar{\varepsilon}_{int} - \bar{\varepsilon}_{imp}}{\bar{\varepsilon}_{int}} = 1 - \frac{\bar{\varepsilon}_{imp}}{\bar{\varepsilon}_{int}} = 1 - \frac{1}{b} \tag{24}$$

The stem altering biomechanics is defined by Equation (23) for values of "a" and "b" $\in \mathbb{R}$, and as a result, the shielding is:

$$SS = 1 - \frac{\bar{\varepsilon}_{imp}}{\bar{\varepsilon}_{int}} = \frac{b-1}{b} + \frac{a}{b\bar{\varepsilon}_{int}} \tag{25}$$

High values of $\bar{\varepsilon}_{int}$ result in a strain shielding equal to Equation (24). Therefore, this expression was used to approximate the shielding caused by stems that do not restore the femoral biomechanics. In the regression graph (Figure 12), the ideal stem is represented by the dashed black line, the stem that restores biomechanics by the blue line and the stem that modifies the mechanical response by the red line. We defined the types of stems from the assumption $\bar{\varepsilon}_{int} = f(\bar{\varepsilon}_{imp})$, and it is necessary to bound the dead zone within the graph:

$$\bar{\varepsilon}_{imp} = (1+s)\bar{\varepsilon}_{int} \Leftrightarrow \bar{\varepsilon}_{imp} = 1.6\bar{\varepsilon}_{int} \Leftrightarrow \bar{\varepsilon}_{int} = 0.625\bar{\varepsilon}_{imp} \tag{26}$$

$$\bar{\varepsilon}_{imp} = (1-s)\bar{\varepsilon}_{int} \Leftrightarrow \bar{\varepsilon}_{imp} = 0.4\bar{\varepsilon}_{int} \Leftrightarrow \bar{\varepsilon}_{int} = 2.5\bar{\varepsilon}_{imp} \tag{27}$$

These lines limit the dead zone, which corresponds to the gray area in Figure 12. From this zone, another two are defined: the purple and green indicate the loss and increase of bone mass, respectively. The adjusted R-Square was used to evaluate how good the linear fit of the equivalent strain of the elements is. The definition of this statistical metric is the proportion of the variance of the dependent variable that can be explained by the independent variable or how well the linear fit is able to model the dependent variable from the independent variable, so it is an indirect measure of how dispersed the points are around the fit line.

The results obtained from the simulations were used for the analysis, with the equivalent strains of the elements of both regions, lateral and medial. The linear adjustment was performed to obtain the "a" and "b" coefficients, the adjusted R-Square and the SS, and to evaluate the response of each femur to the implantation of the customized stems and the influence of the material from which it is made. Then, using scatter plots by regions, areas of the femoral stem that can be optimized in a following work to mitigate shielding were visualized. Finally, equivalent strain maps extracted from NX® were obtained for the intact and implanted femur with the selected stem and material to verify if the analysis in the lateral and medial zone is representative for both femurs.

3.2. Analysis

The linear fits between the equivalent strain of the intact and implanted femur with each stem (V1, V2 and V3) were performed for GC1 using both loading states (ISO and jogging) and materials (Ti6V4Al and Ti21S), whose metrics are summarized in Table 5. Figure 13 shows the strain shielding and the adjusted R-Square produced by each stem graphically. In addition, the SS_{avg} was calculated for comparison with the SS obtained from the coefficient "b" of the regression.

Table 5. Results for GC1.

		Ti6Al4V			Ti21S		
		V1	V2	V3	V1	V2	V3
ISO	Adjusted R^2	0.759	0.78	0.78	0.793	0.804	0.804
	Constant (a)	0.024	0.023	0.023	0.017	0.017	0.017
	Coefficient (b)	1.511	1.41	1.412	1.468	1.397	1.399
	SS	0.338	0.291	0.292	0.319	0.284	0.285
	SS_{avg}	0.574	0.535	0.533	0.496	0.473	0.47
Jogging	Adjusted R^2	0.642	0.683	0.683	0.677	0.701	0.701
	Constant (a)	0.013	0.013	0.013	0.008	0.008	0.008
	Coefficient (b)	1.792	1.706	1.708	1.735	1.663	1.664
	SS	0.442	0.414	0.415	0.424	0.399	0.399
	SS_{avg}	0.605	0.58	0.578	0.525	0.506	0.504

The designer is looking for the stem to be as close as possible to the ideal model, with zero shielding, so the implant with the lowest value should be selected. However, as explained in the previous section, the adjusted R^2 is a statistic that evaluates the goodness of the linear fit, so when it is closer to the unit, it is deduced that the points of the curve present a linear trend and are close to the line, which in turn validates the SS obtained. Therefore, there should be a compromise between the adjusted R^2 and the shielding. Figure 13 shows that the lowest SS and highest adjusted R^2 occur when Ti21S is used; regarding implant geometry, the V2 and V3 stems have a very similar mechanical response, with V3 being superior in the SS by thousandths. Then, to select which of the two implants is the indicated one, its volume was evaluated, because the prosthesis with greater volume is heavier, limits the gait and causes patient discomfort. The V2 stem has a volume of 33.25 cm^3 and V3, 32.368 cm^3; because V3 is lighter and has metrics similar to those of V2, it is the ideal implant for GC1.

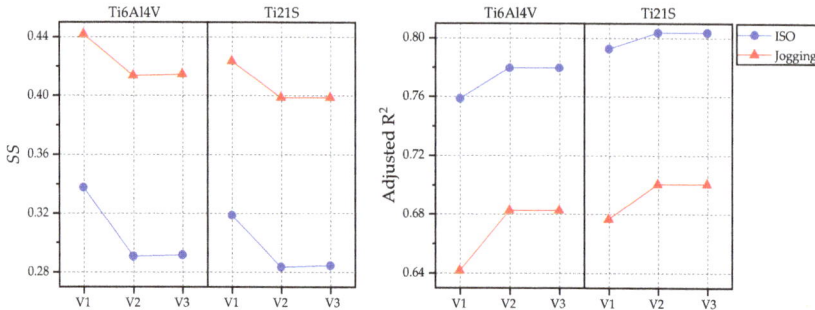

Figure 13. Graphs of the SS and adjusted R^2 for GC1.

The influence of the load has not been mentioned, because evaluating either leads to the same conclusion; therefore, to further understand its effect, we plot the response of GC1 to the insertion of the selected stem (V3) when the femur is subjected to the ISO and jogging loads (Figure 14).

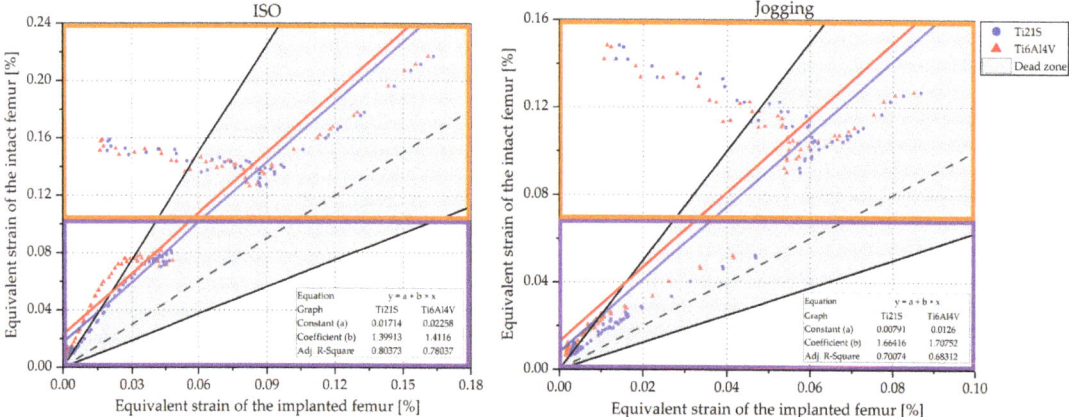

Figure 14. Regression graph of GC1 under ISO and jogging loads.

Examining the range of the axes, it is perceived that jogging loads the proximal femur less in comparison to the ISO force; this is due to its mechanical nature. The femoral neck fracture is caused by high energy mechanisms such as an axial load on the femur; for this reason, ISO overloads it more. Graphically, the ISO force distributes the load better along the femur; for this reason, the points of its regression graph are more concentrated and follow a linear pattern. On the contrary, the jogging load disperses the points more and causes them not to adapt to the regression; as a result, the adjusted R^2 is low (Figure 13). Nevertheless, the conclusions obtained by analyzing any of the two load states do not change, i.e., whether examining the femur under ISO or jogging, the same geometry and material is selected. From this perspective, since the use of the ISO force facilitates the testing of prototypes and allows comparison of the experimental results with the finite element analysis, its use is recommended for the evaluation of femoral implants.

Regression graphs show the influence of the material on the femoral mechanical response. For a closer analysis of its effect, the purple-colored area of the ISO graph in Figure 14 was evaluated.

Young's modulus is related to strain shielding. Figure 15 shows that Ti6Al4V, a material with high stiffness compared to the femur, causes greater shielding, and consequently exposes the points to the bone resorption area.

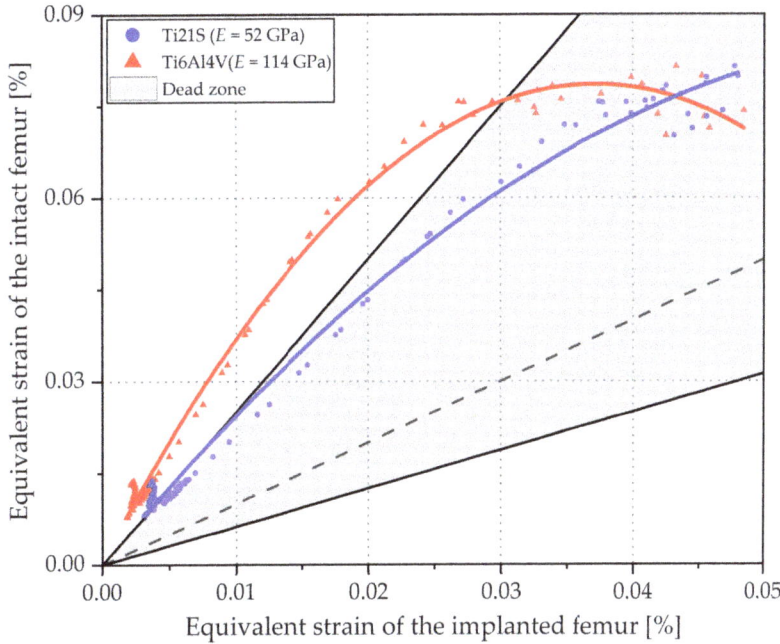

Figure 15. Influence of the modulus of elasticity of the stem material.

In addition, due to its quadratic tendency, it alters the femoral biomechanics since it moves away from the linear behavior of the ideal stem and its adjusted R^2 is lower (Figure 13). In contrast, Ti21S, having lower stiffness, approaches the linear response of the stem that preserves, in a scaled form, the strain of the proximal femur anterior to the THR and maintains the points within the dead zone. In this way, it is verified that, in spite of being the same stem (V3), the selected material originates different mechanical responses; therefore, having a stiffness closer to that of the femur allows the designer to evaluate the behavior originated by the geometry and distinguish it from that caused by the mechanical properties of the material.

The regression graphs were divided by orange and purple squares enclosing the medial and lateral zones, respectively Figure 14. The medial shows a set of points that follows a negative slope and is outside of the dead zone; the stem is made of either Ti6Al4V or Ti21S. To analyze this behavior in depth, Figure 16 shows the scatter plots of the equivalent strain of the intact and implanted femur subjected to both loading states.

The plots of the equivalent strain with respect to the Z-coordinate show that the red and blue curves of the implanted femur mimic the black curve that corresponds to the strain of the intact femur, but in the medial part, from Z = −10, the curves of the implanted femur diverge due to the geometry of V3; this section originates the set of points with the negative slope mentioned above.

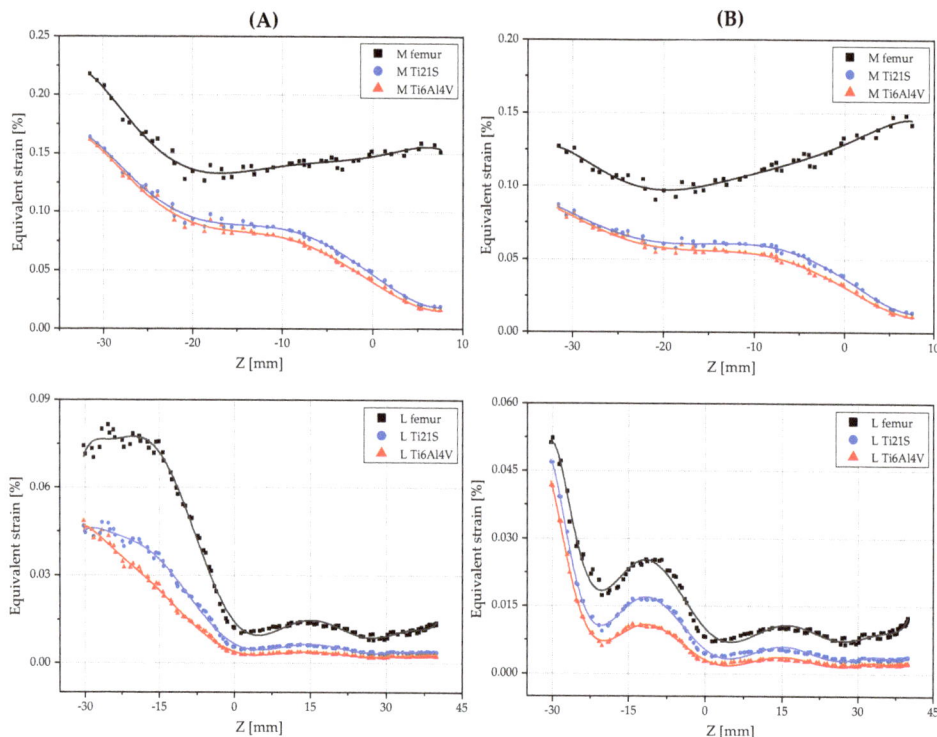

Figure 16. Scatter plots of the equivalent strain of GC1 in the medial and lateral sides under (**A**) ISO and (**B**) jogging loads.

The scatter plots complement the results obtained from the regression graphs. These plots confirm that the material with the lower modulus of elasticity not only reduces the difference between the strain of the intact and implanted femur, but also preserves the femoral biomechanics. Furthermore, it certifies that any of the loads is useful to select the geometry and material of the stem; further proof of this is that both exhibit the alteration of the medial curve of the implanted femur from Z = −10 onwards. For this reason and due to the above advantages, to study the customized stems of GC2, the femur subjected to the ISO force was evaluated by performing the same analysis of GC1.

The metrics of the GC2 linear fits are summarized in Table 6, and Figure 17 exposes the SS and adjusted R^2 produced by each stem graphically.

Table 6. Results for GC2.

	Ti6Al4V			Ti21S		
	V1	V2	V3	V1	V2	V3
Adjusted R^2	0.617	0.668	0.667	0.703	0.72	0.719
Constant (*a*)	0.097	0.09	0.089	0.073	0.073	0.072
Coefficient (*b*)	1.213	1.052	1.043	1.26	1.082	1.079
SS	0.176	0.049	0.041	0.206	0.076	0.073
SS_{avg}	0.611	0.512	0.505	0.521	0.443	0.437

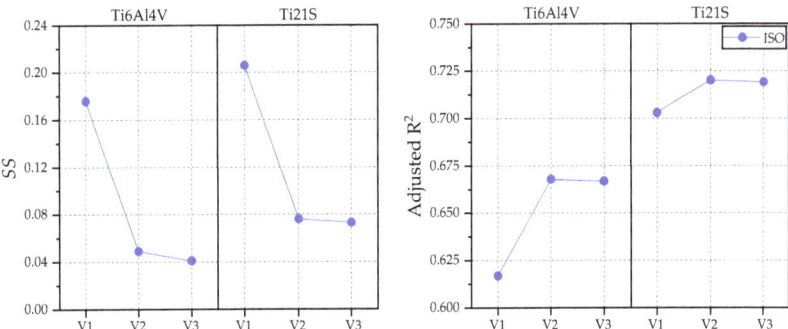

Figure 17. Graphs of the SS and adjusted R^2 for GC2.

Figure 17 shows that the shielding caused by the Ti21S stem is higher compared to those manufactured with Ti6Al4V, which is contradictory to the deductions obtained from the previous analysis. However, the adjusted R^2 of the Ti21S stem is much higher and, because the shielding is a result of the linear fit, Ti6Al4V cannot be reliably selected as a material in this case. Regarding geometry, again, V2 and V3 have very similar metrics, with the smaller volume being the reason that the V3 stem is preferred. When the metrics do not allow correct selection of the material, the visual method is used. Figure 18 shows the regression graph generated by the chosen geometry, produced with both materials.

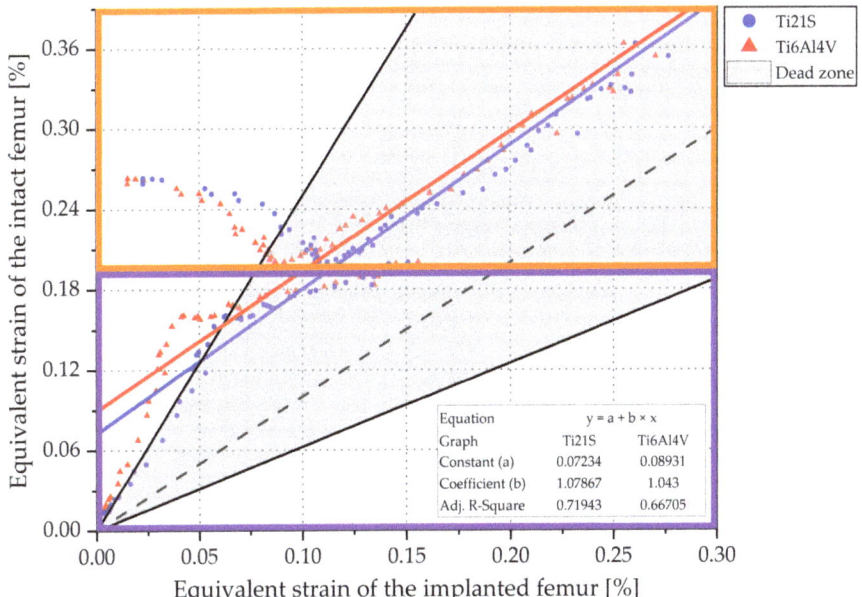

Figure 18. Regression graph of GC2.

The lateral area of the graph (purple box) shows the influence of material stiffness on the restoration of the femoral biomechanics. It is evident that Ti6Al4V locates a greater number of points outside the dead zone; therefore, its shielding is greater, and the deductions based on the theory and the previous analysis are not contradicted by the information shown in the figure. In short, Ti21S is the ideal material for the fabrication of the customized stem.

The value of the independent term (*a*) indicates whether the implant deviates from the ideal behavior and alters the load distribution along the femur, which, in this case, is quantified by the equivalent strain; therefore, the farther the regression is from the center of coordinates ($|a| > 0$), the more the implanted femur strain diverges with respect to the intact one, modifying the load received by the bone and increasing the shielding.

The independent term for GC2 defines that Ti6Al4V more strongly alters the mechanical response of the femur, because the red line is more distant from the center of coordinates. Graphically, the Ti6Al4V regression is above the Ti21S line, cutting the Y-axis at point 0.089.

Figure 18 shows, in the orange box, corresponding to the medial zone, a set of points outside the dead zone and with a negative slope, and, in the purple box, corresponding to the lateral zone, a series of points moving away from the linear trend. The scatter plot of GC2 (Figure 19) supports the choice of material and allows us to identify in which specific regions the geometry of the selected stem should be optimized. In the medial region, it should be improved from $Z = -15$ onwards, and, in the lateral region, from $Z = -30$ to $Z = -15$ because, in these ranges, the strain of the implanted femur, with the stem made of either Ti6Al4V or Ti21S, diverges from the strain of the intact femur, with this effect resulting from the geometry of the V3.

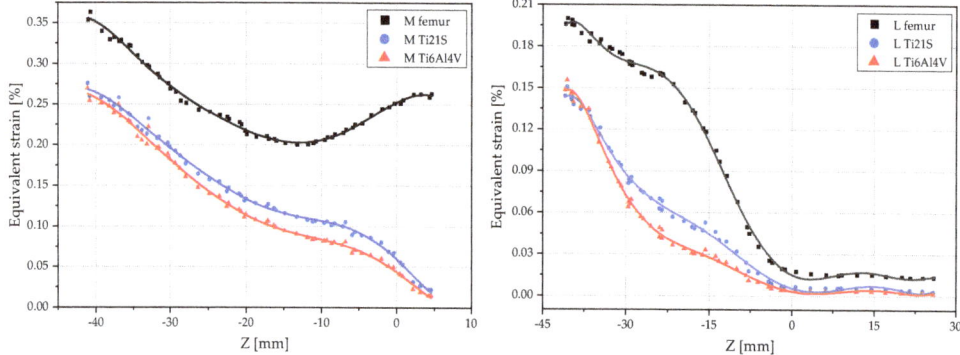

Figure 19. Scatter plots of the equivalent strain of GC2 in the medial and lateral sides.

Once the geometry and the material of the customized stem have been selected, it is necessary to verify whether the SS obtained through the proposed method is better than the SS_{avg}. For this purpose, we resort to the strain maps, which provide equivalent information to performing photoelastic tests. All the maps of the intact femur of both geometric cases distribute the color scale in the range from 0 to 0.4, while the range goes from 0 to 0.29 and from 0 to 0.21 for the maps of the implanting femur of GC1, and from 0 to 0.37 and from 0 to 0.23 for the maps of GC2 (Figure 20), both calculated from the SS and SS_{avg}, respectively, a consequence of the insertion of the V3 stem made of Ti21S.

From the maps, it is verified that the SS adequately quantifies the strain shielding in comparison to the SS_{avg}, because of the similarity between the strain maps of the intact and implanted femur when this metric is used. As the femur is mostly within the dead zone, it favors bone ingrowth, which can be supplemented with osteoconductive liners, benefiting the secondary stability, prolonging its lifespan and improving cementless fixation. Likewise, the strain maps certify that the shear planes used for postprocessing (Figure 11), which allow us to study the mechanical behavior and shielding in the lateral and medial part, are a representative sample of the mechanical response of the entire proximal femur. This plane was obtained from the Y-coordinate of the elliptical adjustment of the implantation section, and it reflects another use of the application that aids not only in design, but also in custom stem analysis and selection. The orange and purple boxes show, similar to how the traumatologist evaluates the shielding radiologically, the decrease in the color scale of the medial and lateral region, respectively, which translates into the loss of bone mass of the femur as a natural response to the removal of the neck. This contrasts with the analysis of the scatter plots of each geometric case (Figures 16 and 19).

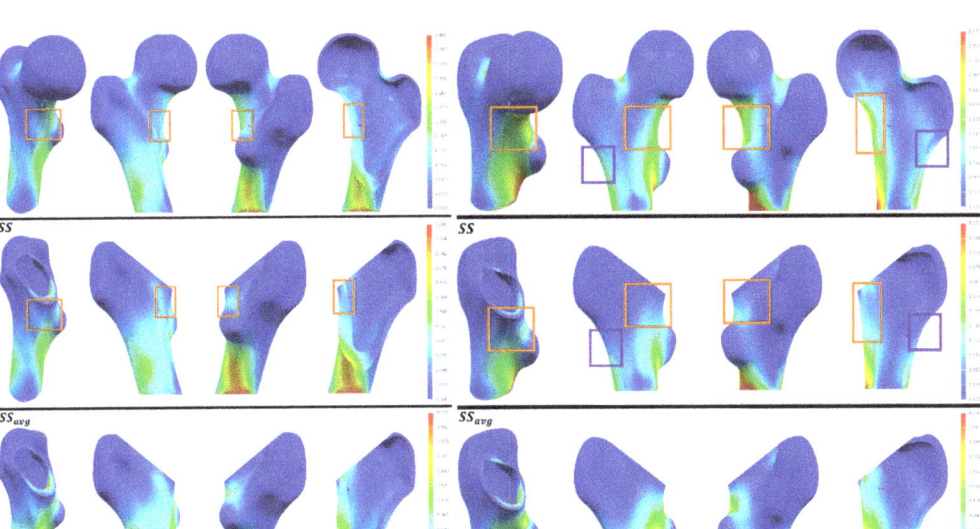

Figure 20. Equivalent strain map of the intact and implanted femur of (**A**) GC1 and (**B**) GC2.

The study performed by Yan et al. [80], whose boundary conditions are similar to this research, on the shielding caused by two commercial stems—one of a conventional type and the other short calcar-loaded—concludes that the SS in the proximal femur caused by the conventional stem is 0.93 and that by the calcar loading stem is 0.82 approximately. Therefore, both commercial implants place the femur outside the dead zone of the bone remodeling curve, so there will be a bone resorption that, in the long-term, will cause the implant to loosen and a revision surgery will be necessary to replace it. Yamako et al. [81], using strain gauges, quantified, through the equivalent strain, that the shielding in the proximal femur caused by a conventional implant made with Ti6Al4V was 0.61, being positioned at the limit of the dead zone. The shielding resulting from the insertion of the V3 stem made of Ti21S was 0.285 and 0.073 for GC1 and GC2, respectively. Therefore, customization is beneficial in the mechanical response of the proximal femur; this is mainly due to the restoration of the parameters of the patient's anatomy (neck–shaft angle, anteversion, offset and femoral cavity) and to the selection of a material that has a modulus of elasticity close to the bone.

However, the precise orientation of the implant is crucial in order not to alter these parameters and consequently its biomechanics; this depends on the surgeon's expertise, but, to avoid human error in the process, technological assistance is becoming more and more common.

Since Ti21S is an isotropic and ductile material, the Von Mises criteria were used. The V3 stem for both geometric cases, subjected to the ISO force, has an average safety factor of 6.055, which guarantees that the implant does not yield to the load. Analyzing the Von Mises stress map of V3 (Figure 21), it is observed that the area with the highest concentration of stresses is the receiving taper; this is beneficial because the stresses generate the compression of the cone walls with the articulating sphere, causing an interference fit and a cold weld between them [82].

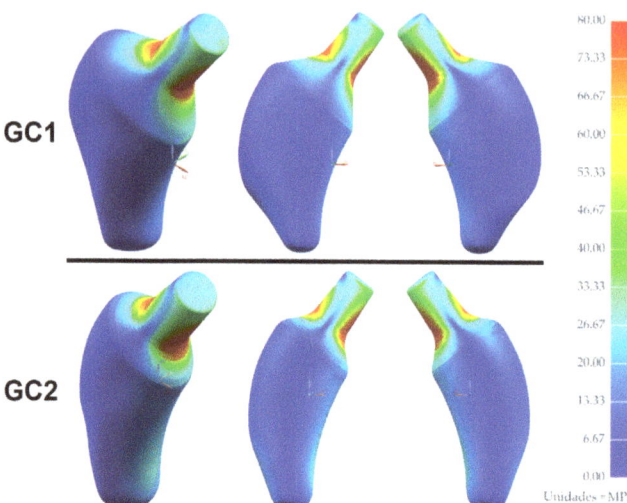

Figure 21. Von Mises stress map of the V3 stem.

To verify the implantability of the prosthesis, PLA prototypes of the V3 stem of both geometric cases were made using fused material deposition printing. With the cortical part of the osteotomy already performed, which was used in the *Implantability* (Section 2.4.3), the "round the corner" technique (Figure 6) performed by the traumatologist was imitated when inserting the stem into the canal, verifying that the implant enters normally (Figure 22).

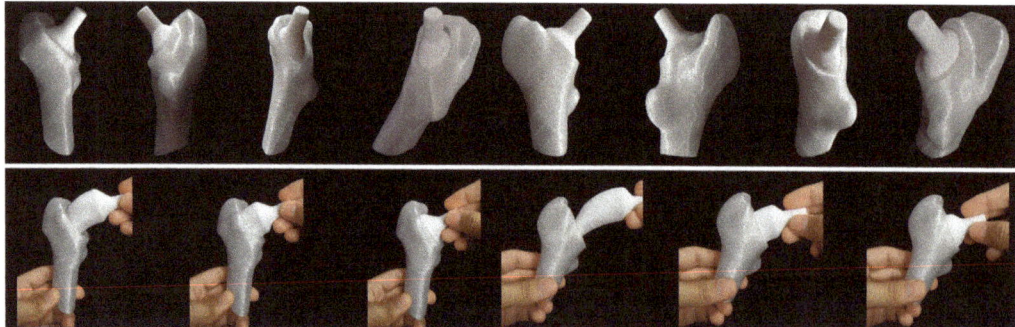

Figure 22. Imitation of the "round the corner" technique for GC1 and GC2.

4. Limitations and Future Proposals
4.1. Limitations of the Study

The authors would like to point out that the study, in its present form, is mainly theoretical and simply aims to present a set of CAD and FEM resources for a more straightforward personalization and in silico evaluation of short stem hip prostheses. Furthermore, the printed prototypes are merely conceptual test probes, in no case intended to be implanted or in vivo tested yet. Several improvements should be performed before considering the proposed designs viable, and the final manufacturing technologies would be completely different to those employed here for a preliminary evaluation of implantability. Probably, for a hypothetical personalized implant with a design such as those presented here, additive manufacturing of metallic alloys (selective laser sintering/melting) would be a good choice, as well as lithography-based ceramic manufacturing using biomedical ceramics, although, in all cases, a final postprocess (sand blasting, PVD-CVD coating) would be beneficial for enhancing osseointegration. The 3D printing of conceptual prototypes helped to initially assess implantability, although design improvements should of course require medical support. In fact, according to the EU Medical Device Regulation 2017/745 (and to most medical regulations worldwide), customized implants cannot reach patients without the prescription and implication in the design procedure of physicians and surgeons. This study provides the basic point of view of biomechanical engineers.

Collaboration with surgeons would be fundamental for improving the design and for considering challenging issues that can be encountered in real-life surgery, including: (1) the need for rasping and for personalized rasps, which could be designed by downscaling the personalized implants and printing both the customized implants and the supporting tools; (2) the occurrence of unexpected collisions during surgery, which could be alleviated by printing two or three models of the customized design, with slightly different surface finishes or scales, and (3) the potential contraindications when abnormal morphologies are present. In the authors' opinion, contraindications for the proposed customized designs would be similar to those applicable to short stems in general: presence of hip dysplasia, severe osteoporosis and previous hip osteotomies.

In any case, before the presented designs can be considered successful solutions, systematic in vitro and in vivo evaluations (with test benches and animal models following the three R principles and applicable regulations) guided by physicians and surgeons are needed.

4.2. Future Research Proposals

Future work, related to the elliptical adjustment application, consists of improving the code and integrating it with a clustering algorithm so that the atypical points of the bone section do not affect the fitting performed by the app. Furthermore, one could consider integrating the program with a computer vision library so that the adjustment is performed only with an image of the cavity (Figure 5B,C), making the extraction of points unnecessary.

For the finite element analysis, one option would be to use the open-source program Bonemat®, whose purpose is to define the elastic properties of each element according to the CT information, thus creating a fully anisotropic mesh that will allow a more accurate evaluation of the mechanical response of the femur before and after stem insertion. Using this program, it is no longer necessary to distinguish between cortical and trabecular bone because the mechanical properties are related to the HU information of each CT voxel, thus simplifying the simulation and making it more personalized.

The stem can improve its geometry by manually regulating its oblique sections, the V3 sections being a limiting condition because, if they are exceeded, the new prosthesis will not be implantable. However, it is possible to program and train a machine learning algorithm that, based on an optimization process, determines the best section that preserves the femoral biomechanics and reduces shielding.

Moreover, topological optimization is an interesting tool that allows a reduction in the weight of the implant and ensures an optimal distribution of the material, as well the optimal load, and it is possible to manufacture it using AM. In fact, the surgeon has the availability of a wide number of prosthesis micro-architectures, thus needing adequate guidelines for the choice of the best one to be implanted in a patient-specific anatomic region [83]. Thus, using strain maps, the designer can improve the stem by mimicking the architecture of the trabecular bone, whose porosity reduces stiffness, decreasing shielding and favoring bone ingrowth, ensuring secondary stability.

A relative micro-displacement analysis should be performed to verify the primary stability, and estimate the secondary stability, of the short prosthesis at the bone–implant interface. In addition, the possibility of designing a short stem that allows the introduction of necessary medications at the postoperative stage should be studied, with the benefits of requiring fewer doses and being applied directly, improving the patient's recovery and reducing the probability of infection. Likewise, the clinical evaluation of the implanted stems should be extended using surgical assistants such as ROBODOC, because it guarantees the correct cutting and reaming, which allow the precise location of the implant according to the design; in addition, it favors primary stability due to the fact that the tight insertion inside the femoral canal restricts relative displacements, favoring the formation of bone tissue.

5. Conclusions

This research work has proposed new tools and concepts that facilitate the custom design of short femoral stems. The application of elliptical adjustment has proven to be very useful for studying femoral morphology, assessing implantability and designing and selecting the customized implant. This instrument can be used in the design of conventional femoral stems or other prostheses in arthroplasty because human error is eliminated when trying to empirically fit an ellipse to the bone cavity.

The implantability has been defined, which is based on the integration of the anatomical parameters with the factors related to surgery, to verify that the stems designed adapt to the femoral cavity and are implantable. Consequently, three geometries for each case study have been designed and evaluated using the finite element method. To analyze the results of the simulations, a methodology based on regression graphs, scatter plots and strain maps that integrate the study of shielding, bone remodeling and femoral biomechanics has been proposed and proven to be more effective and provide more information to the designer compared to the conventional methodology. Based on this, the V3 stem has been selected for having low shielding, keeping the femur within the dead zone and being light in relation to the other geometries, and the Ti21S material, because it restores femoral biomechanics, reduces shielding and does not present the adverse effects of Ti6Al4V related to Alzheimer's disease and cytotoxicity caused by vanadium.

It has been proven, through analysis, that customized implants restore the patient's functional mobility, improving their quality of life, because they reproduce the physiological distribution of the femur, in a scaled form, but subsequent optimization is necessary to resemble more closely the mechanical response before surgery. Likewise, it has been shown that the ISO force, being a high energy mechanism, better distributing the load along the femur and facilitating prototype testing and analysis by the finite element method, should be the load used to evaluate the mechanical response of the femur to the insertion of short prostheses.

Author Contributions: Conceptualization, methodology and formal analysis, W.S.-R., C.O., A.D.L.; design and simulation, W.S.-R.; writing—original draft, W.S.-R.; writing—review and editing, C.O., A.D.L.; supervision, C.O., A.D.L. All authors have read and agreed to the published version of the manuscript.

Funding: This research was funded by Fondo Nacional de Desarrollo Científico, Tecnológico y de Innovación Tecnológica (FONDECYT) of Peru by contract N°316-2019, which afforded W.S.R. the opportunity to remain in the Product Development Laboratory at Universidad Politécnica de Madrid for three months.

Institutional Review Board Statement: Not applicable.

Informed Consent Statement: The Cancer Imaging Archive (TCIA) is a service that de-identifies and hosts a large archive of medical images of cancer, accessible for public download after informed consent from patients. The data are organized as "collections"—typically, patients' imaging related by a common disease (e.g., lung cancer), image modality or type (MRI, CT, digital histopathology, etc.) or research focus. DICOM is the primary file format used by TCIA for radiology imaging. Supporting data related to the images, such as patient outcomes, treatment details, genomics and expert analyses, are also provided when available. In this study, two samples of hip images from TCIA were used as input for the design and simulation studies.

Data Availability Statement: To promote collaborative work and encourage the development and improvement of the proposed design methodology, the STL files of both geometric cases, including their cortical and osteotomized bone, and the short stems are openly available in https://www.thingiverse.com/thing:5187570 (accessed on 3 January 2022).

Acknowledgments: The authors express their gratitude to the members of the Grupo de Biomecánica at Universidad de Piura and División de Ingeniería de Máquinas at Universidad Politécnica de Madrid, especially Maritza Requejo, Mar Cogollo, Pedro Ortego, Isabel Moscol, Gustavo Grosso and Javier Tuesta, for inspiring and supporting us to conduct research that benefits society. The authors acknowledge the advice of the reviewers and their relevant questions, which helped us to present a more detailed and consistent paper, and to better describe the limitations and current challenges.

Conflicts of Interest: The authors declare no conflict of interest.

References

1. Ojeda, C. Estudio de La Influencia de Estabilidad Primaria En El Diseño de Vástagos de Prótesis Femorales Personalizadas: Aplicación a Paciente Específico. Ph.D. Thesis, Universidad Politécnica de Madrid, Madrid, Spain, 2009.
2. Santori, N.; Lucidi, M.; Santori, F.S. Proximal Load Transfer with a Stemless Uncemented Femoral Implant. *J. Orthop. Traumatol.* **2006**, *7*, 154–160. [CrossRef]
3. Fetto, J.F.; Bettinger, P.; Austin, K. Reexamination of Hip Biomechanics during Unilateral Stance. *Am. J. Orthop. Belle Mead NJ* **1995**, *24*, 605–612. [PubMed]
4. Duque Morán, J.; Navarro Navarro, R.; Navarro García, R.; Ruiz Caballero, J.A. Biomecánica de La Protesis Total de Cadera: Cementadas y No Cementadas. *Canar. Médica Quirúrgica* **2011**, *9*, 32–47.
5. Ridzwan, M.I.Z.; Shuib, S.; Hassan, A.Y.; Shokri, A.A.; Mohamad Ib, M.N. Problem of Stress Shielding and Improvement to the Hip Implant Designs: A Review. *J. Med. Sci.* **2007**, *7*, 460–467. [CrossRef]
6. Swiontkowski, M.F.; Winquist, R.A.; Hansen, S.T. Fractures of the Femoral Neck in Patients between the Ages of Twelve and Forty-Nine Years. *J. Bone Jt. Surg.* **1984**, *66*, 837–846. [CrossRef]
7. Sornay-Rendu, E.; Boutroy, S.; Munoz, F.; Delmas, P.D. Alterations of Cortical and Trabecular Architecture Are Associated with Fractures in Postmenopausal Women, Partially Independent of Decreased BMD Measured by DXA: The OFELY Study. *J. Bone Miner. Res.* **2007**, *22*, 425–433. [CrossRef]
8. Garden, R.S. Low-Angle Fixation in Fractures of the Femoral Neck. *J. Bone Jt. Surg. Br.* **1961**, *43-B*, 647–663. [CrossRef]
9. Santori, N.; Albanese, C.V.; Learmonth, I.D.; Santori, F.S. Bone Preservation with a Conservative Metaphyseal Loading Implant. *Hip Int.* **2006**, *16*, 16–21. [CrossRef]
10. Jasty, M.; Krushell, R.; Zalenski, E.; O'Connor, D.; Sedlacek, R.; Harris, W. The Contribution of the Nonporous Distal Stem to the Stability of Proximally Porous-Coated Canine Femoral Components. *J. Arthroplast.* **1993**, *8*, 33–41. [CrossRef]
11. Zuley, M.L.; Jarosz, R.; Drake, B.F.; Rancilio, D.; Klim, A.; Rieger-Christ, K.; Lemmerman, J. Radiology Data from The Cancer Genome Atlas Prostate Adenocarcinoma [TCGA-PRAD] Collection. *Cancer Imaging Arch.* **2016**. [CrossRef]
12. Yorke, A.A.; McDonald, G.C.; Solis, D.; Guerrero, T. Pelvic Reference Data. *Cancer Imaging Arch.* **2019**. [CrossRef]
13. Rinaldi, G.; Capitani, D.; Maspero, F.; Scita, V. Mid-Term Results with a Neck-Preserving Femoral Stem for Total Hip Arthroplasty. *HIP Int.* **2018**, *28*, 28–34. [CrossRef]
14. Sabatini, A.L.; Goswami, T. Hip Implants VII: Finite Element Analysis and Optimization of Cross-Sections. *Mater. Des.* **2008**, *29*, 1438–1446. [CrossRef]
15. Fitzgibbon, A.; Pilu, M.; Fisher, R.B. Direct Least Square Fitting of Ellipses. *IEEE Trans. Pattern Anal. Mach. Intell.* **1999**, *21*, 476–480. [CrossRef]
16. Paton, K. Conic Sections in Chromosome Analysis. *Pattern Recognit.* **1970**, *2*, 39–51. [CrossRef]
17. Cruz-Díaz, C. Ajuste Robusto de Múltiples Elipses Usando Algoritmos Genéticos. Master's Thesis, Instituto Politécnico Nacional, Mexico City, México, 2012.
18. Solórzano, W.; Ojeda, C.; Diaz Lantada, A. Biomechanical Study of Proximal Femur for Designing Stems for Total Hip Replacement. *Appl. Sci.* **2020**, *10*, 4208. [CrossRef]
19. Carter, L.W.; Stovall, D.O.; Young, T.R. Determination of Accuracy of Preoperative Templating of Noncemented Femoral Prostheses. *J. Arthroplast.* **1995**, *10*, 507–513. [CrossRef]
20. Crooijmans, H.J.A.; Laumen, A.M.R.P.; van Pul, C.; van Mourik, J.B.A. A New Digital Preoperative Planning Method for Total Hip Arthroplasties. *Clin. Orthop.* **2009**, *467*, 909–916. [CrossRef]
21. Fottner, A.; Peter, C.V.; Schmidutz, F.; Wanke-Jellinek, L.; Schröder, C.; Mazoochian, F.; Jansson, V. Biomechanical Evaluation of Different Offset Versions of a Cementless Hip Prosthesis by 3-Dimensional Measurement of Micromotions. *Clin. Biomech.* **2011**, *26*, 830–835. [CrossRef] [PubMed]
22. Viceconti, M.; Lattanzi, R.; Antonietti, B.; Paderni, S.; Olmi, R.; Sudanese, A.; Toni, A. CT-Based Surgical Planning Software Improves the Accuracy of Total Hip Replacement Preoperative Planning. *Med. Eng. Phys.* **2003**, *25*, 371–377. [CrossRef]
23. Toth, K.; Sohar, G. Short-Stem Hip Arthroplasty. In *Arthroplasty—Update*; Kinov, P., Ed.; InTech: London, UK, 2013; ISBN 978-953-51-0995-2.
24. Wen-ming, X.; Ai-min, W.; Qi, W.; Chang-Hua, L.; Jian-fei, Z.; Fang-fang, X. An Integrated CAD/CAM/Robotic Milling Method for Custom Cementless Femoral Prostheses. *Med. Eng. Phys.* **2015**, *37*, 911–915. [CrossRef] [PubMed]
25. Baharuddin, M.Y.; Salleh, S.-H.; Zulkifly, A.H.; Lee, M.H.; Mohd Noor, A. Morphological Study of the Newly Designed Cementless Femoral Stem. *BioMed Res. Int.* **2014**, *2014*, 692328. [CrossRef] [PubMed]

26. Wang, Z.; Li, H.; Zhou, Y.; Deng, W. Three-Dimensional Femoral Morphology in Hartofilakidis Type C Developmental Dysplastic Hips and the Implications for Total Hip Arthroplasty. *Int. Orthop.* **2020**, *44*, 1935–1942. [CrossRef]
27. Zhang, R.-Y.; Su, X.-Y.; Zhao, J.-X.; Li, J.-T.; Zhang, L.-C.; Tang, P.-F. Three-Dimensional Morphological Analysis of the Femoral Neck Torsion Angle—An Anatomical Study. *J. Orthop. Surg.* **2020**, *15*, 192. [CrossRef] [PubMed]
28. Gilligan, I.; Chandraphak, S.; Mahakkanukrauh, P. Femoral Neck-Shaft Angle in Humans: Variation Relating to Climate, Clothing, Lifestyle, Sex, Age and Side. *J. Anat.* **2013**, *223*, 133–151. [CrossRef]
29. Gusmão, L.C.B.D.; Sousa Rodrigues, C.F.D.; Martins, J.S.; Silva, A.J.D. Ángulo de Inclinación Del Fémur En El Hombre y Su Relación Con La Coxa Vara y La Coxa Valga. *Int. J. Morphol.* **2011**, *29*, 389–392. [CrossRef]
30. Houcke, J.; Khanduja, V.; Pattyn, C.; Audenaert, E. The History of Biomechanics in Total Hip Arthroplasty. *Indian J. Orthop.* **2017**, *51*, 359–367. [CrossRef]
31. Charles, M.N.; Bourne, R.B.; Davey, J.R.; Greenwald, A.S.; Morrey, B.F.; Rorabeck, C.H. Soft-Tissue Balancing of the Hip: The Role of Femoral Offset Restoration. *Instr. Course Lect.* **2005**, *54*, 131–141. [CrossRef]
32. Widmer, K.-H.; Majewski, M. The Impact of the CCD-Angle on Range of Motion and Cup Positioning in Total Hip Arthroplasty. *Clin. Biomech.* **2005**, *20*, 723–728. [CrossRef]
33. Yadav, P.; Shefelbine, S.J.; Gutierrez-Farewik, E.M. Effect of Growth Plate Geometry and Growth Direction on Prediction of Proximal Femoral Morphology. *J. Biomech.* **2016**, *49*, 1613–1619. [CrossRef]
34. Lewinnek, G.E.; Lewis, J.L.; Tarr, R.; Compere, C.L.; Zimmerman, J.R. Dislocations after Total Hip-Replacement Arthroplasties. *J. Bone Jt. Surg.* **1978**, *60*, 217–220. [CrossRef]
35. Matovinović, D.; Nemec, B.; Gulan, G.; Sestan, B.; Ravlić-Gulan, J. Comparison in Regression of Femoral Neck Anteversion in Children with Normal, Intoeing and Outtoeing Gait—Prospective Study. *Coll. Antropol.* **1998**, *22*, 525–532.
36. Svenningsen, S.; Apalset, K.; Terjesen, T.; Anda, S. Regression of Femoral Anteversion: A Prospective Study of Intoeing Children. *Acta Orthop. Scand.* **1989**, *60*, 170–173. [CrossRef]
37. Bergmann, G.; Graichen, F.; Rohlmann, A. Hip Joint Loading during Walking and Running, Measured in Two Patients. *J. Biomech.* **1993**, *26*, 969–990. [CrossRef]
38. Hauptfleisch, J.; Glyn-Jones, S.; Beard, D.J.; Gill, H.S.; Murray, D.W. The Premature Failure of the Charnley Elite-Plus Stem: A CONFIRMATION OF RSA PREDICTIONS. *J. Bone Jt. Surg.* **2006**, *88-B*, 179–183. [CrossRef]
39. Heller, M.O.; Bergmann, G.; Deuretzbacher, G.; Claes, L.; Haas, N.P.; Duda, G.N. Influence of Femoral Anteversion on Proximal Femoral Loading: Measurement and Simulation in Four Patients. *Clin. Biomech.* **2001**, *16*, 644–649. [CrossRef]
40. Widmer, K.-H.; Zurfluh, B. Compliant Positioning of Total Hip Components for Optimal Range of Motion. *J. Orthop. Res.* **2004**, *22*, 815–821. [CrossRef] [PubMed]
41. Dorr, L.D.; Malik, A.; Dastane, M.; Wan, Z. Combined Anteversion Technique for Total Hip Arthroplasty. *Clin. Orthop.* **2009**, *467*, 119–127. [CrossRef]
42. Matsushita, A.; Nakashima, Y.; Jingushi, S.; Yamamoto, T.; Kuraoka, A.; Iwamoto, Y. Effects of the Femoral Offset and the Head Size on the Safe Range of Motion in Total Hip Arthroplasty. *J. Arthroplast.* **2009**, *24*, 646–651. [CrossRef] [PubMed]
43. Chandler, D.R.; Glousman, R.; Hull, D.; Mcguire, P.J.; Clarke, I.C.; Sarmiento, A. Prosthetic Hip Range of Motion and Impingement: The Effects of Head and Neck Geometry. *Clin. Orthop.* **1982**, *166*, 284–291. [CrossRef]
44. Berstock, J.R.; Hughes, A.M.; Lindh, A.M.; Smith, E.J. A Radiographic Comparison of Femoral Offset after Cemented and Cementless Total Hip Arthroplasty. *HIP Int.* **2014**, *24*, 582–586. [CrossRef]
45. Sakalkale, D.P.; Sharkey, P.F.; Eng, K.; Hozack, W.J.; Rothman, R.H. Effect of Femoral Component Offset on Polyethylene Wear in Total Hip Arthroplasty. *Clin. Orthop.* **2001**, *388*, 125–134. [CrossRef]
46. Kheir, M.M.; Drayer, N.J.; Chen, A.F. An Update on Cementless Femoral Fixation in Total Hip Arthroplasty. *J. Bone Jt. Surg.* **2020**, *102*, 1646–1661. [CrossRef]
47. Gómez-García, F.; Fernández-Fairen, M.; Espinosa-Mendoza, R.L. A Proposal for the Study of Cementless Short-Stem Hip Prostheses. *Acta Ortop. Mex.* **2016**, *30*, 204–215.
48. Dimitriou, D.; Tsai, T.-Y.; Kwon, Y.-M. The Effect of Femoral Neck Osteotomy on Femoral Component Position of a Primary Cementless Total Hip Arthroplasty. *Int. Orthop.* **2015**, *39*, 2315–2321. [CrossRef]
49. Michel, M.C.; Witschger, P. MicroHip: A Minimally Invasive Procedure for Total Hip Replacement Surgery Using a Modified Smith-Peterson Approach. *Ortop. Traumatol. Rehabil.* **2007**, *9*, 46–51. [PubMed]
50. Gombár, C.; Janositz, G.; Friebert, G.; Sisák, K. The DePuy Proxima™ Short Stem for Total Hip Arthroplasty—Excellent Outcome at a Minimum of 7 Years. *J. Orthop. Surg.* **2019**, *27*, 2309499019838668. [CrossRef] [PubMed]
51. Morlock, M.M.; Hube, R.; Wassilew, G.; Prange, F.; Huber, G.; Perka, C. Taper Corrosion: A Complication of Total Hip Arthroplasty. *EFORT Open Rev.* **2020**, *5*, 776–784. [CrossRef] [PubMed]
52. Wang, S.; Zhou, X.; Liu, L.; Shi, Z.; Hao, Y. On the Design and Properties of Porous Femoral Stems with Adjustable Stiffness Gradient. *Med. Eng. Phys.* **2020**, *81*, 30–38. [CrossRef]
53. Brown, T.D.; Ferguson, A.B. Mechanical Property Distributions in the Cancellous Bone of the Human Proximal Femur. *Acta Orthop. Scand.* **1980**, *51*, 429–437. [CrossRef]
54. Rho, J.Y.; Hobatho, M.C.; Ashman, R.B. Relations of Mechanical Properties to Density and CT Numbers in Human Bone. *Med. Eng. Phys.* **1995**, *17*, 347–355. [CrossRef]

55. Keyak, J.H.; Kaneko, T.S.; Tehranzadeh, J.; Skinner, H.B. Predicting Proximal Femoral Strength Using Structural Engineering Models. *Clin. Orthop. Relat. Res.* **2005**, *437*, 219–228. [CrossRef]
56. Schileo, E.; Dall'Ara, E.; Taddei, F.; Malandrino, A.; Schotkamp, T.; Baleani, M.; Viceconti, M. An Accurate Estimation of Bone Density Improves the Accuracy of Subject-Specific Finite Element Models. *J. Biomech.* **2008**, *41*, 2483–2491. [CrossRef]
57. Pithioux, M. Lois de Comportement et Modèles de Rupture des os Longs en Accidentologie. Ph.D. Thesis, Université de la Méditerranée, Marseille, Francia, 2000.
58. Peng, L.; Bai, J.; Zeng, X.; Zhou, Y. Comparison of Isotropic and Orthotropic Material Property Assignments on Femoral Finite Element Models under Two Loading Conditions. *Med. Eng. Phys.* **2006**, *28*, 227–233. [CrossRef]
59. Hernandez, C.J. Chapter A2 Cancellous Bone. In *Handbook of Biomaterial Properties*; Murphy, W., Black, J., Hastings, G., Eds.; Springer: New York, NY, USA, 2016; pp. 15–21. ISBN 978-1-4939-3303-7.
60. Wirtz, D.C.; Schiffers, N.; Pandorf, T.; Radermacher, K.; Weichert, D.; Forst, R. Critical Evaluation of Known Bone Material Properties to Realize Anisotropic FE-Simulation of the Proximal Femur. *J. Biomech.* **2000**, *33*, 1325–1330. [CrossRef]
61. Yang, H.-S.; Guo, T.-T.; Wu, J.-H.; Ma, X. Inhomogeneous Material Property Assignment and Orientation Definition of Transverse Isotropy of Femur. *J. Biomed. Sci. Eng.* **2009**, *2*, 419–424. [CrossRef]
62. Rickert, D.; Lendlein, A.; Peters, I.; Moses, M.A.; Franke, R.-P. Biocompatibility Testing of Novel Multifunctional Polymeric Biomaterials for Tissue Engineering Applications in Head and Neck Surgery: An Overview. *Eur. Arch. Otorhinolaryngol.* **2006**, *263*, 215–222. [CrossRef] [PubMed]
63. Becker, E.L.; Landau, I. International Dictionary of Medicine and Biology. *J. Clin. Eng.* **1986**, *11*, 134. [CrossRef]
64. Sam Froes, F.H. Titanium for Medical and Dental Applications—An Introduction. In *Titanium in Medical and Dental Applications*; Elsevier: Amsterdam, The Netherlands, 2018; pp. 3–21. ISBN 978-0-12-812456-7.
65. Choroszyński, M.; Choroszyński, M.R.; Skrzypek, S.J. Biomaterials for Hip Implants—Important Considerations Relating to the Choice of Materials. *Bio-Algorithms Med-Syst.* **2017**, *13*, 133–145. [CrossRef]
66. Okazaki, Y.; Gotoh, E. Comparison of Metal Release from Various Metallic Biomaterials in Vitro. *Biomaterials* **2005**, *26*, 11–21. [CrossRef] [PubMed]
67. Browne, M.; Gregson, P.J. Surface Modification of Titanium Alloy Implants. *Biomaterials* **1994**, *15*, 894–898. [CrossRef]
68. Pellizzari, M.; Jam, A.; Tschon, M.; Fini, M.; Lora, C.; Benedetti, M. A 3D-Printed Ultra-Low Young's Modulus β-Ti Alloy for Biomedical Applications. *Materials* **2020**, *13*, 2792. [CrossRef] [PubMed]
69. Petrovic, V.; Vicente Haro Gonzalez, J.; Jordá Ferrando, O.; Delgado Gordillo, J.; Ramón Blasco Puchades, J.; Portolés Griñan, L. Additive Layered Manufacturing: Sectors of Industrial Application Shown through Case Studies. *Int. J. Prod. Res.* **2011**, *49*, 1061–1079. [CrossRef]
70. Horn, T.J.; Harrysson, O.L.A. Overview of Current Additive Manufacturing Technologies and Selected Applications. *Sci. Prog.* **2012**, *95*, 255–282. [CrossRef] [PubMed]
71. Markwardt, J.; Friedrichs, J.; Werner, C.; Davids, A.; Weise, H.; Lesche, R.; Weber, A.; Range, U.; Meißner, H.; Lauer, G.; et al. Experimental Study on the Behavior of Primary Human Osteoblasts on Laser-Cused Pure Titanium Surfaces: Human Osteoblasts on Laser—Cused Titanium. *J. Biomed. Mater. Res. A* **2014**, *102*, 1422–1430. [CrossRef]
72. Ponader, S.; Von Wilmowsky, C.; Widenmayer, M.; Lutz, R.; Heinl, P.; Körner, C.; Singer, R.F.; Nkenke, E.; Neukam, F.W.; Schlegel, K.A. In Vivo Performance of Selective Electron Beam-Melted Ti-6Al-4V Structures. *J. Biomed. Mater. Res. Part A* **2010**, *92*, 56–62. [CrossRef]
73. ASTM International. *ASTM International ASTM F136-08: Standard Specification for Wrought Titanium-6 Aluminum-4 Vanadium ELI (Extra Low Interstitial) Alloy for Surgical Implant Applications (UNS R56401)*; ASTM International: West Conshohocken, PA, USA, 2008.
74. Bergmann, G.; Bender, A.; Dymke, J.; Duda, G.; Damm, P. Standardized Loads Acting in Hip Implants. *PLoS ONE* **2016**, *11*, e0155612. [CrossRef]
75. International Organization for Standardization (ISO). *International Standard ISO 7206-4 Implants for Surgery-Partial and Total Hip Joint Prostheses, Part 4: Determination of Endurance Properties and Performance of Stemmed Femoral Components*; International Organization for Standardization (ISO): Geneva, Switzerland, 2010.
76. Wolff, J. Das Gesetz der Transformation der Knochen. *DMW—Dtsch. Med. Wochenschr.* **1893**, *19*, 1222–1224. [CrossRef]
77. Frost, H.M. Bone's Mechanostat: A 2003 Update. *Anat. Rec.* **2003**, *275A*, 1081–1101. [CrossRef]
78. Mikić, B.; Carter, D.R. Bone Strain Gage Data and Theoretical Models of Functional Adaptation. *J. Biomech.* **1995**, *28*, 465–469. [CrossRef]
79. Turner, A.W.L.; Gillies, R.M.; Sekel, R.; Morris, P.; Bruce, W.; Walsh, W.R. Computational Bone Remodelling Simulations and Comparisons with DEXA Results. *J. Orthop. Res.* **2005**, *23*, 705–712. [CrossRef]
80. Yan, S.G.; Chevalier, Y.; Liu, F.; Hua, X.; Schreiner, A.; Jansson, V.; Schmidutz, F. Metaphyseal Anchoring Short Stem Hip Arthroplasty Provides a More Physiological Load Transfer: A Comparative Finite Element Analysis Study. *J. Orthop. Surg.* **2020**, *15*, 498. [CrossRef] [PubMed]
81. Yamako, G.; Chosa, E.; Totoribe, K.; Hanada, S.; Masahashi, N.; Yamada, N.; Itoi, E. In-Vitro Biomechanical Evaluation of Stress Shielding and Initial Stability of a Low-Modulus Hip Stem Made of β Type Ti-33.6Nb-4Sn Alloy. *Med. Eng. Phys.* **2014**, *36*, 1665–1671. [CrossRef] [PubMed]

82. Srinivasan, A.; Jung, E.; Levine, B.R. Modularity of the Femoral Component in Total Hip Arthroplasty. *J. Am. Acad. Orthop. Surg.* **2012**, *20*, 214–222. [CrossRef] [PubMed]
83. Boccaccio, A.; Uva, A.E.; Fiorentino, M.; Monno, G.; Ballini, A.; Desiate, A. Optimal Load for Bone Tissue Scaffolds with an Assigned Geometry. *Int. J. Med. Sci.* **2018**, *15*, 16–22. [CrossRef] [PubMed]

Article

The Influence of Mathematical Definitions on Patellar Kinematics Representations

Adrian Sauer [1,2,*], Maeruan Kebbach [3], Allan Maas [1,2], William M. Mihalko [4] and Thomas M. Grupp [1,2]

1. Research and Development, Aesculap AG, 78532 Tuttlingen, Germany; Allan.Maas@aesculap.de (A.M.); Thomas.Grupp@aesculap.de (T.M.G.)
2. Department of Orthopaedic and Trauma Surgery, Musculoskeletal University Center Munich (MUM), Campus Grosshadern, Ludwig Maximilians University Munich, 81377 Munich, Germany
3. Department of Orthopaedics, Rostock University Medical Center, 18057 Rostock, Germany; Maeruan.Kebbach@med.uni-rostock.de
4. Campbell Clinic Department of Orthopaedic Surgery and Biomedical Engineering, University of Tennessee Health Science Center, Memphis, TN 38163, USA; wmihalko@campbellclinic.com
* Correspondence: Adrian.Sauer@aesculap.de

Abstract: A correlation between patellar kinematics and anterior knee pain is widely accepted. However, there is no consensus on how they are connected or what profile of patellar kinematics would minimize anterior knee pain. Nevertheless, answering this question by merging existing studies is further complicated by the variety of ways to describe patellar kinematics. Therefore, this study describes the most frequently used conventions for defining patellar kinematics, focusing on the rotations. The similarities and differences between the Cardan sequences and angles calculated by projecting axes are analyzed. Additionally, a tool is provided to enable the conversion of kinematic data between definitions in different studies. The choice of convention has a considerable impact on the absolute values and the clinical characteristics of the patello-femoral angles. In fact, the angles that result from using different mathematical conventions to describe a given patello-femoral rotation from our analyses differ up to a Root Mean Squared Error of 111.49° for patellar flexion, 55.72° for patellar spin and 35.39° for patellar tilt. To compare clinical kinematic patello-femoral results, every dataset must follow the same convention. Furthermore, researchers should be aware of the used convention's implications to ensure reproducibility when interpreting and comparing such data.

Keywords: knee joint; patello-femoral joint; kinematics; cardan sequence; euler angles; conversion

Citation: Sauer, A.; Kebbach, M.; Maas, A.; Mihalko, W.M.; Grupp, T.M. The Influence of Mathematical Definitions on Patellar Kinematics Representations. *Materials* **2021**, *14*, 7644. https://doi.org/10.3390/ma14247644

Academic Editor: Oskar Sachenkov

Received: 30 September 2021
Accepted: 9 December 2021
Published: 11 December 2021

Publisher's Note: MDPI stays neutral with regard to jurisdictional claims in published maps and institutional affiliations.

Copyright: © 2021 by the authors. Licensee MDPI, Basel, Switzerland. This article is an open access article distributed under the terms and conditions of the Creative Commons Attribution (CC BY) license (https://creativecommons.org/licenses/by/4.0/).

1. Introduction

Patello-femoral pain has a prevalence of more than 20% in the general population [1] and a high percentage of unsatisfied patients after total knee arthroplasty complain of anterior knee pain [2,3]. Even if satisfactory patellar tracking and kinematics seem to be evident on physical exam [4,5], it is still not entirely clear, what defines good patellar tracking and how healthy patellar kinematics can be quantified [6].

Although previous studies have investigated patellar kinematics [6,7], it remains unclear what ideal patellar kinematics encompass. What we know about patellar kinematics is that the patellar native facets or prosthetic button should be centered in the trochlear groove without subluxation or tilt throughout range of motion [6,7]. Even with descriptions of how the patella tracks from full extension to flexion, the issue of kinematic reference frame and rotations to describe motion has not been standardized.

The difficulty of converting between the multiple existing mathematical definitions of patellar kinematics or even understanding them properly is one of the major challenges for answering these questions. When we consider rotations, the way they are described has many implications. Researchers should be aware of these implications when they work with kinematic data concerning the patello-femoral joint. The choice of coordinate systems and the mathematical definition for describing patello-femoral kinematics can lead to substantial differences in the resulting curves and how one interprets whether normal kinematics have been established. Therefore, studies without a detailed description of the underlying definitions [8–12] are of limited value for researchers and clinicians.

There is a recommendation on definitions of joint coordinate systems for various joints from the International Society of Biomechanics. However, they do not provide a recommended definition for the knee joint, especially not for the patello-femoral joint [13–15]. However, since the publication of these recommendations, a lot of research has been conducted for the patello-femoral joint kinematics. To describe these kinematics, a definition using a floating axis for patellar spin [16], which has been recommended by Bull et al. [17], has been increasingly used in the past few years. This definition follows the same principle used by Grood and Suntay [18] for the tibio-femoral joint.

Nevertheless, there are still lots of studies that use more uncommon conventions for patello-femoral rotations which can lead to fundamental changes in the values for patellar flexion, spin and tilt [6,17]. Therefore, for interpretation of patello-femoral kinematics data a clear understanding of the underlying conventions is needed. If data with different underlying definitions should be compared, the ability to transform between several conventions is very helpful, but rarely described in literature.

The aim of this article is to give an overview of the different methods that can be used to describe patellar kinematics, with a particular focus on rotations. In contrast to previous publications, this article will consider the implications associated with each definition in greater detail. Additionally, the mathematics behind the patellar rotations will be given, including ways to convert data between some of the most common definitions. Therefore, the purpose is to enable researchers to choose the definition of patellar kinematics that is most suitable to them, as well as allow them to easily perform any conversions necessary to be able to compare the outcomes of different studies. As Supplementary Material, a Matlab template is provided to perform the most common conversions easily.

2. Materials and Methods
2.1. Coordinates and Definitions

In the following description, two reference frames will mainly be used: one attached to the femur and another to the patella, with the directions x, y and z pointing laterally, anteriorly and proximally, respectively, in full extension. If the femur or patella rotate and translate during flexion, the attached coordinate systems will follow the bones' movements. Figure 1 shows a possible system of axes for the patellar kinematics. Since the exact orientation depends on the chosen definition, it is neither described nor shown in more detail. The question of how these directions can be identified, i.e., which landmarks can be used, is highly dependent on the data and the measurement methods of every single study and will therefore not be deeply discussed. A possibility to define a reference frame for the patella is presented and validated by Innocenti et al. [19].

The most clinically relevant kinematic parameters are patellar shift, patellar flexion (α), spin (β) and tilt (γ) [5,17,20–22]. The patella's resulting movements are relative to the femur as follows: patellar flexion is defined by the relative rotation of the patella around the medio-lateral flexion axis (x-axis). In most cases, the *femoral* flexion axis is used as the rotation axis for patellar flexion. In a clinical context, patellar spin and tilt are represented by the rotations around the local *patellar* y- and z-axes, respectively. Alternatively, it is also common to use a floating axis for patellar tilt [23,24] or patellar spin [25–27]. The patellar shift is the translation of the patella in the medio-lateral direction with respect to the femur or the trochlear groove. Usually, the patellar shift is given as the patellar movement in the direction of the *femoral* x-axis, which can be calculated by projecting the vector of the relative patellar translations on the normalized direction of the femoral flexion axis. The values of shift, flexion, spin and tilt for a given instance of patellar motion can differ considerably from this intuitive understanding if different definitions are used.

Figure 1 shows the described components of the patellar kinematics schematically without giving precise axes and signs because these are dependent on the chosen definition.

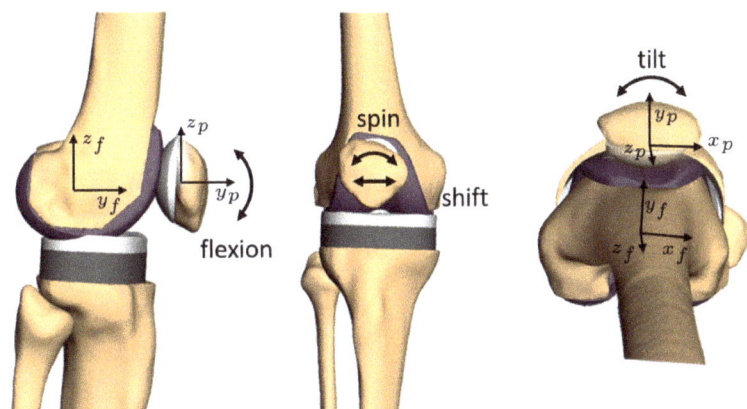

Figure 1. Overview on patellar kinematics and underlying coordinate systems. (**Left**): patellar flexion in a sagittal view, (**center**): medio-lateral movement (shift) and patellar spin in an anterior view, (**right**): patellar tilt in a proximal view. The exact location and orientation of the rotation axes depend on the chosen definition and is therefore not shown in detail. The patellar axes are labelled with x_p, y_p, z_p and the femoral axes with x_f, y_f, z_f.

The movements of the patella in the sagittal plane are highly dependent on the definition of the origin of the femoral coordinate system. If individual anatomical landmarks are used to define this origin, the comparison between different subjects can be problematic. However, since these movements are of secondary clinical interest [17] and are controlled mainly by joint geometry [28], they are not discussed here.

2.2. How Rotations Can Be Described

There are several ways to describe the rotational state of an object in a coordinate system. For example, well-known methods include rotational matrices, quaternions, projected angles, helical axes, and extrinsic and intrinsic rotation sequences. In biomechanics, representations allowing the comparison of rotations around the individual joint axes are needed. Therefore, rotation matrices, quaternions and helical axes are only suitable to process data, but not to interpret the results of studies according to patellar kinematics.

First, a fundamental property of rotations should be noted: rotations are not commutative. This means that—regardless of the mathematical description—the resulting pose for a sequence of rotations about multiple axes is generally not equal to the same rotations in another sequence. This statement remains valid for both multiple rotations in time and rotation sequences to describe a pose at one point in time.

To describe any rotation in three-dimensional (3D) space, a rotation sequence consisting of three consecutive rotations around three axes can be used. If the orientation of the axes is changed by each elemental rotation, the sequence is referred to as *intrinsic*. For initial coordinate axes ξ, η, ζ, the axes after the first rotation are written as ξ', η', ζ' and after the second rotation as ξ'', η'', ζ''. If all rotations occur around the axes as they are in their initial orientation, independent of previous rotation steps, the rotation sequence is called an *extrinsic* rotation.

In biomechanical literature, sometimes an intrinsic sequence of rotations around *three different axes* is called an Euler rotation, and the associated angles are termed *Euler angles* [29,30]. The correct term for these angles is in fact *Cardan angles* or *Tait-Bryan-angles*. Actual Euler angles are an intrinsic rotation sequence, which uses the same axis for the first and the third rotation (e.g., rotations around the axes ξ, η', ξ'') [31]. While Cardan angles can be used for a biomechanical description of patello-femoral rotations, Euler angles are difficult to relate to the clinical terms of patellar flexion, spin and tilt. Figure 2 shows three different rotation sequences which lead to a particular patellar orientation.

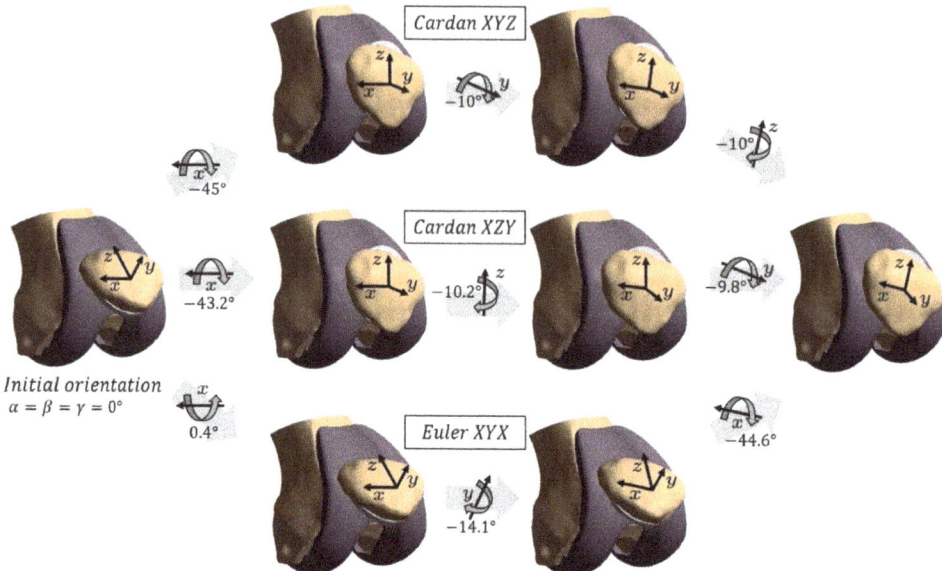

Figure 2. Illustration of three exemplary rotation sequences representing one particular patellar orientation. All sequences start in a patellar orientation where the patellar axes are parallel to the femoral axes. The upper and middle paths show the Cardan sequences XYZ and XZY, respectively, and the lower path represents the Euler sequence XYX.

Two disadvantages are occasionally associated in connection with Cardan angles. The first, sequence dependency, can simply be solved by appropriate standardization of the sequence [32]. The second is termed gimbal lock, which occurs when the second angle of the sequence is ±90° [33]. In this case, the axes of the first and third rotation are collinear, and one degree of freedom is lost. Since the patellar spin and tilt do not reach absolute values of 90° [6,7] sequences with either of these in the middle position are free from this problem for the description of patellar kinematics.

If femoral axes are used to describe the patello-femoral rotations [34], the interpretation of the rotations around the femoral anterior-posterior and proximo-distal axes as patellar spin and patellar tilt, respectively, are in general no longer correct. For a patellar flexion near 0° the deviation remains small, but for one of 90° the femoral tilt axis is parallel to the patellar spin axis in a sagittal view. The same applies to the femoral rotation axis and the patellar tilt axis. In this case, the calculated values for tilt and spin of the patella are switched, in addition to a possible inversion of their sign, relative to the common clinical interpretation [17] (see also Figure 3). Researchers should be aware that Cardan sequences ending with patellar flexion (ZYX and YZX) show the same pattern (e.g., [35]).

For knee kinematics, it seems that there exists another valuable way to describe the rotations. It is called the *three-cylinder open-chain representation*. For the tibio-femoral joint, it was first mentioned by Grood and Suntay in 1983 [18]. It was suggested in 2009 by Merican and Amis [36] and later on proven in 2014 by MacWilliams and Davis [37] that the Grood and Suntay definition is equivalent to a Cardan XYZ-sequence. This sequence was regardless found suitable to describe the tibio-femoral kinematics [38,39].

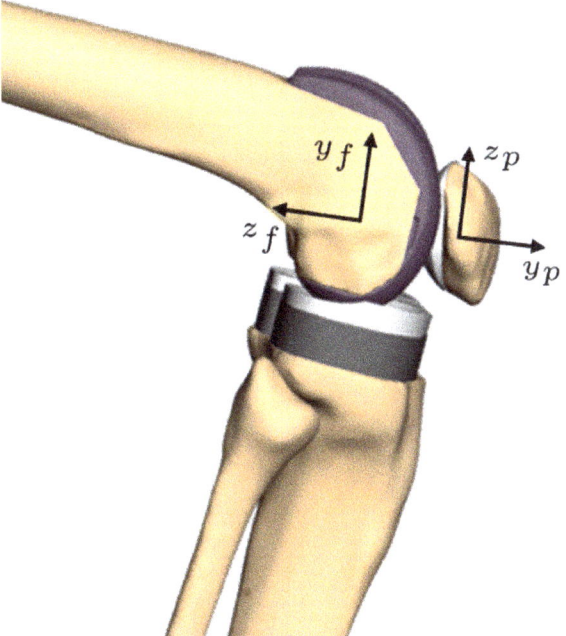

Figure 3. Illustration of the switch of meaning of patellar spin and tilt for an absolute patellar flexion angle of 90°, if described with respect to the femoral coordinate system. Rotating the patella around the anatomical patellar proximo-distal axis (z_p) equals a rotation around the femoral anterior-posterior axis (y_f) for 90° of patello-femoral flexion. The z_f- and y_p-axes show a similar pattern.

For the patello-femoral joint, the three-cylinder open-chain representation was first applied in 1992 by Hefzy et al. [16]. Since this way to describe patellar kinematics was recommended in a method paper [17], it is frequently used. The proof of the equivalence of the Grood and Suntay rotations and a Cardan sequence for the tibio-femoral joint [37] can be carried out analogously for the patello-femoral joint. Therefore, the three-cylinder open-chain representation of the patello-femoral kinematics is—from the rotations point of view—equivalent to the use of the Cardan sequence XYZ. Since Bull et al. [17] do not describe translations in a sagittal plane, their recommendation is to measure the patellar shift relative to the femoral medio-lateral axis and describe the patellar rotations with respect to the femoral axes as a Cardan XYZ-sequence.

The methods, which use a sequence of rotations around rotated or not rotated co-ordinate axes, lead to the inherent possibility to receive a description of the pose of the associated body very straightforward by executing the rotations one after the other. Nevertheless, some authors [40,41] determine the rotations of the patella by projecting the patellar coordinate vectors on the planes of the femoral coordinate system and calculate the angle between these projected vectors and the associated vectors from the femoral system. For instance, the patellar flexion can be calculated by projecting the y-vector of the patella onto the sagittal plane of the femoral coordinate system and calculating the angle between this projection and the y-vector of the femoral system. If a projection is carried out onto the patellar coordinate planes, a rotation relative to the patellar axes can be calculated. This method in general does not lead to values for patellar flexion, spin and tilt which can be executed in sequence to acquire the full rotation of the patella relative to the femur. For every projection plane, there are two vectors that can be projected to calculate the rotation around the axis perpendicular to this plane. As the results of this paper will show, the calculated angles for these two projected directions differ a lot for

3D rotations. For example, the calculation of the patellar flexion as the angle between the patellar *y*-axis projected onto the sagittal plane of the femoral system differs from the same calculation with the *z*-axes. Therefore, the angles are highly dependent on the choice of planes and axes.

2.3. Conversions

A method to convert a rotation from one definition to another is to calculate the rotation matrix representing the rotation and use this matrix to calculate the parameters of the desired definition. To that end, methods will be given to convert the common definitions into matrix form and back. Figure 4 shows an overview of the conversion paths introduced in this paper. The conversions are first given for intrinsic and extrinsic rotation sequences.

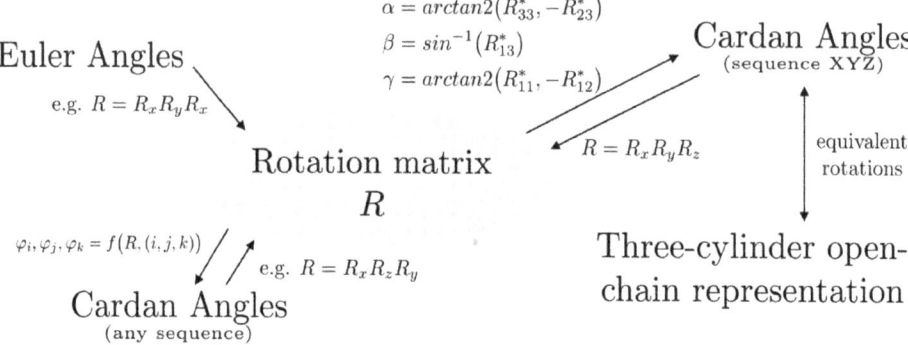

Figure 4. Overview of transformation paths between different descriptions of patellar rotations.

2.3.1. Rotation Sequences

With the standard uniaxial 3D rotation matrices $R_x(\alpha)$, $R_y(\beta)$ and $R_z(\gamma)$ for the rotations around x, y and z, the rotation matrix R of the full 3D rotation can be calculated by matrix multiplication. For an extrinsic rotation sequence, the resulting matrix R is the matrix product of the corresponding uniaxial rotation matrices, expressed in the opposite sequence order from left to right. For an intrinsic rotation sequence, the order of the uniaxial rotation matrices is inverted. For example, the rotation matrix $R = R_x(\alpha) \cdot R_y(\beta) \cdot R_z(\gamma)$ represents the extrinsic rotation sequence ZYX and the intrinsic sequence XYZ. It is important to note that this method works the same for Euler angles.

The formulas for calculating the angles α, β and γ for a specific sequence from a given rotation matrix R^* with the entry R_{st}^* in line s and column t are obviously dependent on the sequence. Rotating around a particular axis twice within one sequence leads to angles that are hard to interpret clinically. Only the equations for Cardan angles are therefore given here.

For certain sequences, the following method is often shown in literature (e.g., intrinsic XYZ-sequence [32,42]). Here, a general formulation which can be used for any sequence, will be introduced.

Let (i,j,k) be a tuple of three different indices with (i,j,k) ∈ {1,2,3}. Then every (i,j,k)-tuple represents a Cardan sequence (e.g., (i,j,k) = (1,3,2) stands for the sequence XZY). To calculate the angles of the patello-femoral rotations, the sign sgn_{ijk} of a sequence tuple is needed. The sign is equal to 1 if the tuple (i,j,k) can be created from the tuple (1,2,3) by an even number of transpositions. If the number of transpositions is odd, sgn_{ijk} is equal to −1 [43]. The equations

$$\varphi_i = \arctan2\,(R_{kk}{}^*, -sgn_{ijk} \cdot R_{jk}{}^*) \qquad (1)$$

$$\varphi_j = \sin^{-1}(sgn_{ijk} \cdot R_{ik}{}^*) \text{ and} \qquad (2)$$

$$\varphi_k = \arctan2\left(R_{ii}{}^*,\, -\mathrm{sgn}_{ijk}\cdot R_{ij}{}^*\right) \tag{3}$$

directly lead to the sought angles for flexion ($\alpha = \varphi_1$), spin ($\beta = \varphi_2$) and tilt ($\gamma = \varphi_3$). The proof for Equations (1)–(3) is given in the Supplementary Materials.

2.3.2. Projected Angles

For the reconstruction of the 3D orientation from projected angles, giving a closed description of all 64 possible axis-plane projection combinations is not feasible in this article. Only a brief overview of the derivation of the equations will be given here. Every given projected angle can be used to determine a semicircle containing all the unit vectors that would lead to this angle if projected. If the j-th axis is projected onto the femoral (or patellar) coordinate plane, the parametrized semicircle can be set to equal the j-th column (or row, in the case of the patellar plane) vector of the rotation matrix. The orthogonality of rotation matrices (pairwise orthogonality of rows and columns, Euclidian norm of every row and column equals 1, third row/column is cross product of first and second row/column) gives the conditions to determine the parameters and the missing entries of R.

2.3.3. Helical Axes

Helical axes are a way to represent a 3D-movement using only one rotation axis for every time step. The movement is described as a translation along this axis and a rotation around it. This method is sometimes used to analyze the actual joint rotation axis, especially for the tibio-femoral joint [44,45]. This description has also been sometimes used for the patello-femoral joint [46,47]. Consequently, the ideas for the conversion of the rotations are discussed here briefly.

The direction of the rotation axis is given by the helical axis. Thus, the Rodrigues' rotation formula [48] can be used to rotate a vector around any given rotation axis. Matrix representation of a helical axis rotation can be achieved by applying this formula to the unit vectors. The rotated unit vectors give the columns of the associated rotation matrix. For the opposite conversion, the direction of this axis is the eigenvector to the eigenvalue of 1, while the angle can be obtained from the other eigenvalues [49,50]. For more details [51] is recommended.

2.3.4. Three-Cylinder Open-Chain Representation

As stated previously, the angles in the three-cylinder open-chain representation [16,17] are equivalent to the angles of the intrinsic XYZ rotation sequence. Therefore, the methods used for this sequence can also be applied to convert rotations from or into this convention.

2.4. Data Processing and Validation

To compare the definitions described earlier in this paper, kinematic data from a previously published and validated musculoskeletal model of the lower right extremity simulating a squat motion [20] was used. This simulation model was implemented in the software SIMPACK (V9.7, Dassault Systèmes Deutschland GmbH, Gilching, Germany). The rotations of the patella relative to the femur were evaluated for a squat motion in all possible Cardan sequences directly in the software. These were used as reference data for verifying the implementation of the previous formulas in MATLAB (R2018a, Mathworks, Natick, Massachusetts, USA). The deviations between the curves were quantified by calculating the Root Mean Squared Error (RMSE).

3. Results

The dataset for a squat includes tibio-femoral flexion angles from $0°$ to $90°$. The discrepancy between the angles according to the different definitions increases with the tibio-femoral flexion angles. Based on the data for the Cardan XYZ-sequence, the angles for the five remaining Cardan sequences and the projected angles were calculated. In order to validate the conversion process, the Cardan sequences for the same dataset were evaluated directly from the general multibody software SIMPACK as reference. The deviation of these results to the values from converting the XYZ data do not exceed 0.0001° (RMSE < $(6.13 \times 10^{-5})°$ for all sequences and angles).

3.1. Cardan Angles

Figure 5 shows the rotation angles for all possible Cardan sequences. The choice of definition has a minor effect on the patellar flexion values. All Cardan sequences except for ZXY have very similar flexion angles. The maximum RMSE within this group is 0.79° (see also Table 1). The patellar flexion angle of the ZXY-sequence differs up to 7.83° from the others (RMSE $\geq 3.71°$).

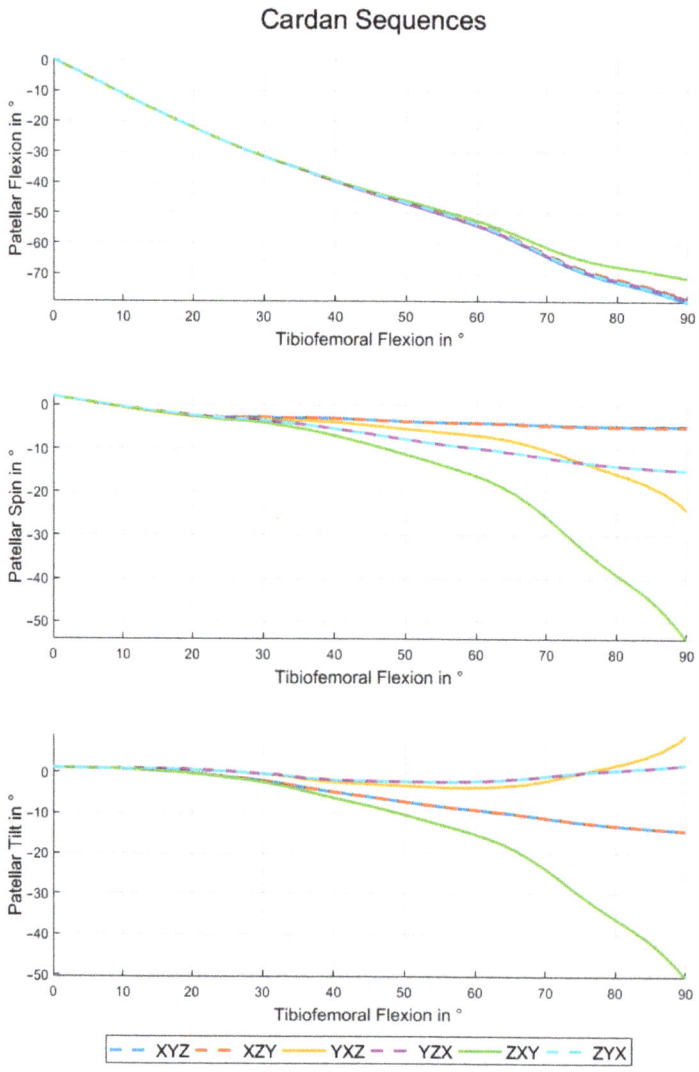

Figure 5. Patellar kinematics for the Cardan sequences. Patellar flexion (**top**), spin (**center**) and tilt (**below**) during a squat motion for all Cardan sequences.

Table 1. RMSE of patellar flexion curves for Cardan sequences and projected angles in °.

	XYZ	XZY	YXZ	YZX	ZXY	ZYX	Y on Patella	Z on Patella	Y on Femur	Z on Femur
XYZ	0	-	-	-	-	-	-	-	-	-
XZY	0.79	0	-	-	-	-	-	-	-	-
YXZ	0.46	0.36	0	-	-	-	-	-	-	-
YZX	0.43	0.4	0.05	0	-	-	-	-	-	-
ZXY	3.71	2.96	3.25	3.28	0	-	-	-	-	-
ZYX	0.39	0.44	0.26	0.25	3.4	0	-	-	-	-
Y on patella	111.48	110.71	111.06	111.09	108.11	111.1	0	-	-	-
Z on patella	111.49	110.72	111.07	111.1	108.12	111.12	0.25	0	-	-
Y on femur	0.79	0	0.36	0.4	2.96	0.44	110.71	110.72	0	-
Z on femur	0	0.79	0.46	0.43	3.71	0.39	111.48	111.49	0.79	0

In terms of patellar spin and patellar tilt the Cardan XYZ- and XZY-sequences are very close (spin: RMSE = 0.09° and tilt: RMSE = 0.03°) as Tables 2 and 3 show. The same is true for YZX and ZYX (spin: RMSE = 0.004° and tilt: RMSE = 0.03°). The sequence ZXY leads to curves which differ considerably from those of other sequences (RMSE up to 27.72°). These deviations are particularly large and grow increasingly for higher tibio-femoral flexion angles, and reach a maximum of 49.22° for the patellar spin and 59.91° for the patellar tilt.

Table 2. RMSE of patellar spin curves for Cardan sequences and projected angles in °.

	XYZ	XZY	YXZ	YZX	ZXY	ZYX	X on Patella	Z on Patella	X on Femur	Z on Femur
XYZ	0	-	-	-	-	-	-	-	-	-
XZY	0.09	0	-	-	-	-	-	-	-	-
YXZ	8.27	8.18	0	-	-	-	-	-	-	-
YZX	6.56	6.48	3.09	0	-	-	-	-	-	-
ZXY	24.37	24.28	16.19	18.06	0	-	-	-	-	-
ZYX	6.56	6.48	3.09	0.004	18.06	0	-	-	-	-
X on patella	7.85	7.93	15.47	14.24	31.53	14.24	0	-	-	-
Z on patella	31.44	31.53	39.56	37.94	55.72	37.94	24.28	0	-	-
X on femur	6.56	6.48	3.09	0	18.06	0.004	14.24	37.94	0	-
Z on femur	8.27	8.18	0	3.09	16.19	3.09	15.47	39.56	3.09	0

Table 3. RMSE of patellar tilt curves for Cardan sequences and projected angles in °.

	XYZ	XZY	YXZ	YZX	ZXY	ZYX	X on Patella	Y on Patella	X on Femur	Y on Femur
XYZ	0	-	-	-	-	-	-	-	-	-
XZY	0.03	0	-	-	-	-	-	-	-	-
YXZ	11.07	11.04	0	-	-	-	-	-	-	-
YZX	9.64	9.61	2.27	0	-	-	-	-	-	-
ZXY	16.69	16.72	27.72	26.14	0	-	-	-	-	-
ZYX	9.65	9.62	2.24	0.03	26.15	0	-	-	-	-
X on patella	19.36	19.33	9.38	9.92	35.39	9.91	0	-	-	-
Y on patella	9.38	9.35	6.79	4.67	24.45	4.7	11.07	0	-	-
X on femur	9.65	9.62	2.24	0.03	26.15	0	9.91	4.7	0	-
Y on femur	16.69	16.72	27.72	26.14	0	26.15	35.39	24.45	26.15	0

3.2. Projected Angles

The projected angles can be divided into two groups. Projecting the patellar reference frame vectors onto the femoral planes gives patello-femoral rotations with respect to the femoral system. Therefore, the angles for patellar flexion are close to the negative of the angles obtained by projecting onto the patellar planes (see also Figure 6, top). The flexion angles calculated by projecting the y- (or z-) axis on the femoral standard plane are equal to the patellar flexion angles from the Cardan XZY- (or XYZ- for the z-axis) sequence as the RMSE of $0°$ for these combinations shows in Table 1. The same pattern is visible for the patellar spin and tilt if the projections on femoral planes and Cardan sequences starting with y- and z-axes are analyzed (see also Tables 2 and 3).

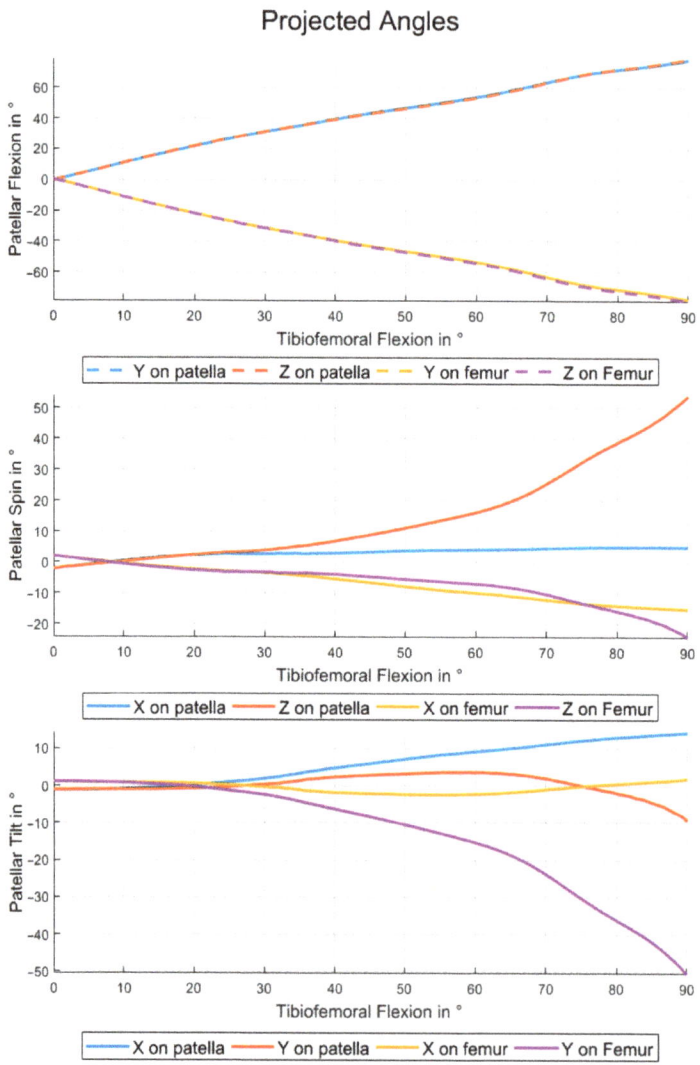

Figure 6. Patellar kinematics using projected angles. Patellar flexion (**top**), spin (**center**) and tilt (**below**) during a squat motion for all possible ways to calculate the angles by projection.

4. Discussion

The first objective of this study was to show the considerable impact of different mathematical definitions on how patello-femoral kinematics are conveyed. A framework was thereby developed to enable researchers to choose the definitions for the patellar kinematics which best fit their needs and to convert patellar kinematics between the different conventions. In this way, comparability and interchangeability between different studies may be facilitated.

The results obtained from the conversions we have presented using the described formulas are in excellent agreement with those obtained from the evaluation using a general multibody software package (RMSE < $(6.13 \times 10^{-5})°$ for all sequences and angles). Therefore, the derived equations and their implementation were verified.

The presented results indicate clearly that the chosen definition for patello-femoral angles has an impact on the description of patello-femoral kinematics. The curves for patellar flexion, spin and tilt deviate not only in magnitude but also in their characteristics (see Figures 5 and 6). Therefore a correct interpretation of patello-femoral kinematic data seems to be impossible if the details of the convention used are not clear.

Deviations between the results with different definitions only occur if rotations around at least two standard directions differ from zero. This condition is based on the fact that the differences are caused by changes in rotation axes for subsequent rotations. For increasing tibio-femoral flexion angles the absolute patellar flexion angle reaches values close to 80°, while the absolute values of patellar spin and tilt stay relatively small if the rotations are considered around the patellar axes. Therefore, the rotation around the x-axis and its position in the Cardan sequence has a special impact on the data for high tibio-femoral and patello-femoral flexion angles. In the case of small patellar spin and tilt, it can be assumed that the rotation axes that occur before the patellar flexion in the Cardan sequence almost agree with the associated femoral coordinate system axes. The rotations placed after the flexion in the sequence can be interpreted as rotations around associated patellar body axes. For the first and last rotations of a Cardan sequence these statements are exactly true.

For the patello-femoral kinematics this means that the first rotation of a sequence is always around the associated axis of the femoral coordinate system and the last rotation is around a patella-fixed axis, as previously shown in Figure 2. The floating axis between these two is close to the associated patellar axis if the sequence starts with the patellar flexion and is close to the associated femoral axis otherwise. In accordance with the previous explanations the two Cardan sequences starting and ending with the patellar flexion (XYZ/XZY and YZX/ZYX) show very small differences (see Figure 5, and Tables 1–3).

For 90° of absolute patellar flexion angle patellar spin and tilt would swap its values if the patellar flexion would be changed from the first position (i.e., tilt and spin close to patellar body fixed rotations) to the third position (i.e., tilt and spin close to femoral body fixed rotation). This is caused by the inherent change from patellar to femoral coordinate system. For the squat cycle showed here, only 80° of absolute patellar flexion are reached. Nevertheless, the described effect can be seen, if the maximum patellar tilt for the sequences XYZ and XZY (15.0°) is compared to the patellar spin for YZX and ZYX (15.7°).

One of the main difficulties in using Cardan sequences to describe rotational poses is the gimbal lock, which occurs if the second rotation of the sequence is equal to 90°. In this case, the axis of the third rotation is parallel to the direction of the first one (or its negative) and the rotation loses one degree of freedom. The absolute values for spin and tilt are far from 90° for healthy knees [6,7]. Therefore, this situation only exists if the patellar flexion is the second rotation of the sequence. Even if the maximum absolute values for patellar flexion in the shown results are less than 80°, an increase of spin and tilt can be mentioned for the YXZ sequence and a comparably larger one for the ZXY sequence. (see Tables 2 and 3). The reason why ZXY differs more from the sequences without the x-rotation on second position than YXZ is that the patello-femoral rotation around the femoral z-axis is bigger than around the femoral y-axis. These effects are making the interpretation of the patellar kinematics in the ZXY and YXZ sequence quite laborious and this description differs a lot from a clinical understanding for spin and tilt. Therefore these sequences are not recommended.

The projected angles also show big differences dependent on the choice of which axis is projected onto which coordinate system. Figure 6 indicates clearly that this method is very sensitive to the choice of projection axis and that this effect increases with growing angles in the other two dimensions. The flexion angles are robust against switching from y- to z-axis for projection, while comparable changes for spin and tilt cause completely different curves due to the relatively higher patellar flexion angles.

For small angles of spin and tilt the projected angles for flexion on the patellar (or femoral) planes can be interpreted as flexion with respect to the patellar (or femoral) system. Due to the high flexion angles, the same interpretation cannot be applied to spin and tilt. For spin and tilt the values are also highly depending on the choice of axis for projection.

It was shown that the deviations between the curves for patellar flexion, spin and tilt for the different kinematic definitions can completely change in both, magnitude and characteristics of the curves. Nevertheless, the differences are not big enough in every case that inadvertently comparing results based on different definitions would be obvious at first glance. Therefore the problem explored by this study should always be considered when dealing with patellar kinematics.

Patellar maltracking is known as possible cause of anterior knee pain [4,5,52–55], but it still remains unknown, how critical patellar kinematics can be distinguished from others [6]. To increase the biomechanical knowledge about how anterior knee pain can be prevented or treated, it will help to bring the available data from the literature together. Therefore, it is essential to properly understand the underlying definitions of every single study and to carefully transform the data into one representation, that can be compared across the available literature. For most of the studies available in literature the given transformations from this study can be utilized.

If researchers are free to choose a convention for their own study, the use of Cardan sequences is recommended due to the available straight forward methods for calculation, conversion and interpretation. Even if all Cardan sequences are theoretically suitable to represent the patello-femoral kinematics in a correct way, some of them are better for intuitive interpretation from a biomechanical side of view. The definitions which are closest to the clinical understanding of patellar flexion, tilt and spin are the two Cardan sequences beginning with patellar flexion (XYZ and XZY), where patellar flexion is given around the *femoral* flexion axis. The last rotation of the sequence is given exactly around the associated *patellar* axis while the second rotation of the sequence is executed around the floating axis perpendicular to the other two. Therefore, the rotation (out of spin and tilt) with the most importance for a certain study should be chosen as last rotation of the sequence. The higher clinical relevance of tilt compared to patellar spin will qualify the Cardan sequence XYZ as recommendation for most studies.

5. Conclusions

In this study the most common definitions for patello-femoral rotations and the most important conversions between them were described. Using kinematic data of a validated squat motion based on motion capture, it was shown that the angles describing the patellar kinematics are highly dependent on the underlying convention.

If researchers are not aware of the described deviations, misinterpretation of results is very likely, which is critical for clinically relevant studies. Additionally, the comparison of different studies should only be executed if equivalence of the used conventions can be ensured or if the data are carefully transformed. Therefore, the methods from this study will help to uncover the complex relationship between patellar kinematics and anterior knee pain.

Supplementary Materials: The following are available online at https://www.mdpi.com/article/10.3390/ma14247644/s1, S1: Matlab-template to transform patellar rotations between the described conventions, S2: Instructions for use of the template (S1), S3: Dataset used in this paper in XYZ-Cardan sequence form, S4: Proof for the conversion formulas (1)–(3).

Author Contributions: Conceptualization, A.S., M.K., A.M., W.M.M. and T.M.G.; methodology, A.S. and M.K.; software, A.S.; validation, A.S. and M.K.; formal analysis, A.S.; investigation, A.S. and M.K.; resources, A.S., M.K., A.M. and T.M.G.; data curation, A.S. and M.K.; writing—original draft preparation, A.S.; writing—review and editing, A.S., M.K., A.M., W.M.M. and T.M.G.; visualization, A.S. and M.K.; supervision, A.M. and T.M.G.; project administration, A.S., A.M. and T.M.G.; funding acquisition, A.M. and T.M.G. All authors have read and agreed to the published version of the manuscript.

Funding: Three of the authors (A.S., A.M., T.M.G.) were funded by B.Braun Aesculap AG, Tuttlingen, Germany. The funder provided support in the form of salaries for authors A.S., A.M. and T.M.G. The funders had no role in study design, data collection and analysis, decision to publish, or preparation of the manuscript.

Institutional Review Board Statement: Not applicable.

Informed Consent Statement: Not applicable.

Data Availability Statement: The dataset used for this study is available as Supplementary Material (S3).

Conflicts of Interest: Three of the authors (A.S., A.M., T.M.G.) are employees of B.Braun Aesculap AG, Tuttlingen, Germany. W.M.M. is a paid consultant at B. Braun Aesculap AG, but did not receive any reimbursement for the current study. M.K. declares no conflicts of interest.

References

1. Smith, B.E.; Selfe, J.; Thacker, D.; Hendrick, P.; Bateman, M.; Moffatt, F.; Rathleff, M.S.; Smith, T.O.; Logan, P. Incidence and prevalence of patellofemoral pain: A systematic review and meta-analysis. *PLoS ONE* **2018**, *13*, e0190892. [CrossRef]
2. Parvizi, J.; Rapuri, V.R.; Saleh, K.J.; Kuskowski, M.A.; Sharkey, P.F.; Mont, M.A. Failure to resurface the patella during total knee arthroplasty may result in more knee pain and secondary surgery. *Clin. Orthop. Relat. Res.* **2005**, *438*, 191–196. [CrossRef]
3. Steinbruck, A.; Schröder, C.; Woiczinski, M.; Muller, T.; Muller, P.E.; Jansson, V.; Fottner, A. Influence of tibial rotation in total knee arthroplasty on knee kinematics and retropatellar pressure: An in vitro study. *Knee Surg. Sports Traumatol. Arthrosc.* **2016**, *24*, 2395–2401. [CrossRef]
4. Powers, C.M.; Witvrouw, E.; Davis, I.S.; Crossley, K.M. Evidence-based framework for a pathomechanical model of patellofemoral pain: 2017 patellofemoral pain consensus statement from the 4th International Patellofemoral Pain Research Retreat, Manchester, UK: Part 3. *Br. J. Sports Med.* **2017**, *51*, 1713–1723. [CrossRef] [PubMed]
5. Wheatley, M.G.A.; Rainbow, M.J.; Clouthier, A.L. Patellofemoral Mechanics: A Review of Pathomechanics and Research Approaches. *Curr. Rev. Musculoskelet. Med.* **2020**, *13*, 326–337. [CrossRef] [PubMed]
6. Yu, Z.; Yao, J.; Wang, X.; Xin, X.; Zhang, K.; Cai, H.; Fan, Y.; Yang, B. Research Methods and Progress of Patellofemoral Joint Kinematics: A Review. *J. Healthc. Eng.* **2019**, *2019*, 9159267. [CrossRef]
7. Katchburian, M.V.; Bull, A.M.J.; Shih, Y.-F.; Heatley, F.W.; Amis, A.A. Measurement of patellar tracking: Assessment and analysis of the literature. *Clin. Orthop. Relat. Res.* **2003**, *412*, 241–259. [CrossRef] [PubMed]
8. Chew, J.T.; Stewart, N.J.; Hanssen, A.D.; Luo, Z.-P.; Rand, J.A.; An, K.-N. Differences in Patellar Tracking and Knee Kinematics Among Three Different Total Knee Designs. *Clin. Orthop. Relat. Res.* **1997**, *345*, 87–98. [CrossRef]
9. Nakamura, S.; Tanaka, Y.; Kuriyama, S.; Nishitani, K.; Ito, H.; Furu, M.; Matsuda, S. Superior-inferior position of patellar component affects patellofemoral kinematics and contact forces in computer simulation. *Clin. Biomech.* **2017**, *45*, 19–24. [CrossRef]
10. Sakai, N.; Luo, Z.-P.; Rand, J.A.; An, K.-N. The effects of tibial rotation on patellar position. *Knee* **1994**, *1*, 133–138. [CrossRef]
11. Sakai, N.; Luo, Z.-P.; Rand, J.A.; An, K.-N. Quadriceps forces and patellar motion in the anatomical model of the patellofemoral joint. *Knee* **1996**, *3*, 1–7. [CrossRef]
12. Sakai, N.; Luo, Z.-P.; Rand, J.A.; An, K.-N. The influence of weakness in the vastus medialis oblique muscle on the patellofemoral joint: An in vitro biomechanical study. *Clin. Biomech.* **2000**, *15*, 335–339. [CrossRef]
13. Wu, G.; Cavanagh, P.R. ISB recommendations for standardization in the reporting of kinematic data. *J. Biomech.* **1995**, *28*, 1257–1261. [CrossRef]
14. Wu, G.; Siegler, S.; Allard, P.; Kirtley, C.; Leardini, A.; Rosenbaum, D.; Whittle, M.; D'Lima, D.D.; Christofolini, L.; Witte, H.; et al. ISB recommendation on definitions of joint coordinate system of various joints for the reporting of human joint motion: Part I: Ankle, hip, and spine. *J. Biomech.* **2002**, *35*, 543–548. [CrossRef]
15. Wu, G.; van der Helm, F.C.T.; Veeger, H.E.J.D.; Makhsous, M.; van Roy, P.; Anglin, C.; Nagels, J.; Karduna, A.R.; McQuade, K.; Wang, X.; et al. ISB recommendation on definitions of joint coordinate systems of various joints for the reporting of human joint motion–Part II: Shoulder, elbow, wrist and hand. *J. Biomech.* **2005**, *38*, 981–992. [CrossRef]
16. Hefzy, M.S.; Jackson, W.T.; Saddemi, S.R.; Hsieh, Y.-F. Effects of tibial rotations on patellar tracking and patello-femoral contact areas. *J. Biomed. Eng.* **1992**, *14*, 329–343. [CrossRef]
17. Bull, A.M.J.; Katchburian, M.V.; Shih, Y.-F.; Amis, A.A. Standardisation of the description of patellofemoral motion and comparison between different techniques. *Knee Surg. Sports Traumatol. Arthrosc.* **2002**, *10*, 184–193. [CrossRef] [PubMed]
18. Grood, E.S.; Suntay, W.J. A joint coordinate system for the clinical description of three-dimensional motions: Application to the knee. *J. Biomech. Eng.* **1983**, *105*, 136–144. [CrossRef]

19. Innocenti, B.; Bori, E.; Piccolo, S. Development and validation of a robust patellar reference coordinate system for biomechanical and clinical studies. *Knee* **2020**, *27*, 81–88. [CrossRef]
20. Kebbach, M.; Darowski, M.; Krueger, S.; Schilling, C.; Grupp, T.M.; Bader, R.; Geier, A. Musculoskeletal Multibody Simulation Analysis on the Impact of Patellar Component Design and Positioning on Joint Dynamics after Unconstrained Total Knee Arthroplasty. *Materials* **2020**, *13*, 2365. [CrossRef]
21. Tischer, T.; Geier, A.; Lenz, R.; Woernle, C.; Bader, R. Impact of the patella height on the strain pattern of the medial patellofemoral ligament after reconstruction: A computer model-based study. *Knee Surg. Sports Traumatol. Arthrosc.* **2017**, *25*, 3123–3133. [CrossRef]
22. Woiczinski, M.; Steinbruck, A.; Weber, P.; Muller, P.E.; Jansson, V.; Schröder, C. Development and validation of a weight-bearing finite element model for total knee replacement. *Comput. Methods Biomech. Biomed. Engin.* **2016**, *19*, 1033–1045. [CrossRef]
23. Ahmed, A.M.; Duncan, N.A.; Tanzer, M. In Vitro Measurement of the Tracking Pattern of the Human Patella. *J. Biomech. Eng.* **1999**, *121*, 222–228. [CrossRef]
24. Hirokawa, S. Three-dimensional mathematical model analysis of the patellofemoral joint. *J. Biomech.* **1991**, *24*, 659–671. [CrossRef]
25. Hsu, H.-C.; Luo, Z.-P.; Rand, J.A.; An, K.-N. Influence of lateral release on patellar tracking and patellofemoral contact characteristics after total knee arthroplasty. *J. Arthroplasty* **1997**, *12*, 74–83. [CrossRef]
26. Koh, T.J.; Grabiner, M.D.; Swart, R.J.de. In vivo tracking of the human patella. *J. Biomech.* **1992**, *25*, 637–643. [CrossRef]
27. Lin, F.; Makhsous, M.; Chang, A.H.; Hendrix, R.W.; Zhang, L.-Q. In vivo and noninvasive six degrees of freedom patellar tracking during voluntary knee movement. *Clin. Biomech.* **2003**, *18*, 401–409. [CrossRef]
28. Anglin, C.; Ho, K.C.T.; Briard, J.-L.; de Lambilly, C.; Plaskos, C.; Nodwell, E.; Stindel, E. In vivo patellar kinematics during total knee arthroplasty. *Comput. Aided Surg.* **2008**, *13*, 377–391. [CrossRef]
29. Ali, A.A.; Mannen, E.M.; Rullkoetter, P.J.; Shelburne, K.B. In vivo comparison of medialized dome and anatomic patellofemoral geometries using subject-specific computational modeling. *J. Orthop. Res.* **2018**, *36*, 1910–1918. [CrossRef] [PubMed]
30. Wang, L.; Wang, C.J. Influence of patellar implantation on the patellofemoral joint of an anatomic customised total knee replacement implant: A case study. *Proc. Inst. Mech. Eng. H* **2020**, *234*, 1370–1383. [CrossRef] [PubMed]
31. Woernle, C. *Mehrkörpersysteme*; Springer: Berlin, Heidelberg, 2011; ISBN 978-3-642-15981-7.
32. Nigg, B.M. *Biomechanics of the Musculo-Skeletal System*, 2nd ed.; Wiley: Chichester, UK, 2002; ISBN 9780471978183.
33. Hemingway, E.G.; O'Reilly, O.M. Perspectives on Euler angle singularities, gimbal lock, and the orthogonality of applied forces and applied moments. *Multibody Syst. Dyn.* **2018**, *44*, 31–56. [CrossRef]
34. Cheung, R.T.H.; Mok, N.W.; Chung, P.Y.M.; Ng, G.Y.F. Non-invasive measurement of the patellofemoral movements during knee extension-flexion: A validation study. *Knee* **2013**, *20*, 213–217. [CrossRef] [PubMed]
35. Dagneaux, L.; Thoreux, P.; Eustache, B.; Canovas, F.; Skalli, W. Sequential 3D analysis of patellofemoral kinematics from biplanar x-rays: In vitro validation protocol. *Orthop. Traumatol. Surg. Res.* **2015**, *101*, 811–818. [CrossRef] [PubMed]
36. Merican, A.M.; Amis, A.A. Iliotibial band tension affects patellofemoral and tibiofemoral kinematics. *J. Biomech.* **2009**, *42*, 1539–1546. [CrossRef] [PubMed]
37. MacWilliams, B.A.; Davis, R.B. Addressing some misperceptions of the joint coordinate system. *J. Biomech. Eng.* **2013**, *135*, 54506. [CrossRef]
38. Lees, A.; Barton, G.; Robinson, M. The influence of Cardan rotation sequence on angular orientation data for the lower limb in the soccer kick. *J. Sports Sci.* **2010**, *28*, 445–450. [CrossRef]
39. Sinclair, J.; Hebron, J.; Hurst, H.; Taylor, P. The influence of different Cardan sequences on three-dimensional cycling kinematics. *Hum. Mov.* **2013**, *14*, 334–339. [CrossRef]
40. Ahmad, C.S.; Kwak, S.D.; Ateshian, G.A.; Warden, W.H.; Steadman, J.R.; Mow, V.C. Effects of patellar tendon adhesion to the anterior tibia on knee mechanics. *Am. J. Sports Med.* **1998**, *26*, 715–724. [CrossRef] [PubMed]
41. Suzuki, T.; Hosseini, A.; Li, J.-S.; Gill, T.J.; Li, G. In vivo patellar tracking and patellofemoral cartilage contacts during dynamic stair ascending. *J. Biomech.* **2012**, *45*, 2432–2437. [CrossRef]
42. Yeadon, M.R. The simulation of aerial movement—I. The determination of orientation angles from film data. *J. Biomech.* **1990**, *23*, 59–66. [CrossRef]
43. Sagan, B.E. *The Symmetric Group: Representations, Combinatorial Algorithms, and Symmetric Functions*, 2nd ed.; Springer: New York, NY, USA, 2010; ISBN 9781441928696.
44. Geier, A.; Aschemann, H.; Lima, D.D.; Woernle, C.; Bader, R. Force Closure Mechanism Modeling for Musculoskeletal Multibody Simulation. *IEEE Trans. Biomed. Eng.* **2018**, *65*, 2471–2482. [CrossRef]
45. Gale, T.; Anderst, W. Tibiofemoral helical axis of motion during the full gait cycle measured using biplane radiography. *Med. Eng. Phys.* **2020**, *86*, 65–70. [CrossRef]
46. Kwak, S.D.; Ahmad, C.S.; Gardner, T.R.; Grelsamer, R.P.; Henry, J.H.; Blankevoort, L.; Ateshian, G.A.; Mow, V.C. Hamstrings and iliotibial band forces affect knee kinematics and contact pattern. *J. Orthop. Res.* **2000**, *18*, 101–108. [CrossRef]
47. Yao, J.; Yang, B.; Niu, W.; Zhou, J.; Wang, Y.; Gong, H.; Ma, H.; Tan, R.; Fan, Y. In vivo measurements of patellar tracking and finite helical axis using a static magnetic resonance based methodology. *Med. Eng. Phys.* **2014**, *36*, 1611–1617. [CrossRef] [PubMed]
48. Rodrigues, O. Des lois géométriques qui régissent les déplacements d'unsystème solide dans l'espace, et de la variation des coordonnéesprovenant de ces déplacements considérés indépendammentdes causes qui peuvent les produire. *J. Math. Pures Appl.* **1840**, *5*, 380–440. (In French)

49. Bottema, O.; Roth, B. *Theoretical Kinematics*; Dover Publ: New York, NY, USA, 1990; ISBN 9780486663463.
50. Chirikjian, G.S.; Kyatkin, A.B. *Engineering Applications of Noncommutative Harmonic Analysis: With Emphasis on Rotation and Motion Groups*; CRC Press: Boca Raton, FL, USA, 2001; ISBN 9780849307485.
51. Woltring, H.J. 3-D attitude representation of human joints: A standardization proposal. *J. Biomech.* **1994**, *27*, 1399–1414. [CrossRef]
52. Fick, C.N.; Jiménez-Silva, R.; Sheehan, F.T.; Grant, C. Patellofemoral kinematics in patellofemoral pain syndrome: The influence of demographic factors. *J. Biomech.* **2021**, *130*, 110819. [CrossRef] [PubMed]
53. Ward, S.R.; Powers, C.M. The influence of patella alta on patellofemoral joint stress during normal and fast walking. *Clin. Biomech.* **2004**, *19*, 1040–1047. [CrossRef]
54. Pal, S.; Besier, T.F.; Beaupre, G.S.; Fredericson, M.; Delp, S.L.; Gold, G.E. Patellar maltracking is prevalent among patellofemoral pain subjects with patella alta: An upright, weightbearing MRI study. *J. Orthop. Res.* **2013**, *31*, 448–457. [CrossRef] [PubMed]
55. Shen, A.; Boden, B.P.; Grant, C.; Carlson, V.R.; Alter, K.E.; Sheehan, F.T. Adolescents and adults with patellofemoral pain exhibit distinct patellar maltracking patterns. *Clin. Biomech.* **2021**, *90*, 105481. [CrossRef] [PubMed]

Article

Tresca Stress Simulation of Metal-on-Metal Total Hip Arthroplasty during Normal Walking Activity

Muhammad Imam Ammarullah [1,2,*], Ilham Yustar Afif [1,2], Mohamad Izzur Maula [1,2], Tri Indah Winarni [2,3,4], Mohammad Tauviqirrahman [1], Imam Akbar [5], Hasan Basri [5], Emile van der Heide [6] and J. Jamari [1,2,*]

1. Department of Mechanical Engineering, Faculty of Engineering, Diponegoro University, Tembalang, Semarang 50275, Central Java, Indonesia; iyustar.afif@gmail.com (I.Y.A.); izzurmaula@gmail.com (M.I.M.); mtauviq99@yahoo.com (M.T.)
2. Undip Biomechanics Engineering & Research Center (UBM-ERC), Diponegoro University, Tembalang, Semarang 50275, Central Java, Indonesia; triwinarni@lecturer.undip.ac.id
3. Department of Anatomy, Faculty of Medicine, Diponegoro University, Tembalang, Semarang 50275, Central Java, Indonesia
4. Center for Biomedical Research (CEBIOR), Faculty of Medicine, Diponegoro University, Tembalang, Semarang 50275, Central Java, Indonesia
5. Department of Mechanical Engineering, Faculty of Engineering, Sriwijaya University, Indralaya 30662, South Sumatra, Indonesia; imamakbarrr0502@gmail.com (I.A.); hasan_basri@unsri.ac.id (H.B.)
6. Laboratory for Surface Technology and Tribology, Faculty of Engineering Technology, University of Twente, Postbox 217, 7500 AE Enschede, The Netherlands; e.vanderheide@utwente.nl
* Correspondence: imamammarullah@gmail.com (M.I.A.); j.jamari@gmail.com (J.J.); Tel.: +62-895-3559-22435 (M.I.A.); +62-813-2674-8417 (J.J.)

Abstract: The selection of biomaterials for bearing in total hip arthroplasty is very important to avoid various risks of primary postoperative failure for patients. The current investigation attempts to analyze the Tresca stress of metal-on-metal bearings with three different materials, namely, cobalt chromium molybdenum (CoCrMo), stainless steel 316L (SS 316L), and titanium alloy (Ti6Al4V). We used computational simulations using a 2D axisymmetric finite element model to predict Tresca stresses under physiological conditions of the human hip joint during normal walking. The simulation results show that Ti6Al4V-on-Ti6Al4V has the best performance to reduce Tresca stress by 45.76% and 39.15%, respectively, compared to CoCrMo-on-CoCrMo and SS 316L-on-SS 316L.

Keywords: Tresca stress; metal-on-metal; total hip arthroplasty; normal walking activity

1. Introduction

Metal-on-metal total hip arthroplasty has been increasingly selected in hip replacement surgery for diseased hips, especially for younger patients with higher activity levels [1]. This is due to the high number of failure cases that have been found in the use of metal-on-polyethylene and ceramic-on-polyethylene bearings, where polyethylene wear induces osteolysis and aseptic loosening. Meanwhile, the use of ceramic-on-ceramic is prone to failure due to cracking due to high-intensity activities, generally carried out by younger users.

Looking at the data published by the Australian Orthopedic Association (AOA) in 2020 [2] shows that the total failure cases of hip arthroplasty with metal-on-metal bearings are relatively higher than other options. Even so, metal-on-metal is still widely used in several developing countries, including Indonesia to meet the domestic market's need for hip joint implants independently without having to import [3–5]. This is based on the advantages of metal-on-metal in terms of relatively affordable prices, easily available raw materials, and limited production equipment.

The main problem with using metal-on-metal is the production of metal ions due to metal wear particles that can spread throughout the body from the bloodstream. These metal ions cause poisoning and various negative reactions in the human body system. Efforts that can be made to prevent implants from failing are the further evaluation of the choice of metal material used for metal-on-metal total hip arthroplasty. Several metal materials are commonly used for metal-on-metal bearings, including cobalt chromium molybdenum (CoCrMo), 316 L stainless steel (SS 316L), and titanium alloy (Ti6Al4V) [6].

In evaluating the performance of total hip arthroplasty, several previous studies used the von Mises stress, such as that conducted by Chethan et al. [7] and Carreiras et al. [8]. However, the use of Tresca stress is considered to be better than the von Mises stress because the Tresca failure theory has a smaller safety area than the von Mises failure theory, so it can be said that the use of Tresca is safer than of von Mises [9]. Research related to the evaluation of the performance of artificial joints with Tresca stress has previously been carried out by Usman and Huang [9] and Abdullah et al. [10]. Unfortunately, the evaluation of the use of materials for metal-on-metal total hip arthroplasty with Tresca stresses has not yet been carried out.

To accommodate this problem, the current study focuses on evaluating the Tresca stress in metal-on-metal bearings with different materials. A finite element-based prediction model has been created to solve problems using computational simulations. Gait loading has been used to reflect Tresca stress conditions more accurately in daily activities for simulating metal-on-metal total hip arthroplasty.

2. Materials and Methods

2.1. Geometric Parameters and Material Properties

The bearing geometry used for the current study was adopted from the research conducted by Jamari et al. [3] and is described in Table 1 for both the femoral head and the acetabular cup, which are commonly used in total hip arthroplasty.

Table 1. Geometric parameters for bearing components.

Parameter	Size (mm)
Femoral head diameter	28
Radial clearance	0.05
Acetabular cup thickness	5

The properties of metallic materials evaluated in the current study were based on previous studies: CoCrMo from Jamari et al. [3] and SS 316L and Ti6Al4V from Jiang et al. [11], which are described in Table 2. All simulated materials are assumed to be homogeneous, isotropic, and linear elastic.

Table 2. Material properties for metallic materials.

Material	Young's Modulus (GPA)	Poisson's Ratio (-)
CoCrMo	210	0.3
SS 316L	193	0.3
Ti6Al4V	110	0.3

To consider the surface roughness during articulation in metal-on-metal bearings, the coefficient of friction is given for the current computational simulations described in Table 3. The coefficient values for various metal-on-metal bearings were adopted from the previous literature: CoCrMo-on-CoCrMo from Jamari et al. [3], SS 316L-on-SS 316L from Jin et al. [12], and Ti6Al4V-on-Ti6Al4V from Arash et al. [13].

Table 3. Coefficient of friction for different materials of metal-on-metal.

Bearings	Coefficient of Friction (-)
CoCrMo-on-CoCrMo	0.2
SS 316L-on-SS 316L	0.8
Ti6Al4V-on-Ti6Al4V	1

2.2. Finite Element Modelling

A computational model is established to analyze metal-on-metal bearings is carried out by analyzing two main components, namely, the acetabular cup and femoral head by adopting a 2D axisymmetric ball-in-socket model as shown in Figure 1. Pelvic bones, fixation system, and femoral stem components were not considered in the current study to simplify the computational process but without significantly affecting the results. The definition of contact was made by configuring the femoral head and acetabular cup contact surfaces as master and slave surfaces, respectively. For contact formulation, discretization method, friction formulation, and threshold are defined as surface to surface, penalty, and nodal, respectively.

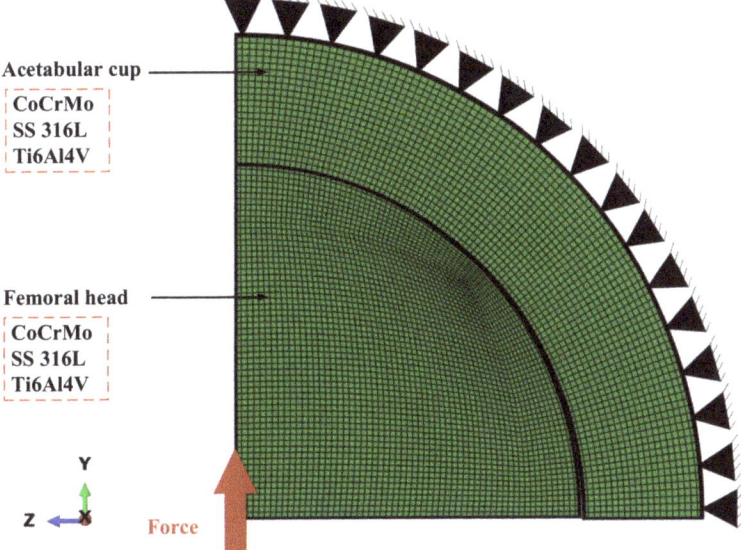

Figure 1. Metal-on-metal total hip arthroplasty model for finite element analysis.

The contact process also ignores micro separation with the femoral head always concentric with the acetabular cup position. The influence of synovial fluid during contact was ignored in the analysis with only considering surface roughness in dry contact. Temperature changes are not analyzed. Then, the acetabular cup component is made immobile by selecting fix constraint set of nodes in the outer acetabular cup area. In addition, the loading condition is given, being the concentrated load on the femoral head with selecting center node of femoral head.

The selection of the number of elements is obtained through a convergence study to find the optimal number of elements using the H-refinement method by finding the least number of elements but with near-accurate results to provide computational time efficiency. A total of 5500 CAX4 elements have been used with details of 2000 and 3500 CAX4 elements for the femoral head and acetabular, respectively, in analyzing Tresca stress on metal-on-metal using ABAQUS/CAE 6.14-1 software. In the present computational simulation, Tresca stress is analyzed in the bulk area of the acetabular cup.

2.3. Gait Cycle

Computational simulation is carried out with physiological loading by adopting normal walking conditions that is the most common activity carried out by patients who have performed total hip replacement surgery. However, in our present computational simulation, we only consider gait loading in the form of resultant force and ignore the range of motion effect for model simplification as conducted by Basri et al. [14]. This was done to focus the present study on the effect of resultant force under normal walking conditions against Tresca stress. The normal walking condition with one full cycle simplified into 32 phases has been adopted from the previous study by Jamari et al. [3] shown in Figure 2. In one loading cycle consisting of the 'stance phase' which is the first 19 phases and continued with the 'swing phase' to completion, the resultant force is highest in the seventh phase of 2326 N that has 2.5–3 times the average human body weight in general. Then, the lowest phase occurs in the 30th phase.

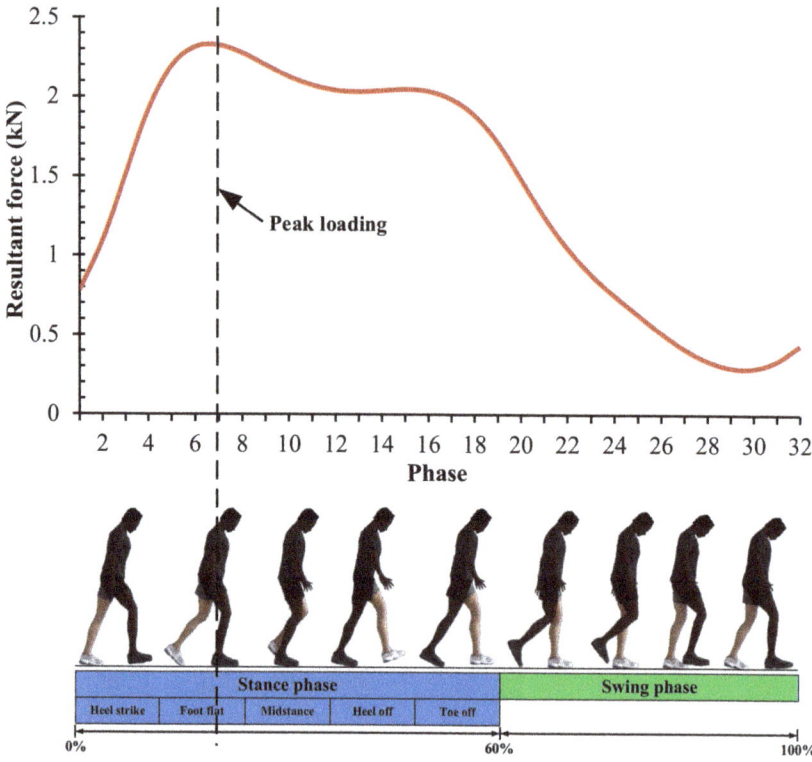

Figure 2. Gair loading based on normal walking condition [3].

3. Results and Discussion

The maximum Tresca stress from metal-on-metal bearing with different metallic materials during normal walking cycle is presented in Figure 3. The comparison of the highest, average, and lowest Tresca stresses of metal-on-metal bearing materials under normal walking conditions are described in Figure 4. The maximum Tresca stress value changes due to the difference in the resultant force applied during loading in the normal walking condition, where the highest maximum Tresca stress value is in the seventh phase for all metal-on-metal bearings in this study.

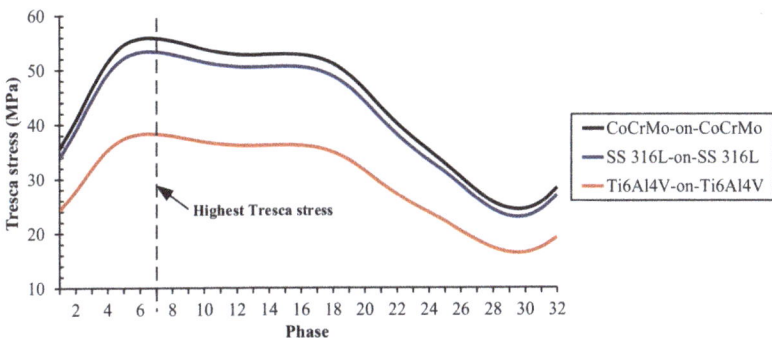

Figure 3. Maximum Tresca stress for different metal-on-metal bearing materials in full cycle.

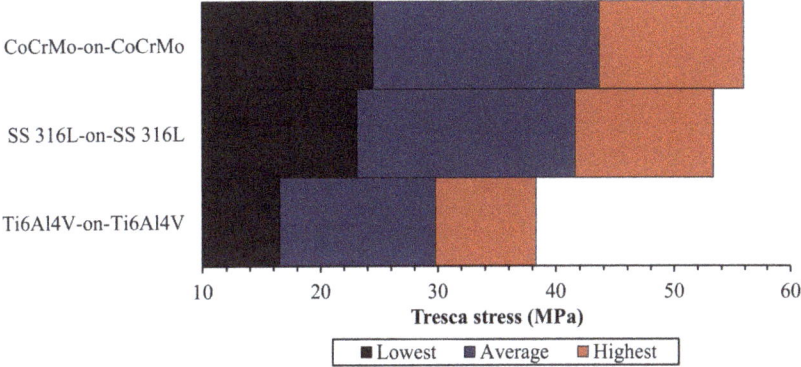

Figure 4. Comparison of highest, average, and lowest Tresca stress of metal-on-metal bearings materials on under normal walking condition.

Ti6Al4V-on-Ti6Al4V has the lowest Tresca stress value of 38.31 MPa. Tresca stress values were found to increase in CoCrMo-on-CoCrMo and SS 316L-on-SS 316L bearings by about 45.76% and 39.15%, respectively, compared to Ti6Al4V-on-Ti6Al4V. Apart from the magnitude of resultant force applied, Young's modulus also plays a significant role in the Tresca stress results. The greater Young's modulus of material under the same loading conditions will give a higher Tresca stress. Therefore, metal-on-metal bearings with CoCrMo material have the highest Tresca stress value under normal walking conditions compared to other bearing materials and vice versa for metal-on-metal bearings using Ti6Al4V material. The maximum Tresca stress values of various metal-on-metal bearings can be seen in Table 4.

Table 4. Maximum Tresca stress for different metal-on-metal bearing materials at peak loading.

Bearing Material	Tresca Stress (MPa)
CoCrMo-on-CoCrMo	55.84
SS 316L-on-SS 316L	53.31
Ti6Al4V-on-Ti6Al4V	38.31

Figure 5 illustrates the distribution contour of Tresca stress from Tresca terminology in ABAQUS [15]. The Tresca stress distribution contour is displayed using 5 selected phases from 32 phases in a normal walking cycle, namely the 1st phase that is the initial cycle, the 7th phase that is the phase with the highest gait loading, the 16th phase that is the middle of the cycle, the 30th phase that is the phase with the lowest gait loading, and the 32nd phase that is the end of the cycle. It can be seen that the distribution contour of the Tresca stress will be wider and the value of the Tresca stress will be higher along with the greater the resultant force applied.

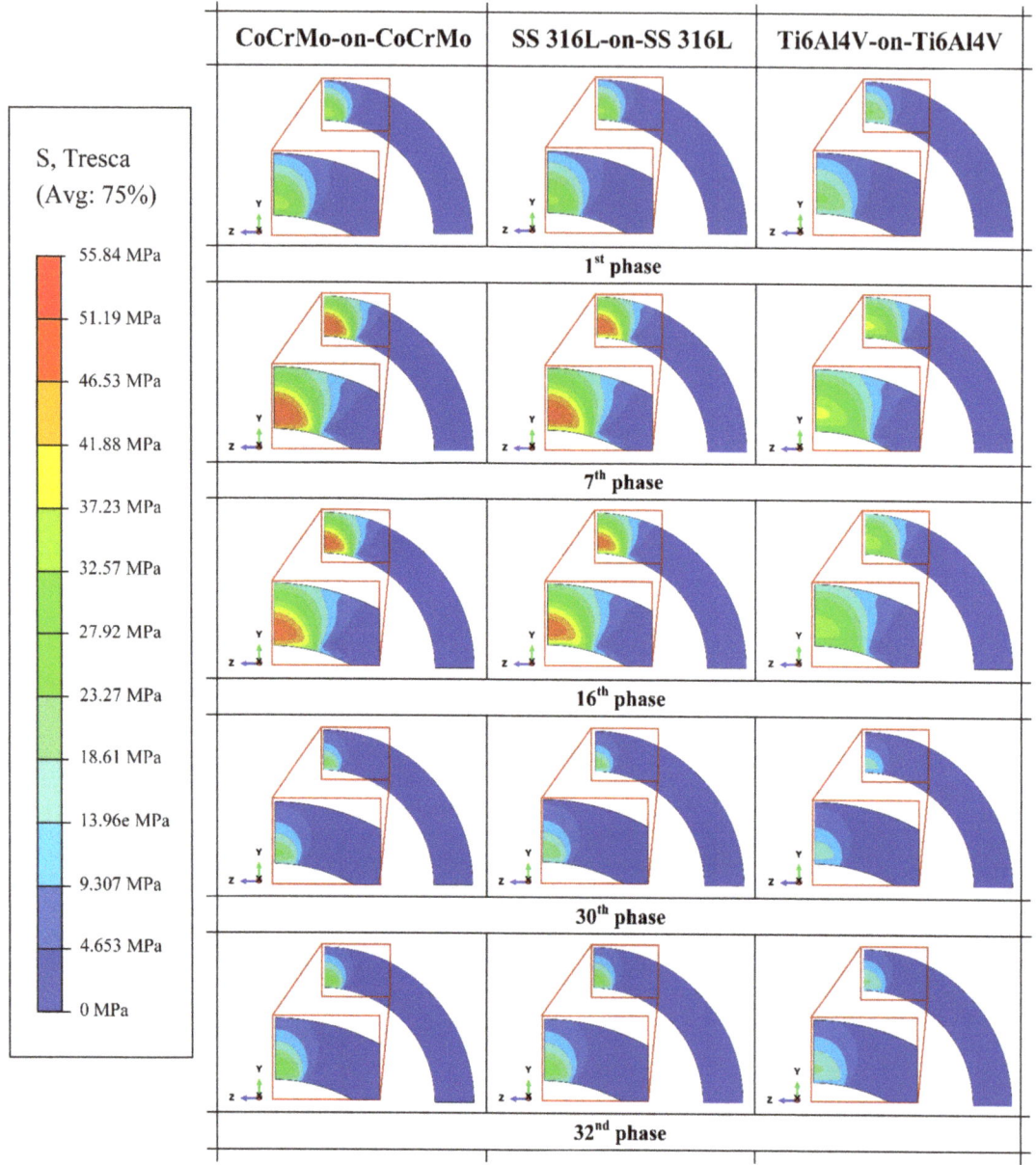

Figure 5. Tresca stress distribution on the acetabular cup for different metal-on-metal bearing materials at selected phases.

The relationship between Tresca stress and acetabular cup thickness on metal-on-metal bearings with different metallic materials during the peak phase and selected phase is illustrated in Figures 6 and 7. The highest Tresca stress does not occur in the surface contact but in the bulk area that is very prone to losing its elastic properties. In addition, the higher

Tresca stress experienced by a material, the higher the probability of failure based on the Tresca failure theory [16]. In the results obtained, it is found that Ti6Al4V-on-Ti6Al4V has the lowest probability of failure with the lowest Tresca stress compared to the other two metal-on-metal bearings.

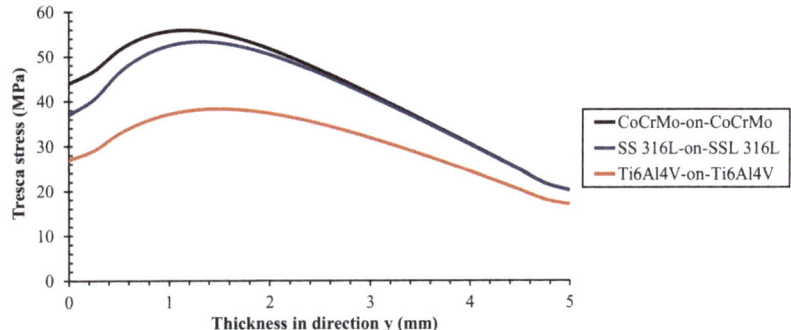

Figure 6. Tresca stress profile as a function of acetabular cup thickness for different metal-on-metal bearing materials at peak loading.

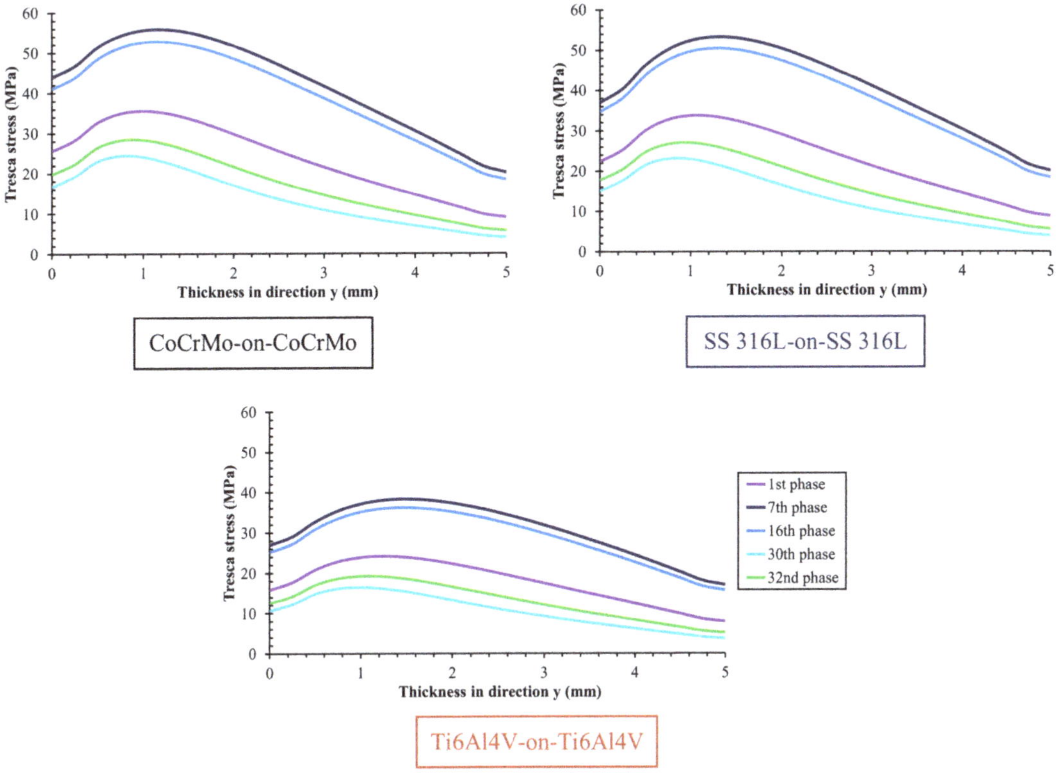

Figure 7. Tresca stress profile as a function of acetabular cup thickness for different metal-on-metal bearing materials at selected phases.

From the Tresca stress results obtained in the current implant design, various development efforts and further studies for metal-on-metal bearings need to be carried out to reduce the probability of failure. Apart from the selection of materials that have been carried out in the current study, several other efforts can be made, namely, examining the geometry parameters [14,17], the application of textured surfaces [18], and the use of coating layers [19]. The technical aspects of the surgeon's total hip replacement surgery with total hip arthroplasty also contributes to the long-term survival of the implant [20].

There are several focuses of attention to be conveyed regarding the limitations in our study. First, the coefficient of friction used is constant to represent the effect of lubrication that occurs, where the coefficient of friction should have a value that varies with time [13]. Second, the finite element model used is simple with a two-dimensional model that can reduce the accrual of the results compared with the results from three-dimensional model that are closer to the actual conditions [3]. Finally, the computational simulation does not take into account the range of motion during a normal walking cycle by only analyzing vertical loads that are irrelevant to actual conditions [11].

4. Conclusions

The evaluation of Tresca stress on metal-on-metal bearings with different metal materials using a 2D axisymmetric finite element prediction model has been successfully presented in our paper. The highest maximum Tresca stress is obtained in the seventh phase, where the resultant force is highest when conditions are normal walking. The distribution of the Tresca stress contour also widens along with the higher Tresca stress values experienced. The Tresca stress values also correlate with failure probability based on the Tresca failure theory, where we found Ti6Al4V-on-Ti6Al4V bearings to be the most superior among other metal-on-metal bearing options for reducing Tresca stresses.

Author Contributions: Conceptualization, M.I.A.; methodology, M.I.A.; software, M.I.A.; validation, M.I.A. and I.A.; formal analysis, M.I.A.; investigation, M.I.A. and I.A.; resources, J.J.; data curation, M.I.A., I.Y.A. and M.I.M.; writing—original draft preparation, M.I.A.; writing—review and editing, T.I.W., M.T., E.v.d.H. and J.J.; visualization, M.I.A.; supervision, T.I.W., M.T., H.B., E.v.d.H. and J.J.; project administration, I.Y.A. and M.I.M.; funding acquisition, H.B. and J.J. All authors have read and agreed to the published version of the manuscript.

Funding: The research was funded by World Class Research UNDIP number 118-23/UN7.6.1/PP/2021 and DIPA of Public Service Agency of Sriwijaya University 2021 SP DIPA-023.17.2.677515/2021. M.I.A. thanks Ministry of Education, Culture, Research and Technology (Kemendikbudristek) and Indonesia Endowment Fund for Education (LPDP) for domestic master degree of Indonesian education scholarships (BPI) under number 1059/J5/KM.01.00/2021.

Institutional Review Board Statement: Not applicable.

Informed Consent Statement: Not applicable.

Data Availability Statement: The data presented in this study are available on request from the corresponding author.

Acknowledgments: We gratefully thank Diponegoro University, Sriwijaya University, and University of Twente as the author's institutions for their strong support in our conducted research.

Conflicts of Interest: The authors declare no conflict of interest.

References

1. Alvarez-Vera, M.; Ortega-Saenz, J.A.; Hernandez-Rodríguez, M.A.L. A study of the wear performance in a hip simulator of a metal-metal Co-Cr alloy with different boron additions. *Wear* **2013**, *301*, 175–181. [CrossRef]
2. *Australian Orthopaedic Association National Joint Replacement Registry. Annual Report 2020*, 2020th ed.; Australian Orthopaedic Association: Adelaide, Australia, 2020.
3. Jamari, J.; Ammarullah, M.I.; Saad, A.P.M.; Syahrom, A.; Uddin, M.; van der Heide, E.; Basri, H. The effect of bottom profile dimples on the femoral head on wear in metal-on-metal total hip arthroplasty. *J. Funct. Biomater.* **2021**, *12*, 38. [CrossRef]

4. Maula, M.I.; Aji, A.L.; Aliyafi, M.B.; Afif, I.Y.; Ammarullah, M.I.; Winarni, T.I.; Jamari, J. The subjective comfort test of autism hug machine portable seat. *J. Intellect. Disabil.-Diagn. Treat.* **2021**, *9*, 182–188. [CrossRef]
5. Jamari, J.; Ammarullah, M.I.; Afif, I.Y.; Ismail, R.; Tauviqirrahman, M.; Bayuseno, A.P. Running-in analysis of transmission gear. *Tribol. Ind.* **2021**, *43*, 434–441. [CrossRef]
6. Mahyudin, F.; Hermawan, H. *Biomaterials and Medical Devices—A Perspective from an Emerging Country*; Springer: Berlin/Heidelberg, Germany, 2016; Volume 58. [CrossRef]
7. Chethan, K.N.; Mohammad, Z.; Shyamasunder Bhat, N.; Satish Shenoy, B. Optimized trapezoidal-shaped hip implant for total hip arthroplasty using finite element analysis. *Cogent Eng.* **2020**, *7*, 1719575. [CrossRef]
8. Carreiras, A.R.; Fonseca, E.M.M.; Martins, D.; Couto, R. The axisymmetric computational study of a femoral component to analysis the effect of titanium alloy and diameter variation. *J. Comput. Appl. Mech.* **2020**, *51*, 403–410. [CrossRef]
9. Huang, S.C. The study of stresses characteristic of contact mechanism in total knee replacement using two-dimensional finite element analysis. *Biomed. Mater. Eng.* **2017**, *28*, 567–578. [CrossRef] [PubMed]
10. Abdullah, A.H.; Asri, M.N.M.; Alias, M.S.; Giha, T. Finite element analysis of cemented hip arthroplasty: Influence of stem tapers. *Proc. Int. Multiconf. Eng. Comput. Sci.* **2010**, *2182*, 2241–2246.
11. Jiang, H.B. Static and Dynamic mechanics analysis on artificial hip joints with different interface designs by the finite element method. *J. Bionic Eng.* **2007**, *4*, 123–131. [CrossRef]
12. Jin, Z.M.; Stone, M.; Ingham, E.; Fisher, J. (v) Biotribology. *Curr. Orthop.* **2006**, *20*, 32–40. [CrossRef]
13. Arash, V.; Anoush, K.; Rabieer, S.M.; Rahmatei, M.; Tavanafar, S. The effects of silver coating on friction coefficient and shear bond strength of steel orthodontic brackets. *Scanning* **2015**, *37*, 294–299. [CrossRef] [PubMed]
14. Basri, H.; Syahrom, A.; Prakoso, A.T.; Wicaksono, D.; Amarullah, M.I.; Ramadhoni, T.S.; Nugraha, R.D. The analysis of dimple geometry on artificial hip joint to the performance of lubrication. *J. Phys. Conf. Ser.* **2019**, *1198*, 042012. [CrossRef]
15. Dassault Systèmes. *Abaqus Analysis User's Guide Volume IV: Elements*; Dassault Systèmes Simulia Corp.: Providence, RI, USA, 2016.
16. Christensen, R.M. *The theory of materials failure. Theory Mater. Fail*; Oxford University Press: Oxford, UK, 2013; Volume 1. [CrossRef]
17. Basri, H.; Syahrom, A.; Ramadhoni, T.S.; Prakoso, A.T.; Ammarullah, M.I.; Vincent. The analysis of the dimple arrangement of the artificial hip joint to the performance of lubrication. *IOP Conf. Ser. Mater. Sci. Eng.* **2019**, *620*, 012116. [CrossRef]
18. Ammarullah, M.I.; Saad, A.P.M.; Syahrom, A.; Basri, H. Contact pressure analysis of acetabular cup surface with dimple addition on total hip arthroplasty using finite element method. *IOP Conf. Ser. Mater. Sci. Eng.* **2021**, *1034*, 012001. [CrossRef]
19. Choudhury, D.; Ay Ching, H.; Mamat, A.B.; Cizek, J.; Abu Osman, N.A.; Vrbka, M.; Hartl, M.; Krupka, I. Fabrication and characterization of DLC coated microdimples on hip prosthesis heads. *J. Biomed. Mater. Res. Part B Appl. Biomater.* **2015**, *103*, 1002–1012. [CrossRef] [PubMed]
20. Giannoudis, P.V. *Practical Procedures in Elective Orthopaedic Surgery: Pelvis and Lower Extremity*, 1st ed.; Springer: Berlin/Heidelberg, Germany, 2013; Volume 9780857298. [CrossRef]

Article

Modeling the Contact Interaction of a Pair of Antagonist Teeth through Individual Protective Mouthguards of Different Geometric Configuration

Anna Kamenskikh [1], Alex G. Kuchumov [1,*] and Inessa Baradina [2]

1 Department of Computational Mathematics, Mechanics and Biomechanics, Perm National Research Polytechnic University, 614990 Perm, Russia; anna_kamenskikh@mail.ru
2 Department of Orthopedic Dentistry and Orthodontics with the Course of Children's Dentistry, Belarusian Medical Academy of Postgraduate Education, 220013 Minsk, Belarus; baradina@yandex.by
* Correspondence: kychymov@inbox.ru; Tel.: +7-(342)-2-39-17-02

Abstract: This study carried out modeling of the contact between a pair of antagonist teeth with/without individual mouthguards with different geometric configurations. Comparisons of the stress–strain state of teeth interacting through a multilayer mouthguard EVA and multilayer mouthguards with an A-silicon interlayer were performed. The influence of the intermediate layer geometry of A-silicone in a multilayer mouthguard with an A-silicon interlayer on the stress–strain state of the human dentition was considered. The teeth geometry was obtained by computed tomography data and patient dental impressions. The contact 2D problem had a constant thickness, frictional contact deformation, and large deformations in the mouthguard. The strain–stress analysis of the biomechanical model was performed by elastoplastic stress–strain theory. Four geometric configurations of the mouthguard were considered within a wide range of functional loads varied from 50 to 300 N. The stress–strain distributions in a teeth pair during contact interaction at different levels of the physiological loads were obtained. The dependences of the maximum level of stress intensity and the plastic deformation intensity were established, and the contact parameters near the occlusion zone were considered. It was found that when using a multilayer mouthguard with an A-silicone interlayer, there is a significant decrease in the stress intensity level in the hard tissues of the teeth, more than eight and four times for the teeth of the upper and lower teeth, respectively.

Keywords: mouthguard; occlusal contact; friction; teeth

Citation: Kamenskikh, A.; Kuchumov, A.G.; Baradina, I. Modeling the Contact Interaction of a Pair of Antagonist Teeth through Individual Protective Mouthguards of Different Geometric Configuration. *Materials* 2021, 14, 7331. https://doi.org/10.3390/ma14237331

Academic Editor: Giovanni Vozzi

Received: 10 November 2021
Accepted: 28 November 2021
Published: 30 November 2021

Publisher's Note: MDPI stays neutral with regard to jurisdictional claims in published maps and institutional affiliations.

Copyright: © 2021 by the authors. Licensee MDPI, Basel, Switzerland. This article is an open access article distributed under the terms and conditions of the Creative Commons Attribution (CC BY) license (https://creativecommons.org/licenses/by/4.0/).

1. Introduction

1.1. Problem Context

The dentofacial system is vital as its elements maintain various physiological processes, such as respiration, digestion and speech [1]. Dental biomechanics issues have increased in number over the past decade. These issues include the study of the heterogeneity of dental tissues [2], modeling the teeth stress–strain state under orthodontic loads [3], orthodontics problems to correct the occlusion [4] and numerical simulation of contact between teeth and dental implants or mouthguards [5]. Furthermore, the condition of the dentofacial system has a significant influence on physiological processes due to tooth injuries during sports and hard physical labor as well psychoemotional stress [6,7].

Today, one of the most effective ways to avoid dental injuries is the use of a mouthguard [8]. As various mouthguard designs exist on the market, there is a need for computer modeling of biomechanical behavior, both of the structures themselves and of the materials from which they are made [9].

Research has been carried out on the following:

- Biomechanical analysis of the effect of the properties of mouthguard materials on the deformation behavior of dental hard tissues [10,11];

- Mathematical modeling of the contact interaction of mouthguards of different geometric configurations with a wide range of physiological loads [12,13];
- A comprehensive interdisciplinary study of the patterns of change in the stress–strain state of the human dentofacial system when using mouthguards [14];
- The influence of the geometric configurations of the mouthguards on the stress–strain state [15], etc.

Currently, there is a particular interest in studying the structural features of mouthguards on the elements of the dentofacial system in order to identify qualitative and quantitative patterns of deformation behavior of teeth.

1.2. Research Objectives

The study objectives were formulated to evaluate the practical application of multilayer Eva mouthguards with an A-silicone interlayer. The research objectives are:

- To solve the problem of deformation of teeth during occlusion for a specific clinical case, with/without mouthguards;
- To model the frictional contact in the area of teeth occlusion;
- To use an elastoplastic model of behavior on the base material of the EVA mouthguards;
- To model various configurations of mouthguards for a clinical case;
- To analyze the influence of the geometry and thickness of the A-silicone layer on the system "mouthguard–teeth".

1.3. Problem Description

Analysis of the influence of the geometric configuration of protective mouthguards on their performance during teeth contact is presented in this study. In this case, the load of the jaw compression is considered as the indentation force with consideration of the friction between the contacting surfaces.

The paper presents a comparative analysis of the deformation behavior of a pair of antagonist teeth during frictional contact interaction through mouthguards of different geometric configurations.

The task was carried out on the basis of the clinical case data. The patients practiced sports professionally. Figure 1 shows the components of a biomechanical unit for one of the clinical cases: the upper jaw cast and CT (Computer tomography) image.

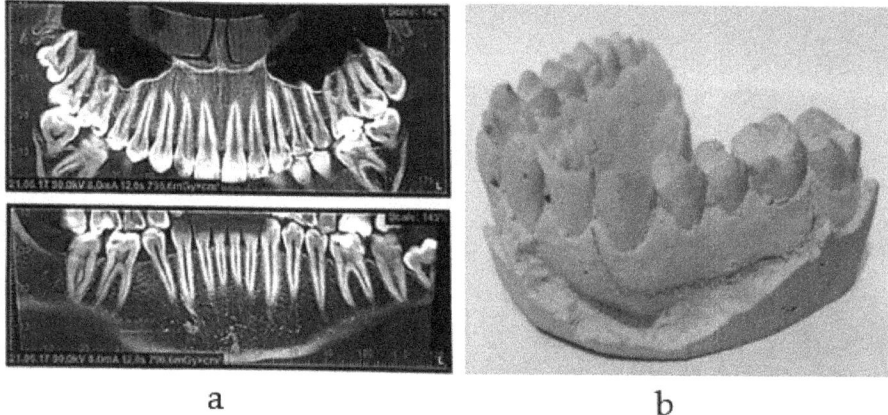

Figure 1. Geometry of the clinical cases: (**a**) CT image; (**b**) the gypsum model of an upper jaw.

Improving the performance of protective mouthguards by introducing additional layers of materials has been considered by many scientific groups [16–18]. In [16], the novel idea of introducing an additional layer of A-silicone into the design of an ethylene–vinyl acetate (EVA) mouthguard was proposed.

The influence of the geometric characteristics of mouthguards on the stress–strain state of the hard tissues was carried out in Reference [19]. When analyzing the results, it was found that the geometric configuration of the A-silicone interlayer of the mouthguard has a significant effect on the deformation behavior of the dentition. However, in that work, the canonical geometries of the teeth were considered in contact through an individually adaptable mouthguard.

2. Materials and Methods

2.1. Design of the Experiment

In this work, an attempt was made to analyze the influence of the geometric configuration and the thickness of the interlayer on the deformation behavior of the elements of the dentition. The peculiarity of the models is the use of data on the geometry of teeth for a real clinical case. To fully evaluate the effectiveness of a multilayer mouthguard with an A-silicone interlayer, the contact of a pair of antagonist teeth was simulated for a clinical case (Figure 1) with and without the multilayer EVA of an individual protective mouthguard.

Five numerical models of the contact of a pair of teeth, with and without protective mouthguards of different geometries (Figure 2), were analyzed.

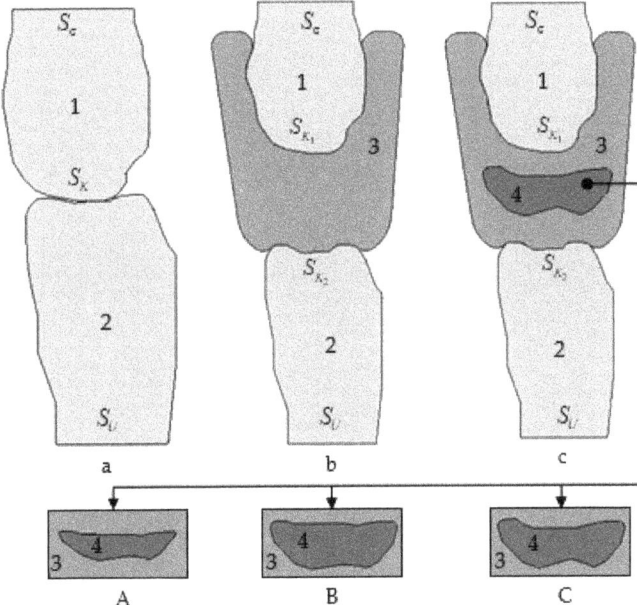

Figure 2. Numerical scheme of the contact between upper (1) and lower (2) teeth with and without a mouthguard: (**a**) case a without mouthguard; (**b**) case b with a multilayer EVA (3) mouthguard; (**c**) case c with a multilayer EVA (3) mouthguard with an A-silicone interlayer (4) of different geometries and thicknesses (**A**–**C**). (S_σ is the boundary where the loading is applied; S_K is the boundary where the contact occurs (for multilayered mouthguards $S_K = S_{K_1} \cup S_{K_2}$); S_U is the boundary where displacements are set).

The geometric configuration of the teeth was based on clinical case data (Figure 1). Mouthguard fit of the teeth geometry and frictional contact were considered. In case a, frictional contact was taken into account in the area of teeth-antagonists occlusion. In cases b, c-A, c-B, c-C, antagonist teeth were in contact with mouthguards.

The maximum thickness of multilayer mouthguards in the area of teeth occlusion was about 7 mm. All multilayer mouthguards were based on EVA layers. Contact interaction between EVA layers was not considered (modeled as a solid). Multilayer mouthguards with a layer of A-silicone were modeled within the framework of the frictional contact between the layers of EVA and A-silicone.

To analyze the influence of mouthguard on the teeth deformation under contact, three variants of the A-silicone interlayer were considered (Figure 2):

- Case c-A—the thickness of the A-silicone interlayer in the occlusion region was 1.6–1.9 mm (22.8–27.1% of the maximum thickness of the mouthguard);
- Case c-B—the thickness of the A-silicone interlayer in the occlusion region was 2.9–3.2 mm (41.4–45.7% of the maximum thickness of the mouthguard);
- Case c-C—the thickness of the A-silicone interlayer in the occlusion region was 2.2–3.2 mm (31.4–45.7% of the maximum thickness of the mouthguard).

For the first two variants of the geometric configuration of the A-silicone interlayer, a slight change in thickness in the occlusion region of 10–15% is characteristic. The third variant of the interlayer has a more significant change in thickness in the occlusion region (more than 30%). It should also be noted that the interlayer thickness was adjusted to the teeth geometry: for the area where the tooth has a smaller contact area, the interlayer was selected to be the thinnest and vice versa.

2.2. Mechanical Properties of the Mouthguard Components

A multilayered mouthguard construction was studied. The mouthguard was from EVA (Drufosoft, Dreve, Germany) with an embedded layer of A-silicone (UfiGelP, Voco, Germany). UfiGelP is a base paste and a catalyst paste, which can be mixed in certain ratios (in this case 1:1). Once cured, UfiGelP is a highly elastic polymer material. The proportion of base and catalyst pastes in the manufacture of the splice tray was selected empirically. Before solidification, the material is soft enough and easily adjusts to the required shape. That allows you to form a layer of A-silicone in the mouthguard. The properties of these materials were obtained from experimental studies performed by the research team from Perm National Research Polytechnic University, Perm State Medical University and Perm State National Research University [16,20]. The Young's modulus E and Poisson's coefficient v are $E = 17.3$ MPa, $v = 0.46$ (EVA) and $E = 0.3$ MPa, $v = 0.49$ (A-silicone). It was shown that EVA exhibits elastoplastic properties. The experimental plots are presented in Reference [20]. To describe the EVA mechanical properties, the deformation theory of elastoplasticity was chosen. A-silicone was shown to be an elastic material.

2.3. Loading and Boundary Conditions

The mathematical problem statement is described in Reference [20] and includes equilibrium equations, physical and geometric relations, as well as contact boundary conditions. When realizing the problem, the mathematical formulation was supplemented by considering the possibility of the appearance of large deformations in the EVA. The task was implemented as a 2D problem. The mathematical formulation of the problem was supplemented by boundary conditions: a constant load varying from 50 to 300 N (indentation force) was applied at the boundary S_σ; at the boundary S_u, movement along the vertical coordinate y was prohibited.

2.4. Numerical Finite Element (FE) Solution and Convergence

The numerical schemes were implemented in the ANSYS software package (version 11.0, ANSYS Inc., Canonsburg, PA, USA), by the finite element method using quadrangular plane finite elements with Lagrangian approximation and two degrees of freedom at each node. The basic procedures for constructing FE models are based on the use of the Galerkin method procedure with the choice of basic functions with a compact support by the finite element method.

The analysis of the convergence of the results of the numerical solution of the problem of contact interaction of a pair of teeth with and without a protective mouthguard from the degree of discretization of the system was carried out in an earlier paper [21]. It was found that a finite element mesh with a gradient concentration of elements to the contact zones provides an optimal solution to the problem in terms of accuracy and computation time: the maximum element size is 0.25 mm, the minimum element size is approximately 4 times less than the maximum one. The finite element subdivision of the biomechanical units considered in this work was performed similarly to the previously selected mesh.

3. Results

The influence of the mouthguard geometry on the stress–strain state and the parameters of the occlusion region was considered in our findings.

Figures 3 and 4 show the stress intensity distributions in the hard tissues of the teeth with an indentation force of 250 N for the upper and lower teeth. As expected, the use of a mouthguard leads to a significant decrease in the level of stress intensity.

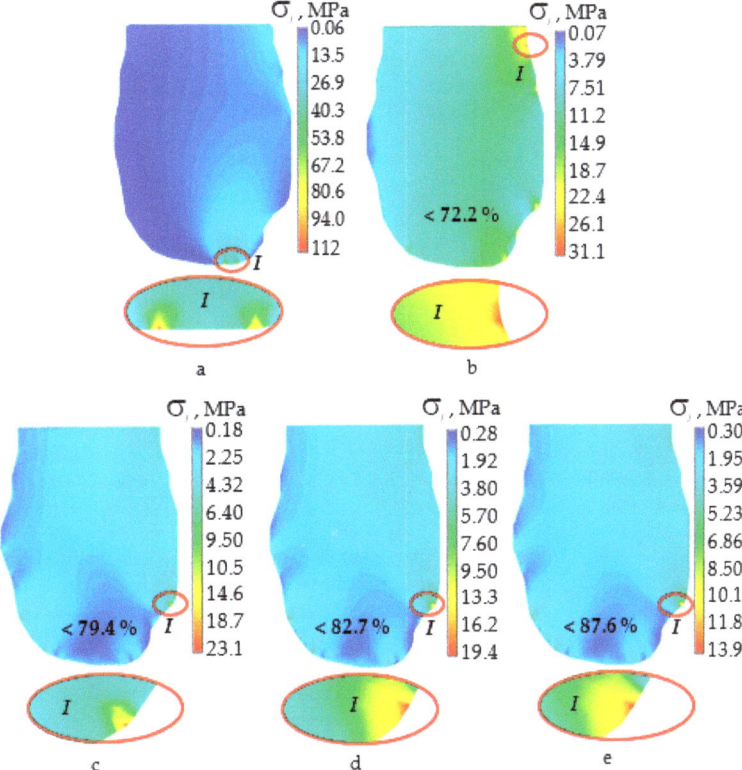

Figure 3. The stress intensity in the upper dentition tooth at 250 N: (**a**) case a; (**b**) case b; (**c**) case c-A; (**d**) case c-B; (**e**) case c-C; I—zone of maximum stress intensity.

In the case of teeth contact without a mouthguard, the maximal stress intensity was observed near the occlusion region. When using mouthguards, a significant decrease in the level of stress intensity was observed. In most cases, the maximum stresses are distributed over a larger area. For the multilayer EVA case, the mouthguard shifted the zone of maximum stresses towards the neck of the tooth.

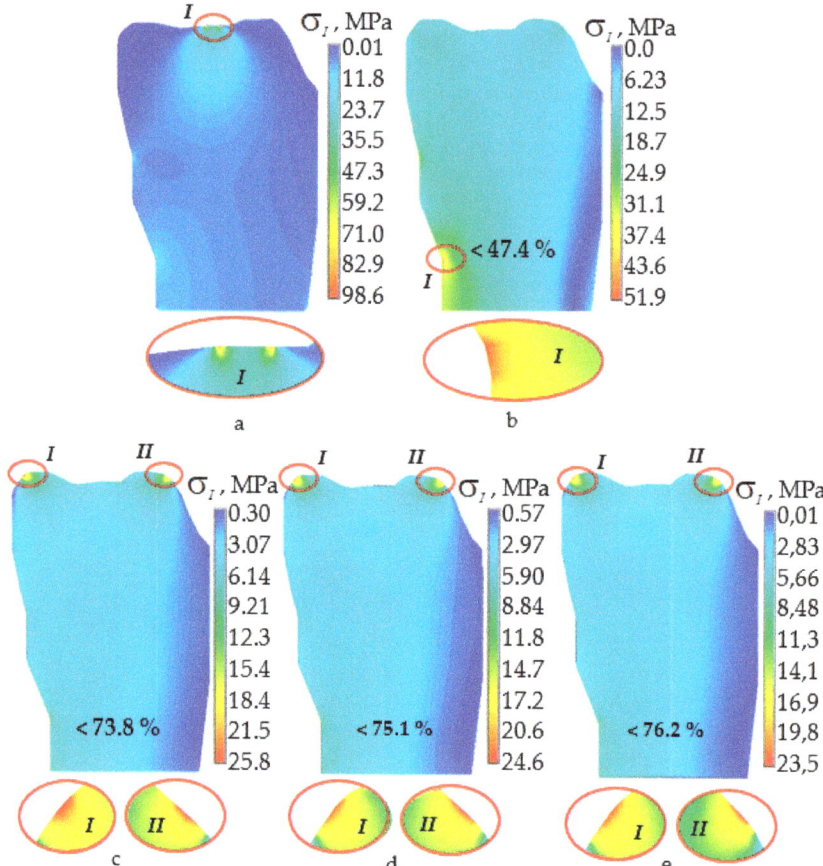

Figure 4. The stress intensity in the lower dentition tooth at 250 N: (**a**) case a; (**b**) case b; (**c**) case c-A; (**d**) case c-B; (**e**) case c-C; I, II—zone of maximum stress intensity.

For multilayer mouthguards with an A-silicon interlayer, the maximum stress intensity was observed in the contact zones, but removed from the tooth occlusion zone. When using individual mouthguards, the intensity of stresses in the tooth of the upper dentition decreased by 3.6 times when using a multilayer EVA mouthguard and on average by 6.2 times when using multilayer mouthguards with an A-silicon interlayer.

The greatest decrease in the level of stress intensity was observed when using a mouthguard with an interlayer adjusted to the geometry of the elements of the dentition (Figure 3e), maximal stress intensity was less by more than eight times.

The decrease in the maximum level of stress intensity in the hard tissues of the tooth of the lower dentition when using a multilayer individual EVA mouthguard was found to be much less than for the tooth of the upper dentition (Figure 4). The maximum stress intensity decreased by only 1.9 times. In this case, the maximum stress intensity when using a multilayer EVA mouthguard was localized near the edge of the contact area with the mouthguard.

When using multilayer mouthguards in the tooth of the lower dentition, the greatest decrease in the maximum level of stress intensity was observed on average four times more without their localization near the contact surface. Of particular interest is the analysis of the influence of the geometric characteristics of the mouthguard on the deformation behavior of the elements of the dentition. Within the framework of a series of numerical experiments, the dependences of the maximum level of stress intensity on the force of indentation in hard tissues of teeth were established for all variants of design schemes (Figure 5). As expected, the dependence of the maximum level of stress intensity on the indentation force between a pair of antagonist teeth without mouthguard usage was linear.

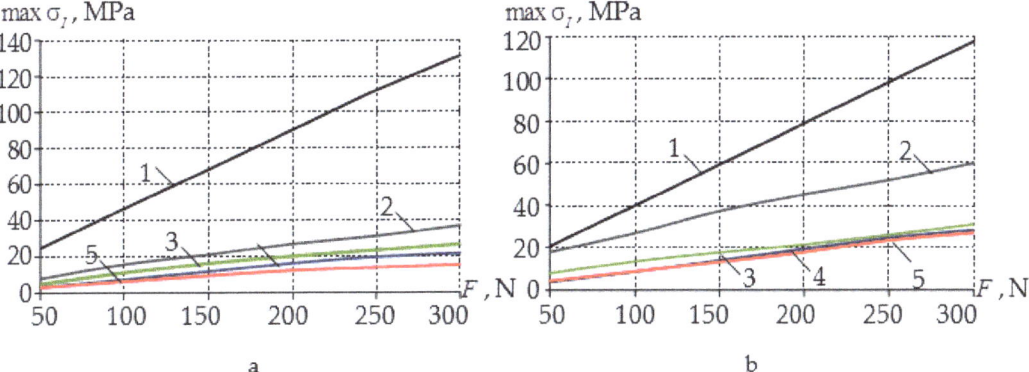

Figure 5. Dependence of maximal stress intensity on indentation force for the teeth of the upper (**a**) and lower (**b**) dentition: 1—case a; 2—case b; 3—case c-A; 4—case c-B; 5—case c-C.

When using a mouthguard, the dependence of the maximal stress intensity on the indentation force was found to be close to linear. With an increase in the force of indentation, an increase in the effect of reducing the intensity of stresses in the hard tissues of the teeth was observed when using all types of mouthguards. In the tooth of the upper dentition, the greatest decrease in the level of stress intensity was observed when using a mouthguard with an interlayer adjusted to the geometry of the elements of the dentition.

The mouthguard use was shown to have a significant impact on the parameters of contact interaction. With the contact of antagonist teeth without a mouthguard, the maximum level of contact pressure and contact shear stress reached 78.6 and 7.19 MPa with an indentation force of 250 N. Figure 6 shows a comparison of the maximum level of the parameters of the contact zones when using individual dental mouthguards of different geometric configurations.

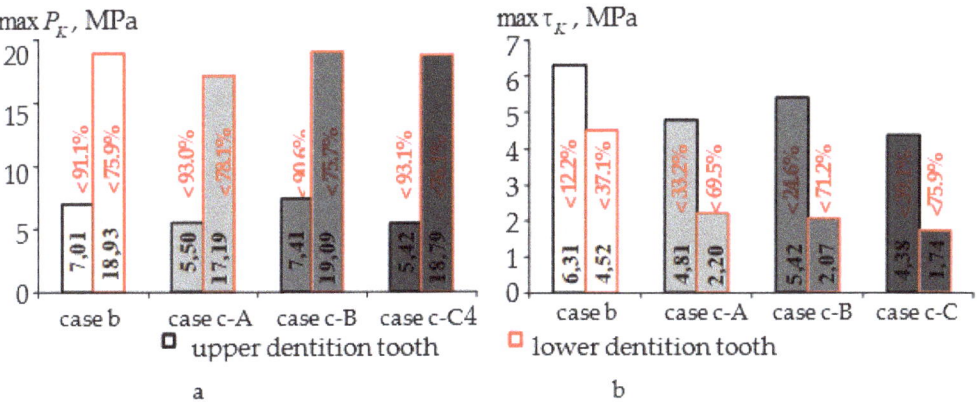

Figure 6. Maximal values of contact parameters: (**a**) max P_K; (**b**) max τ_K.

A significant decrease in the maximum level of contact pressure can be noted for all types of dental mouthguards. The maximum level of contact shear stress decreased less than the contact pressure in the zone of contact with the tooth of the upper dentition.

The maximum contact pressure was higher in the tooth of the lower dentition, and the maximum contact shear stress was in the tooth of the upper dentition (Figure 6). As in the case of stress intensity, the maximum decrease in the level of contact interaction parameters was observed when using a mouthguard with an interlayer adjusted to the geometry of the elements of the dentition (case c-C).

The dependence of the level of plastic deformations in the mouthguard was not linear and did not exceed 30% at a maximum load.

4. Discussion

4.1. Limitation Statement

Within the framework of the study, the models of materials and objects of the study had a number of limitations.

When modeling the EVA behavior model, it is possible to consider the viscosity of the material. Additional studies of the deformation material and further clarification of the constitutive relations are required. Friction coefficients of EVA-tooth material and EVA-A-silicone are constant at 0.3; the coefficient of friction was taken from the reference literature. Investigation of the friction of materials requires specialized equipment and an original test procedure. Considering the experimentally obtained frictional properties of materials will make it possible to obtain a better picture of deformation of the elements of the dentition.

Within the framework of the task, the multilayered teeth were not considered, which introduces an additional error into the model. The tooth is a composite structure. In further studies, it is planned to consider the multilayered tooth with different properties of the materials of the layers.

The 2D FEA problem was solved. The contact between the layers of the mouthguard was considered, but the level of the parameters of the EVA-A-silicon contact zone was a lower order of magnitude than in the occlusion area. No delamination of the interlayer was observed during the simulation. Frictional contact with a previously unknown contact area, and the nature of the distribution of contact state status zones was realized in the zone of teeth closing.

The study of the influence of the geometry of the mouthguards on the deformation of the elements of the dentition was carried out in the first approximation. Researchers have a number of challenges that follow:

- Clarification of the physical, mechanical and frictional properties of the materials of the biomechanical unit;
- Clarification of the models of teeth and analysis of the influence of their multilayerness and the nature of the conjugation of layers on the deformation of the biomechanical unit in flat and axisymmetric formulations;
- Clarification of the level and type of loads acting on the biomechanical unit;
- Transition to three-dimensional models.

Mouthguard thickness has a significant impact on the patient's comfort. Recent studies considered an influence of the mouthguard's thickness on its performance [22–24]. Westerman et al. [22] revealed that the rational thickness for an EVA mouthguard is 4 mm. When a mouthguard's thickness exceeds 4 mm, there are some negative effects on the patient's speech and breath. Bochnig et al. [23] studied various mouthguards with thicknesses ranging from 2 up to 11 mm. It was concluded that thickness increase by insertion of additional layers results in protective properties of the construction. Sarac et al. [24] obtained similar results.

We studied multilayer mouthguards with 7 mm thickness, including harder intermediate layers. The thickness of the mouthguard can have a number of limitations when athletes use them.

4.2. Main Results

Almost all studies in the field of sports dental medicine pay attention to the need to use mouthguards to prevent dentofacial injuries during sport activities with a high and moderate level of injury [17,25,26]. The effectiveness of mouthguards has been proven in practice. The choice of mouthguards is wide [26]: from standard to individual, from single-layer to multilayer, from ordinary to specialized, etc. Standard and thermoplastic designs of protective mouthguards have a number of disadvantages [18]:

1. The loose fit, which leads to the need to hold the mouthguard, difficulty breathing and distorted speech.
2. Adaptation to the dentition through heating and seating is similar to taking a dental impression; this mouthguard fit on the teeth depends on the human factor, which is often not ideal.
3. Fracture and creep can be observed with repeated heat treatment, as well as an increase in the hardness of materials, which leads to its early failure and a negative effect on the athlete's body.

The most effective protective dental constructions are individual mouthguards [18,27]. Sousa et al. [27] consider different designs of mouthguards, highlighting individual mouthguards as more effective, including multilayer ones. The classic single-layer mouthguard design reduces the maximum stress intensity in the teeth by a maximum of 54.4% [14]. The multilayer individual EVA mouthguard reduces the maximum stress intensity in teeth by 72.2%, as shown in this work. The efficiency of the multilayer mouthguard is more than 17% higher than that of the classic design. The use of a multilayer mouthguard from relatively soft material changed the stresses distribution nature in the dentition elements (Figures 3 and 4b). In this case, the maximum stress level was observed in the tooth neck. The stress in the teeth decreased in general but increased in the tooth neck by 2.2–2.3 times.

Rationalization of mouthguard designs has been ongoing. The work is aimed at a number of factors [28,29]: materials, production methods, and geometry, etc. The assumption is considered that by introducing an intermediate layer made of a harder material into a structure, it is possible to achieve a better effect in protecting the teeth [16,29,30]. The intermediate interlayer of A-silicone in mouthguards made it possible to maximally reduce the stress intensity of the teeth by 87.6%, as shown in this work. Nevertheless, the insertion of additional elements or layers into the structure of the mouthguards is not always effective. For example, the mouthguard with an interlayer of a silica-nylon mesh did not show an improvement in the mechanical reaction of teeth [28]. Mouthguards with A-silicone interlayer reduced tooth stress significantly more than adaptive and multilayer EVA mouthguards, the study showed. At the same time, it was established that the interlayer geometry significantly affects the performance of the biomechanical unit. The maximum stress concentration in the upper tooth was observed, with an interlayer thickness of 22.8–27.1% of the total mouthguard thickness, but not in the teeth closing zone. The geometry of the A-silicone interlayer of the mouthguard affects the deformation of the elements of the dentition. An improperly chosen interlayer shape can lead to a stress concentration in the tooth, with further cracking.

Another feature of this work is the use of a nonlinear model of EVA material behavior. Kerr et al. [17] revealed that when choosing and analyzing the operation of mouthguards, it is necessary to consider the physical and mechanical properties of the materials from which they are made. One of the most common materials in mouthguards is the EVA polymer [28]. Currently, there is a significant amount of research devoted to the analysis of the properties of EVA from different manufacturers [29,30]. The behavior EVA model is nonlinear and reflects elastoplastic deformation. Many works consider the EVA material as elastic, for example, Lokhov et al. [16]. Consideration of EVA in terms of elastic deformation distorts the research results. The many effects and patterns were made noticeable due to the nonlinear description of the EVA operation within this study.

4.3. The Mouthguard Thickness Analysis

Standard thickness of EVA mouthguards is 3–4 mm. This work considered a multilayer EVA kappa with a thickness of 7 mm. Of interest is the comparative analysis of the operation of EVA mouthguards of standard thickness and mouthguards with a layer of A-silicone. The stresses in the teeth in contact with EVA mouthguards of 3 and 4 mm thickness of a load of up to 600 N are shown in Figure 7.

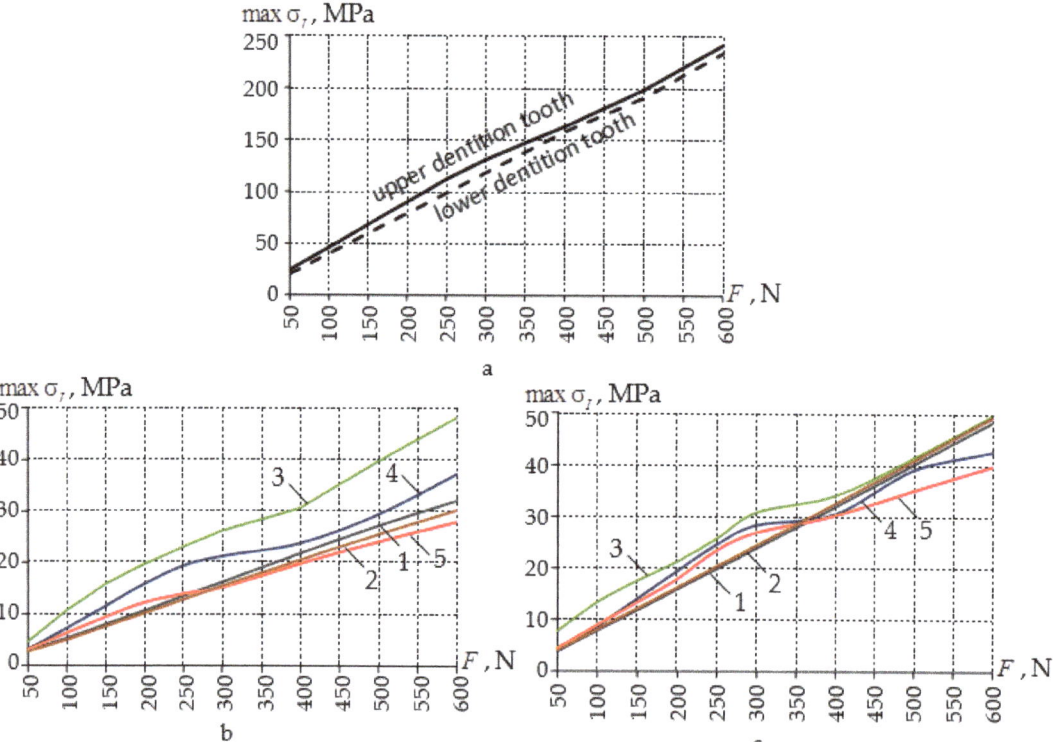

Figure 7. Dependence of maximal stress intensity on indentation force without (**a**) and with (**b**,**c**) a mouthguard (for teeth of the upper (**b**) and lower (**c**) teeth dentition): 1—case b of 4 mm thickness; 2—case b of 3 mm thickness; 3—case c-A; 4—case c-B; 5—case c-C.

It was found that with a standard thickness, the EVA aligner reduces the intensity of stresses in the teeth to a level close to the aligners with a layer of A-silicone. The maximum intensity of stresses in a mouthguard made of EVA with a thickness of 3–4 mm were observed in the neck of the tooth.

At loads less than 250–350 N, EVA splints perform better than multilayer splints with an A-silicone interlayer. Under heavy loads, the mouthguard with an interlayer adjusted to the geometry of the elements of the dentition (case c-C) reduces tooth stress better than all other aligners reviewed.

The geometry of the A-silicone interlayer has a significant effect on the deformation distributions.

5. Conclusions

A comparative analysis was carried out on the deformation behavior of a pair of antagonist teeth during contact with and without mouthguards of different geometric configurations, with a range of functional loads (50–300 N). Data were obtained on the intensity of stresses and deformations, the parameters of the contact zone, as well as the dependence of the maximum level of deformation characteristics on the physiological load.

Analysis of the results of a series of numerical experiments established the following conclusions:

- Mouthguards, with an additional intermediate layer of A-silicone, make it possible to reduce the level of stress intensity in the hard tissues of the teeth by 15–25% more than when using individual multilayer EVA mouthguards.
- There is no pronounced localization of zones of maximum stress intensity in the hard tissues of the teeth, for all considered options for the geometry of the intermediate layer of A-silicone, when using multilayer mouthguards with an A-silicon interlayer.
- The geometric configuration of the A-silicone interlayer has a significant effect on the stress–strain state of a pair of antagonist teeth and a mouthguard.
- The greatest decrease in the level of deformation characteristics of the investigated unit is observed when using a mouthguard with an interlayer, adjusted to the geometry of the elements of the dentition.

The thickness of 7 mm multilayer mouthguards can lead to a number of limitations when used by athletes. Additional practical research is required on the physical and psychoemotional state of patients when they are using it.

Author Contributions: Conceptualization, A.K., A.G.K. and I.B.; methodology, A.K.; software, A.K.; validation, A.K. and A.G.K.; writing—original draft preparation, A.K. and A.G.K.; writing—review and editing, A.K., A.G.K. and I.B.; visualization, A.K.; funding acquisition, A.K. All authors have read and agreed to the published version of the manuscript.

Funding: Anna Kamenskikh and Alex G. Kuchumov acknowledge the financial support of the Ministry of Science and Higher Education of the Russian Federation in the framework of the program of activities of the Perm Scientific and Educational Center "Rational Subsoil Use".

Institutional Review Board Statement: Not applicable.

Informed Consent Statement: Not applicable.

Data Availability Statement: Not applicable.

Conflicts of Interest: The authors declare no conflict of interest.

References

1. Prado, D.G.D.A.; Sovinski, S.R.P.; Nary, H.; Brasolotto, A.G.; Berretin-Felix, G. Oral motor control and orofacial functions in individuals with dentofacial deformity Controle motor oral e funções orofaciais em indivíduos com deformidade dentofacial. *Audiol. Commun. Res.* **2015**, *20*, 76–83. [CrossRef]
2. Liao, Z.; Chen, J.; Zhang, Z.; Li, W.; Swain, M.; Li, Q. Computational modeling of dynamic behaviors of human teeth. *J. Biomech.* **2015**, *48*, 4214–4220. [CrossRef]
3. Fedorova, N.V. The study of the stress-strain state of the dental ceramic implants depending on their shape and bone mineralization degree. *Russ. J. Biomech.* **2019**, *23*, 388–394. [CrossRef]
4. Peck, C.C. Biomechanics of occlusion—Implications for oral rehabilitation. *J. Oral Rehabil.* **2016**, *43*, 205–214. [CrossRef] [PubMed]
5. Carvalho, V.; Soares, P.; Verissimo, C.; Pessoa, R.; Versluis, A.; Soares, C. Mouthguard Biomechanics for Protecting Dental Implants from Impact: Experimental and Finite Element Impact Analysis. *Int. J. Oral Maxillofac. Implant.* **2018**, *33*, 335–343. [CrossRef]
6. Mills, S.; Canal, E. Prevention of Athletic Dental Injuries: The Mouthguard. In *Modern Sports Dentistry*; Spring: Berlin/Heidelberg, Germany, 2018; pp. 111–133. ISBN 9783319444161.
7. Tribst, J.P.M.; Dal Piva, A.M.D.O.; Borges, A.L.S.; Bottino, M.A. Simulation of mouthguard use in preventing dental injuries caused by different impacts in sports activities. *Sport Sci. Health* **2019**, *15*, 85–90. [CrossRef]
8. Khan, S.A.; Fatima, M.; Hassan, M.; Khalid, N.; Iqbal, A.; Raja, A.A.; Annas, M. MOUTHGUARDS. *Prof. Med. J.* **2018**, *25*, 1029–1033. [CrossRef]
9. Fasciglione, D.; Persic, R.; Pohl, Y.; Filippi, A. Dental injuries in inline skating? Level of information and prevention. *Dent. Traumatol.* **2007**, *23*, 143–148. [CrossRef]
10. Westerman, B.; Stringfellow, P.M.; Eccleston, J.A. The effect on energy absorption of hard inserts in laminated EVA mouthguards. *Aust. Dent. J.* **2000**, *45*, 21–23. [CrossRef] [PubMed]
11. Westerman, B.; Stringfellow, P.M.; Eccleston, J.A. Beneficial effects of air inclusions on the performance of ethylene vinyl acetate (EVA) mouthguard material. *Br. J. Sports Med.* **2002**, *36*, 51–53. [CrossRef]

12. Otani, T.; Kobayashi, M.; Nozaki, K.; Gonda, T.; Maeda, Y.; Tanaka, M. Influence of mouthguards and their palatal design on the stress-State of tooth-periodontal ligament-bone complex under static loading. *Dent. Traumatol.* **2018**, *34*, 208–213. [CrossRef]
13. Gialain, I.O.; Coto, N.P.; Driemeier, L.; Noritomi, P.Y.; Dias, R.B.E. A three-dimensional finite element analysis of the sports mouthguard. *Dent. Traumatol.* **2016**, *32*, 409–415. [CrossRef] [PubMed]
14. Kamenskikh, A.A.; Ustjugova, T.N.; Kuchumov, A.G. Modelling of the tooth contact through one-layered mouthguard. *J. Phys. Conf. Ser.* **2018**, *1129*, 012014. [CrossRef]
15. Kamenskih, A.; Ustugova, T.; Kuchumov, A.G.; Taiar, R. Numerical evaluation of sport mouthguard application. *Adv. Intell. Syst. Comput.* **2020**, *1018*, 581–585. [CrossRef]
16. Lokhov, V.A.; Kuchumov, A.G.; Merzlyakov, A.F.; Astashina, N.B.; Ozhgikhina, E.S.; Tropin, V.A. Experimental investigation of materials of novel sport mouthguard design. *Russ. J. Biomech.* **2015**, *19*, 354–364. [CrossRef]
17. Kerr, I.L. Mouth Guards for the Prevention of Injuries in Contact Sports. *Sport. Med.* **1986**, *3*, 415–427. [CrossRef] [PubMed]
18. Ahn, H.-W.; Lee, S.-Y.; Yu, H.; Park, J.-Y.; Kim, K.-A.; Kim, S.-J. Force Distribution of a Novel Core-Reinforced Multilayered Mandibular Advancement Device. *Sensors* **2021**, *21*, 3383. [CrossRef] [PubMed]
19. Kamenskikh, A. The Analysis of the Work of Materials of Mouthguard Designs during Biomechanical Deformation. *Solid State Phenom.* **2018**, *284*, 1355–1360. [CrossRef]
20. Kamenskikh, A.; Astashina, N.B.; Lesnikova, Y.; Sergeeva, E.; Kuchumov, A.G. Numerical and experimental study of the functional loads distribution in the dental system to evaluate the new design of the sports dental splint. *Ser. Biomech.* **2018**, *32*, 3–15.
21. Kamenskikh, A.A.; Ustjugova, T.N.; Kuchumov, A.G. Comparative analysis of mechanical behavior of the tooth pair contacting with different mouthguard configurations. *IOP Conf. Ser. Mater. Sci. Eng.* **2019**, *511*, 012003. [CrossRef]
22. Westerman, B.; Stringfellow, P.M.; Eccleston, J.A. EVA mouthguards: How thick should they be? *Dent. Traumatol.* **2002**, *18*, 24–27. [CrossRef]
23. Bochnig, M.S.; Oh, M.J.; Nagel, T.; Ziegler, F.; Jost-Brinkmann, P.G. Comparison of the shock absorption capacities of di_erent mouthguards. *Dent. Traumatol.* **2017**, *33*, 205–213. [CrossRef] [PubMed]
24. Sarac, R.; Helbig, J.; Dräger, J.; Jost-Brinkmann, P.-G. A Comparative Study of Shock Absorption Capacities of Custom Fabricated Mouthguards Using a Triangulation Sensor. *Materials* **2019**, *12*, 3535. [CrossRef]
25. Ferrari, C.H.; Ferreira De Medeiros, J.M. Dental trauma and level of information: Mouthguard use in different contact sports. *Dent. Traumatol.* **2002**, *18*, 144–147. [CrossRef] [PubMed]
26. Knapik, J.J.; Marshall, S.W.; Lee, R.B.; Darakjy, S.S.; Jones, S.B.; Mitchener, T.A.; delaCruz, G.G.; Jones, B.H. Mouthguards in Sport Activities. *Sport. Med.* **2007**, *37*, 117–144. [CrossRef]
27. Sousa, A.M.; Pinho, A.C.; Messias, A.; Piedade, A.P. Present status in polymeric mouthguards. A future area for additive manufacturing? *Polymers* **2020**, *12*, 1490. [CrossRef]
28. Tribst, J.P.M.; Dal Piva, A.M.D.O.; de Carvalho, P.C.K.; de Queiroz Gonçalves, P.H.P.; Borges, A.L.S.; de Arruda Paes-Junior, T.J. Does silica–nylon mesh improves the biomechanical response of custom-made mouthguards? *Sport Sci. Health* **2020**, *16*, 75–84. [CrossRef]
29. Takeda, T.; Ishigami, K.; Mishima, O.; Karasawa, K.; Kurokawa, K.; Kajima, T.; Nakajima, K. Easy fabrication of a new type of mouthguard incorporating a hard insert and space and offering improved shock absorption ability. *Dent. Traumatol.* **2011**, *27*, 489–495. [CrossRef] [PubMed]
30. Matsuda, Y.; Nakajima, K.; Saitou, M.; Katano, K.; Kanemitsu, A.; Takeda, T.; Fukuda, K. The effect of light-cured resin with a glass fiber net as an intermediate material for Hard & Space mouthguard. *Dent. Traumatol.* **2020**, *36*, 654–661. [CrossRef]

Article

Finite Element Analysis of Customized Acetabular Implant and Bone after Pelvic Tumour Resection throughout the Gait Cycle

Leonid Maslov [1,2,*], Alexey Borovkov [1], Irina Maslova [1], Dmitriy Soloviev [1], Mikhail Zhmaylo [1] and Fedor Tarasenko [1]

[1] Institute for Advanced Manufacturing Technologies, Peter the Great St. Petersburg Polytechnic University, 29 Politekhnicheskaya, 195251 St. Petersburg, Russia; borovkov@compmechlab.com (A.B.); maslova.i@compmechlab.ru (I.M.); Solovyov.d@compmechlab.com (D.S.); zhmaylo@compmechlab.com (M.Z.); tarasenko@compmechlab.com (F.T.)

[2] Department of Theoretical and Applied Mechanics, Ivanovo State Power Engineering University, 34 Rabfakovskaya, 153003 Ivanovo, Russia

* Correspondence: leonid-maslov@mail.ru or tipm@tipm.ispu.ru

Citation: Maslov, L.; Borovkov, A.; Maslova, I.; Soloviev, D.; Zhmaylo, M.; Tarasenko, F. Finite Element Analysis of Customized Acetabular Implant and Bone after Pelvic Tumour Resection throughout the Gait Cycle. *Materials* **2021**, *14*, 7066. https://doi.org/10.3390/ma14227066

Academic Editors: Oskar Sachenkov and Antoniac Iulian

Received: 13 October 2021
Accepted: 18 November 2021
Published: 21 November 2021

Publisher's Note: MDPI stays neutral with regard to jurisdictional claims in published maps and institutional affiliations.

Copyright: © 2021 by the authors. Licensee MDPI, Basel, Switzerland. This article is an open access article distributed under the terms and conditions of the Creative Commons Attribution (CC BY) license (https://creativecommons.org/licenses/by/4.0/).

Abstract: The aim of this paper is to investigate and compare the stress distribution of a reconstructed pelvis under different screw forces in a typical walking pattern. Computer-aided design models of the pelvic bones and sacrum made based on computer tomography images and individually designed implants are the basis for creating finite element models, which are imported into ABAQUS software. The screws provide compression loading and bring the implant and pelvic bones together. The sacrum is fixed at the level of the L5 vertebrae. The variants of strength analyses are carried out with four different screw pretension forces. The loads equivalent to the hip joint reaction forces arising during moderate walking are applied to reference points based on the centres of the acetabulum. According to the results of the performed analyses, the optimal and critical values of screw forces are estimated for the current model. The highest stresses among all the models occurred in the screws and implant. As soon as the screw force increases up to the ultimate value, the bone tissue might be locally destroyed. The results prove that the developed implant design with optimal screw pretension forces should have good biomechanical characteristics.

Keywords: strength; computer simulation; finite element analysis; implant; pelvis; walking

1. Introduction

The human pelvis is a geometrically complex, biomechanical structure that carries the weight of the human body and stabilizes and protects inner organs. The pelvis can be damaged due to problems with the primary implant, infections, accidents, or bone tumours, which usually involve a large area of tissue removal and affect the patients' lives. Due to the complex anatomical structure, the reconstruction of pelvic biomechanics after the loss of bone structure is still a challenge [1]. Various implants are used for different types of pelvis injuries, such as modular pelvis prostheses, saddle prostheses, pedestal cups, and custom-made pelvis prostheses. Among them, custom-made endoprostheses are matched with the patient's bones, which in turn can reduce the risk of infection, dislocation, or failure of the implant [2]. Therefore, a custom prosthesis design is in demand when it is required to treat a complex bone fracture or replace a primary serial implant.

Previous research [3] presented pelvis reconstruction by applying a fibula and a variation of the methods of internal fixation of the implant. In this study, a vertical load of 500 N was applied to the L3 lumbar vertebrae, and the pelvis was considered to be in a bipedal standing position. As a result, the stress concentration in the fibula implant was extremely high, but this effect was minimized by internal fixation, which partially transferred the stresses from the fibula to the screw system. Additionally, a high stress concentration was detected in the implant. Among the four methods of fixation, the best method was a double rod system with an L5-S1 pedicle and iliac screws, which provided the lowest stress concentration and the lowest displacement of the pelvis.

A previous study [4] describes a modular endoprosthesis for the damaged half of the pelvis. In the course of this research, a comparative analysis of the stress distribution between the healthy and reconstructed pelvis was carried out in three static positions: sitting, standing, and standing on the foot of the injured side. The loads and boundary conditions were similar to those described in the study above [3]. In the healthy pelvis, the stress distribution was concentrated in the upper region of the acetabulum, arcuate line, sacroiliac joint, sacral midline, and, in particular, the upper region of the greater sciatic notch. In the reconstructed pelvis, the stress distribution was concentrated on the proximal area of the pubic plate, the top of the acetabulum, the connection between the CS fixator and acetabular cup, and the fixation between the prosthesis and sacroiliac joint. The stress distribution in the reconstructed pelvis was similar to the stress distribution in the healthy pelvis in the three different static positions.

Generally, the clinical efficacy and biomechanical features of the implants used for pelvic injuries should be evaluated through biomechanical experiments in vitro. However, irregular geometry and material heterogeneity of the pelvis often make mechanical experiments challenging [5].

In modern orthopaedic biomechanics, a computational approach was developed for analysing the stress and strain distributions of a hip joint endoprosthesis [6]. The study under consideration is based on the finite element (FE) method to investigate stresses on the bones and implant.

The FE method has proven to be a powerful tool in reducing the cost and time in many biomechanical studies and has become an important tool for understanding overall biomechanical behaviour. Nevertheless, many factors, such as material properties, anatomical geometry, the integrity of the human structure, and boundary conditions, could influence the accuracy of FE results [7].

Thus, the FE method is becoming increasingly popular in pelvis biomechanics research and plays a critical role in failure analysis and revision prosthesis design [2]. Although some FE analyses of custom-made prostheses have been carried out, studies of the influence of the prestress of the screws on the biomechanical performance of a reconstructed pelvis for walking patterns are rarely reported [2].

The aim of this study was to investigate the stress distribution of the pelvis reconstructed by individual endoprostheses with different screw forces and then identify which force value is optimal for tightening the implant and the bone. After that, the stress distribution in the "bone–endoprosthesis" system was obtained for typical walking loads and chosen screw forces.

2. Materials and Methods
2.1. Finite Element Models

Three-dimensional reconstruction was performed for the case of a young patient whose weight was 50 kg. He underwent treatment at the Federal State Budgetary Institution National Medical Research Center of Oncology named after N. N. Blokhin of the Ministry of Health of the Russian Federation (N.N. Blokhin NMRCO). Three-dimensional models of the patient's pelvis were obtained using next-generation, multi-slice computed tomography with high resolution and innovative software [8]. These 3D models were provided by N. N. Blokhin NMRCO (Figure 1). The CAD models consisted of several faceted surface bodies.

The abovementioned CAD models were the basis for developing a computer model of the individual endoprosthesis. The design of the customized endoprosthesis is shown in Figure 2. The main parts of the implant are the cup (1), the bearing flange on the iliac bone (2), the bearing flange on the pubic bone (3), and the bearing flange on the ischiatic bone (4).

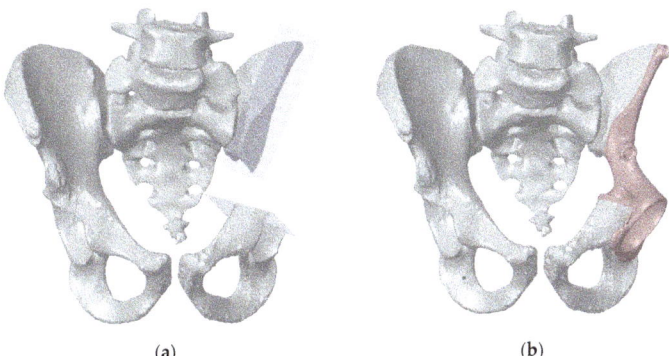

Figure 1. A virtual resection of the pelvic bones for pelvic tumour surgery: (**a**) resection planes; (**b**) the pelvis reconstructed with the individual endoprosthesis.

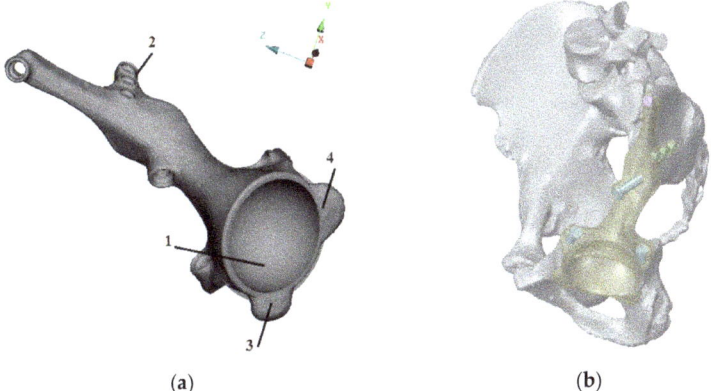

Figure 2. Developed design of the individual endoprosthesis considered in this paper: (**a**) design and main components of the implant: the cup (1), the bearing flange on the iliac bone (2), the bearing flange on the pubic bone (3), and the bearing flange on the ischiatic bone (4). (**b**) Implant position and fixation by seven screws.

    The personalised implant fastened to the damaged parts of the pelvic bones with seven screws (Figure 3). The drilled holes, which are also shown in the figure, have the same numbering as the screws. Screws with numbers from 1 to 3 have a length of 55 mm and diameter of 6.5 mm, and screws from 5 to 7 are 15 mm long and have a diameter of 4.5 mm. Screw 4 has a length of 45 mm and diameter of 5.5 mm.

    The development of models for numerical analysis is a relatively complicated process. Medical researchers are always concerned about verifying analytical models because incorrect assumptions in the FE model might lead to an incorrect stress distribution [9–13]. For this reason, a great amount of time was devoted to the development of the FE model, in particular the choice of the FE types and the mesh grid density.

    Additionally, it was necessary to consider the performance of the computer while defining the parameters of the finite element models. The current study was performed using an ordinary workstation, and the mesh grid parameters were chosen in such a way that the analysis model could be successfully run on the machine. The total number of finite elements in the assembly amounted to 1,203,061 elements of the mesh prepared in ABAQUS/CAE software (Dassault Systems Simulia Corp., Johnston, RI, USA) (Figure 4).

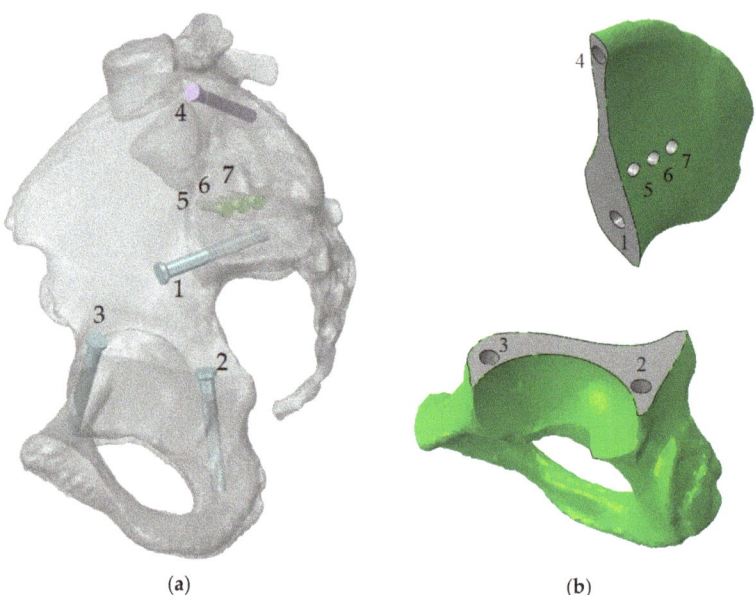

Figure 3. Implant fixation: (**a**) the screws numbered from 1 to 7 and their positions (the implant is hidden); (**b**) the holes numbered from 1 to 7 in the bone parts, the same numeration as this for the screws.

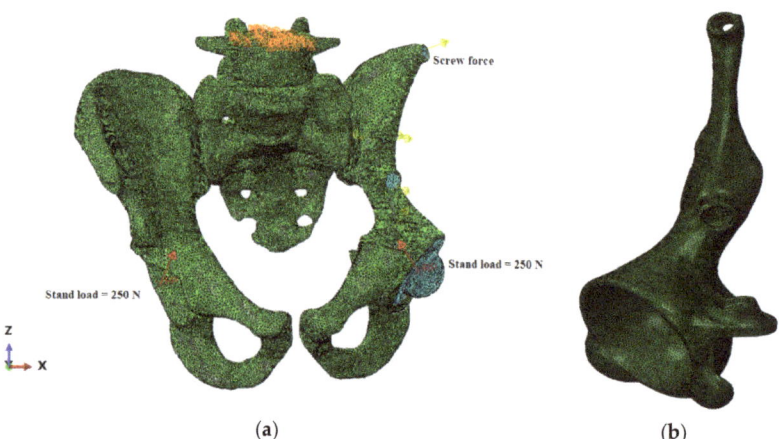

Figure 4. Finite element model of the "bone–endoprosthesis" system: (**a**) finite element mesh on the pelvis; (**b**) finite element mesh on the implant.

The finite element size for the pelvic bones and implant ranged from 0.5 mm to 2 mm. It complied with the conclusion of previous research [2] studying the dependence of the optimal mesh size and obtaining reliable results. It should be mentioned that the cortical layer was modelled as a solid body with a constant thickness by offsetting the outer surface inward by 0.5 mm (Figure 3b). The characteristics of the FE model are shown in Table 1.

Table 1. Characteristics of the finite element model.

Part	Number of Finite Elements	Finite Element Type	FEM Verification		
			Aspect	Skewness	Warping
Top of the damaged half of the pelvic bone	77,392	Four-node linear solid tetrahedral C3D4 type	No violations	2892 elements off	No violations
Bottom of the damaged half of the pelvic bone	122,131			3031 elements off	
Healthy pelvic bone	265,775			8807 elements off	
Sacrum	44,746			2009 elements off	
Implant	329,828			12,615 elements off	

The quality of FE models is verified by three criteria: aspect ratio, skewness, and warping. The verification showed that the general quality of the mesh was relatively high, but some elements did not satisfy the quality criteria, as indicated in Table 1. Note that according to the aspect ratio and warping criteria, no poor-quality elements were found. The poor-quality elements appeared due to the strict skewness criteria. However, at current skewness settings, the maximum error is no more than 3.5% of the total quantity of the finite elements. So, the amount of such elements was quite low, and this fact could be neglected without loss of accuracy.

2.2. Material Properties

Particular attention should be given to the physical and mechanical properties of the bone tissue. The pelvic bone consists mainly of low-density spongy tissue and a thin and dense cortical layer. Most of the load is transferred through the cortical layer, and the spongy tissue works as a support material, preventing the cortical layer from collapsing. Due to age and other reasons that may cause degradation of bone tissue, the mechanical properties can change [14–17]. Bone tissue is anisotropic [12,14,18,19], and the distribution of Young's modulus depends linearly on the density. However, the variation in Young's modulus both for the compact and spongy tissue is negligible for the pelvic bone, as proven previously [9,14]. Many papers devoted to the pelvis finite element study considered the pelvic bone as isotropic and linear elastic [1–8]. After analysing the data from these sources, the mechanical properties are summarised in Table 2.

Table 2. Material properties used in the present study.

Material	Young's Modulus, GPa	Poisson's Ratio	Ultimate Stress, MPa	
			Yield	Fatigue
Cortical tissue [9,18]	17	0.3	80–150 [18]	the same as the yield stress
Spongy tissue [14,20]	0.07	0.2	1.4–2.1 [20]	
Normal Ti-6Al-4V [21]	113.8	0.34	950	310–610 [21]
3D printed Ti-6Al-4V [22,23]	123.4	0.26	910	200–500 [23]
Polyethylene [24]	1	0.35	26	-

It is important to mention that the yield strength reported previously [12–14] for spongy bone seems excessively high. From a clinical point of view [20], the spongy bone has a yield strength of approximately 1%, so some additional evaluation is required to confirm that the spongy bone could be loaded up to 2–3% of strain without any plastic deformation, damage, or fracture. Therefore, a value range of 1.4–2.1 MPa was used in this study as the ultimate strength of the spongy bone. Data reported previously [20] were confirmed by the authors' research on spongy tissues extracted from the femoral head after surgery [15].

The next important point is to study the material characteristics of the other parts of the assembly. The main advantages of titanium alloy Ti-6Al-4V are good corrosion resistance under all conditions and excellent biocompatibility.

The screws and prosthetic head should be made of normal solid titanium alloy Ti-6Al-4V [21], and the considered implant is manufactured by 3D printing [22]. The mechanical properties of 3D printed and solid titanium are slightly different, particularly those characterizing the fatigue behaviour [23].

Finally, the acetabular liner is made of ultrahigh-molecular-weight polyethylene [24]. This polyethylene, reinforced with chemical cross-links, is distinguished by its strength, low friction coefficient, and high biocompatibility. These properties allow using it for artificial joints.

2.3. Loads and Boundary Conditions

Boundary conditions and contact interactions were defined in ABAQUS/CAE software (Dassault Systems Simulia Corp., Johnston, RI, USA).

The solution of the task in ABAQUS included two steps: screw tightening and standing. The region of the contact interaction and boundary conditions were unchanged between the steps. The model had contact interactions with friction in the following pairs: the implant and pelvic bone and the acetabular liner and the artificial femoral head. The coefficient of friction was equal to 0.2 for the titanium–bone pair and 0.15 for the polyethylene–titanium pair. The screw heads and threads are bonded with implant and bone correspondingly, so these contact pairs are considered as linear contacts without friction and separation.

The first step was to bring together the implant and the pelvic bone with the screws. Therefore, a special compressive screw force was applied to each screw. The calculations were conducted with four values of screw forces: 100, 500, 1000, and 1500 N. The upper surface of the sacrum is fully constrained.

The second step simulated the walking cycle as a slow, quasistatic process. The reaction forces corresponding to the patient's weight were applied to the reference points in the centres of the acetabular cups (Figure 5). The reaction force was obtained from the HIP98 software available from the OrthoLoad open resource (https://orthoload.com/test-loads/data-collection-hip98/, accessed on 1 October 2021). The software generates the biomechanical forces based on the special database, which was developed within study [25] using special instrumented implants.

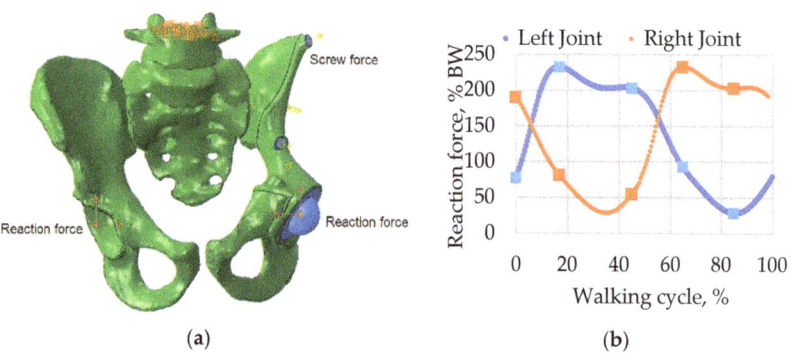

Figure 5. Boundary conditions and applied forces: (**a**) model with boundary conditions and loads including screw pretension and reaction forces applied to the centres of the left and right joints; (**b**) reaction force curves for walking simulation measured as a percentage of body weight (BW).

At the beginning of the second loading step, where the walking process was simulated, the length of the screws was fixed, which occurs at the end of the pretension phase. A detailed strength analysis was further performed for screw force values of 500 N and 1000 N. The general kinematic boundary conditions remained unchanged from the previous step of the considered computer simulation.

2.4. Model Validation

The FE method requires strict validation of the model because an inaccurate model may lead to incorrect and unreliable results [9–11]. Comprehensive experimental validation of the FE model of the reconstructed pelvis is not carried out before the surgery because a definite forecast of the bone stress state is required before performing surgery.

However, the evaluation of the general adequacy of the model was carried out in accordance with several parameters. First, the FE mesh was assessed based on several criteria (aspect ratio, skewness, warping), and mesh quality was evaluated based on element size, type, and shape. Second, the physical and mechanical characteristics of materials ensuring model accuracy were obtained from reliable sources [9,12,14,16,18,19].

Kinematic boundary conditions for the model and contact regions were applied and approved by the medical studies described in other articles [11,26,27]. The loading scheme, which was used in current research and based on the loading of the structure with the hip joint force acting as a reaction force, was the same as used previously [28]. It allowed us to describe the stress-strain state in a more accurate way.

The preliminary frequency analysis proved that the reconstructed pelvis assembly was joint and adjusted correctly, and all the connections were set up properly.

Another validation based on the displacement distribution (Figure 6) showed that the values obtained in the current study were within range close to the ranges described in papers [1,10,29]. The results of the analysis of the stress and strain fields were also close to previous results [1,2,10], proving the principal qualitative and quantitative conformance.

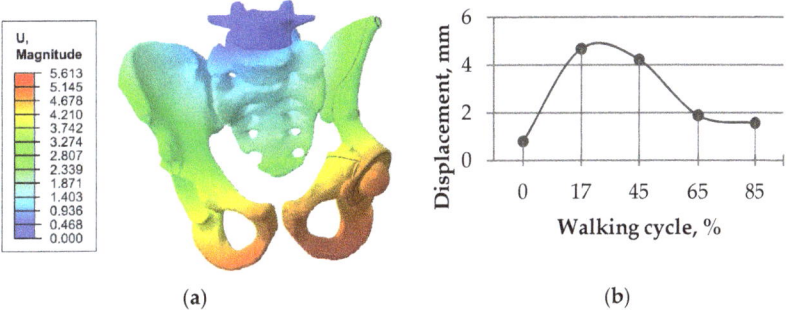

Figure 6. Total displacement in the "bone–endoprosthesis" system for the walking condition and a screw load of 1000 N: (**a**) typical displacement distribution (mm); (**b**) displacement values for the full gait cycle at the spherical center of the left femoral head.

3. Results

In this study, the problem of assessing the screw force effect on overall stress distribution is solved for the four values of pretension load equal to 100, 500, 1000, and 1500 N, and reliable results were obtained. Based on these results, a detailed analysis of the structural strength in the case of slow walking is presented below with emphasis for the screw force values of 500 and 1000 N.

3.1. Results for the Stage of Tightening the Screw Simulation

In the beginning of the study, the initial step of screw pretension was analysed, and the optimal value for the screw force was chosen. Since there was no specific value for the pretension force, it was decided to investigate this point in more detail. For the final assembly, the calculations were carried out with several values of the pretension force from 100 N up to 1500 N, and the force values were considered incrementally.

Figure 7 demonstrates the dependence between the maximum stress in each of the bolts and the applied tension force. The typical stress distribution in the screws and implant itself caused by the screw preload force of 1000 N is shown in Figure 8.

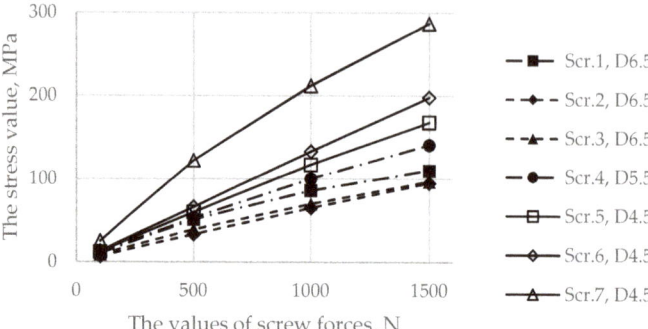

Figure 7. Dependence of the maximum von Mises stress in the screws as a function of the applied preload values.

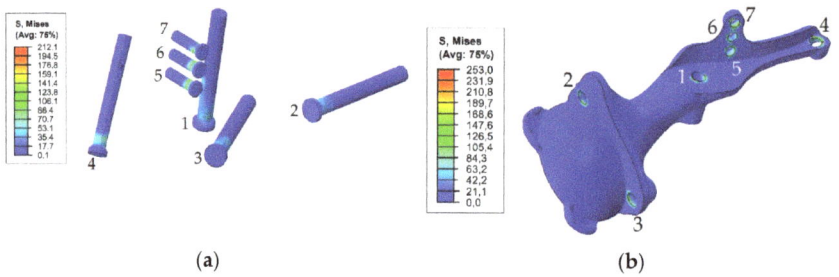

Figure 8. Equivalent von Mises stress (MPa) distribution caused by a pretension load of 1000 N: (**a**) in the screw system; (**b**) in the implant.

Particular attention should be given to the stresses in the pelvic bones because the destruction of the bone might be initiated due to the high value of the screw force. Figure 9 represents comparative graphs of the maximum stress values arising in the spongy (a) and cortical (b) layers in the first row of the finite elements of holes 1, 2, and 3 in the spongy tissue and holes 5, 6, and 7 in the cortical tissue.

Figure 10 shows the typical stress distribution around the holes in the resected pelvic bone after surgery.

The comparison was carried out separately for the spongy and cortical layers because screws with numbers 1, 2, 3, and 4 interacted only with the spongy layer. However, screws with numbers 5, 6, and 7 crossed the cortical layer. The screw with number 4 entered the spongy tissue but was located very close to the cortical layer.

As shown in Figures 8 and 9, the maximum von Mises stress values that occurred in screw 7 and holes 4 and 7 in the case of a pretension force of 1500 N were close to the ultimate stresses according to Table 2. A significant change in the stresses took place near the holes, as shown in Figure 10, while the remaining volume of the bone was almost free from the stresses.

The analysis proved that a pretension force of close to or higher than 1500 N may lead to local bone fracture, and further analysis of the walking simulation made no sense. The optimal screw force was expected to be between 500 N and 1000 N, since it did not cause any bone destruction but provided sufficient contact between the bone and implant. Therefore, the following strength analysis of the "bone–endoprosthesis" system was carried out for a walking condition assuming screw forces equal to 500 N and 1000 N.

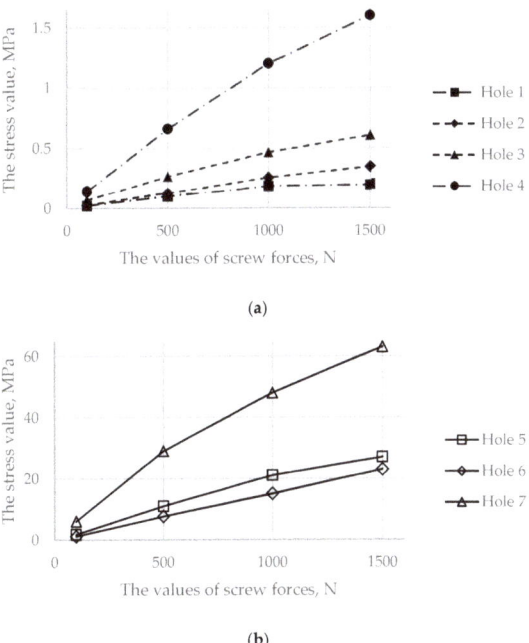

Figure 9. Comparative graphs of the maximum von Mises stresses occurring in the reconstructed pelvis bone on the hole boundaries: (**a**) the spongy tissue; (**b**) the cortical tissue.

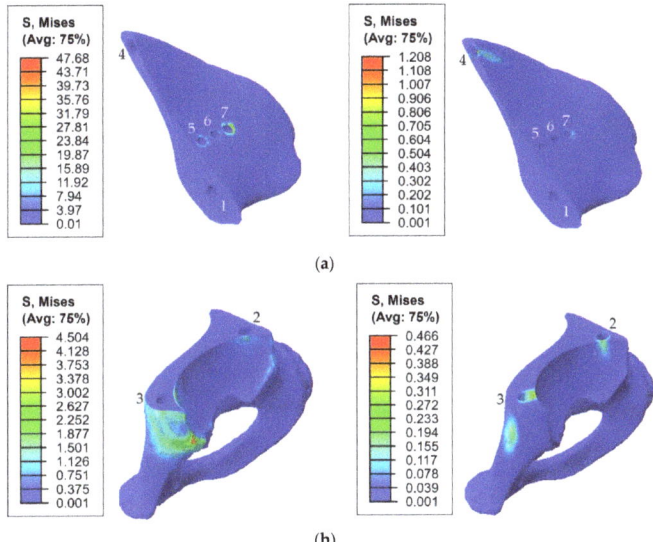

Figure 10. Equivalent von Mises stress (MPa) distribution in the bone tissue caused by a pretension load of 1000 N: (**a**) the upper part of the resected pelvic bone; (**b**) the lower part of the resected pelvic bone. The images on the left show the stresses in the cortical tissue, and the images on the right show the stresses in the spongy tissue.

3.2. Results of the Walking Cycle Simulation

The detailed walking FE simulation was carried out for the prepared model with screw pretension values equal to 500 N and 1000 N as follows from the previous section.

First, Figure 11 shows the von Mises stresses in the screws preloaded with forces of 500 and 1000 N, while the hip joint reaction force altered according to the graph in Figure 5 that simulates the gait cycle.

In addition, for the initial stage of screw pretension, the stress state in the titanium parts of the endoprosthesis in the walking condition was also considered. Particular attention was given to the load case corresponding to 17% of the gait cycle because this phase is the phase of the maximum reaction force applied to the centre of the left joint, where the implant was placed. However, obviously, extremely high stresses in the screws occurred at 45% of the gait cycle (Figure 11).

Similarly, the peak von Mises stresses for the considered walking phases were expected on the hole edges (Figure 12). The highest stresses in the implant took place at 45% of the gait cycle in the case of a 1000 N preload but at 17% and 65% for a 500 N preload.

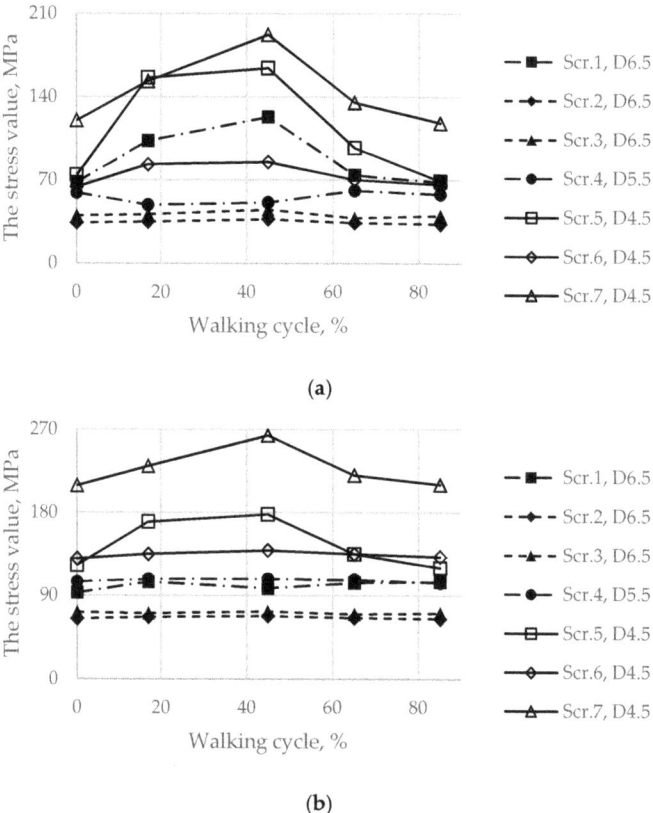

Figure 11. Maximum von Mises stress in each of the screws with pretension in the walking cycle phases: (**a**) pretension force equal to 500 N; (**b**) pretension force equal to 1000 N.

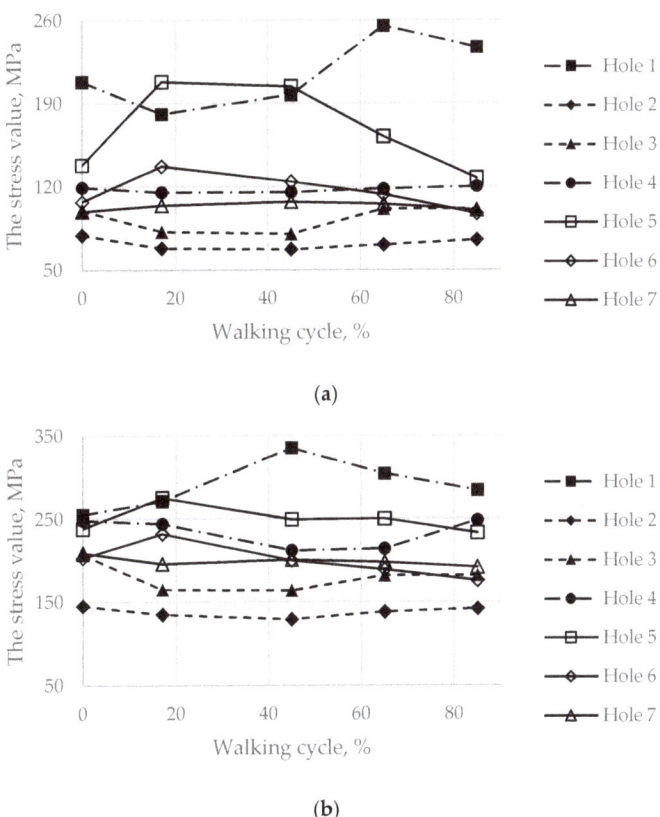

(a)

(b)

Figure 12. Maximum von Mises stress in the implant on the boundaries of the holes in the walking cycle phases: (**a**) pretension force equal to 500 N; (**b**) pretension force equal to 1000 N.

The general view of the typical stress distribution in the endoprosthesis is shown in Figure 13 for a preload of 1000 N and gait cycle phase equal to 45%.

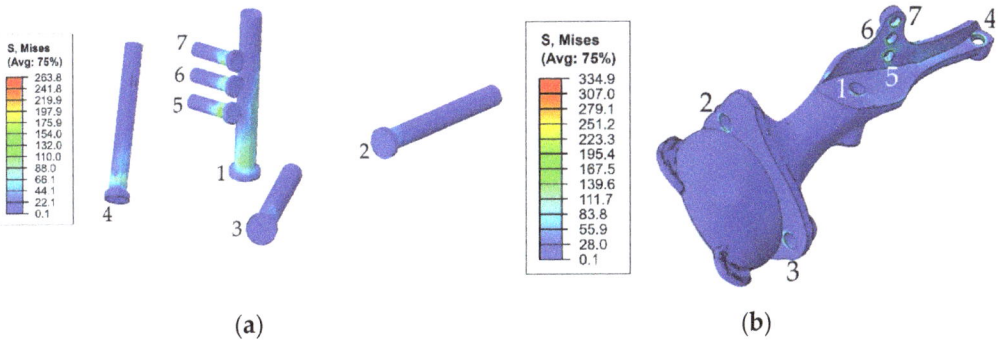

Figure 13. Equivalent von Mises stress (MPa) distribution in the endoprosthesis titanium parts caused by the pretension load of 1000 N and joint reaction forces at 45% of the gait cycle: (**a**) screw system; (**b**) implant body.

Finally, the maximum von Mises stresses evaluated on the boundaries of the holes for the screws at every load point in the walking cycle are shown in Figure 14.

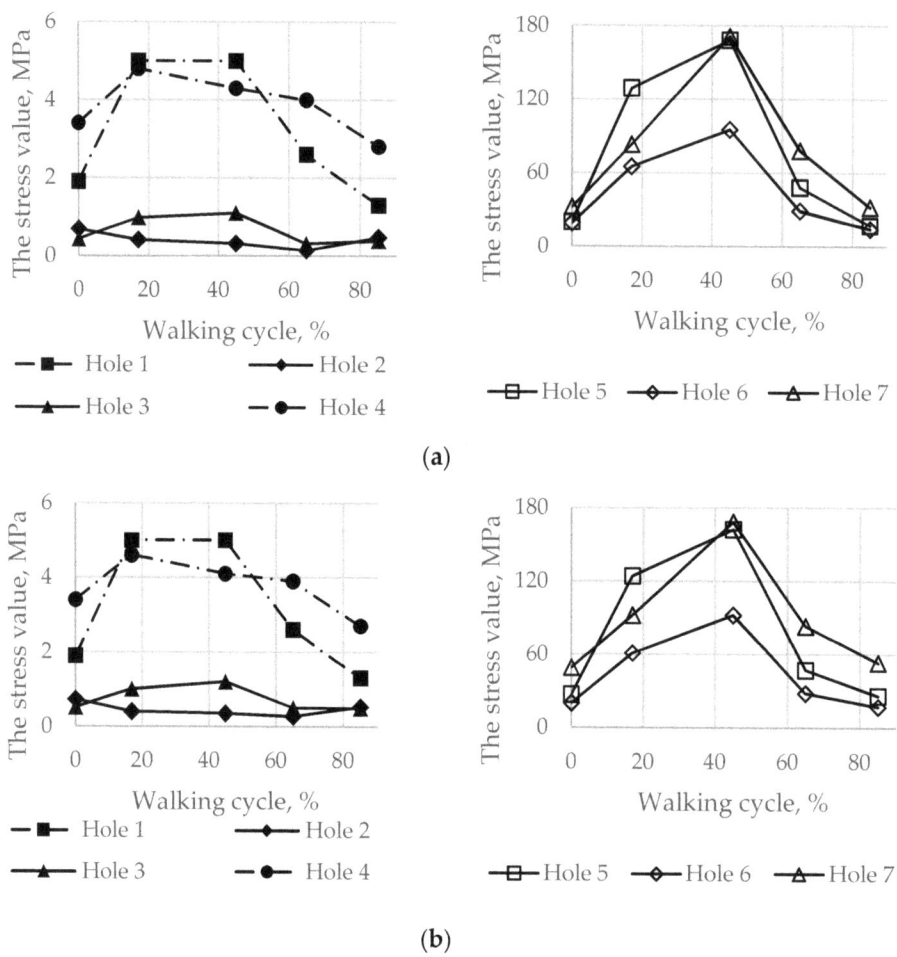

Figure 14. Maximum von Mises stress in the bone tissue on the boundaries of the holes in the walking cycle phases: (**a**) the pretension force equal to 500 N; (**b**) the pretension force equal to 1000 N.

The total stress distribution that occurred in the pelvic bones for the typical gait phase was 45%, and the screw force of 1000 N is presented in Figure 15.

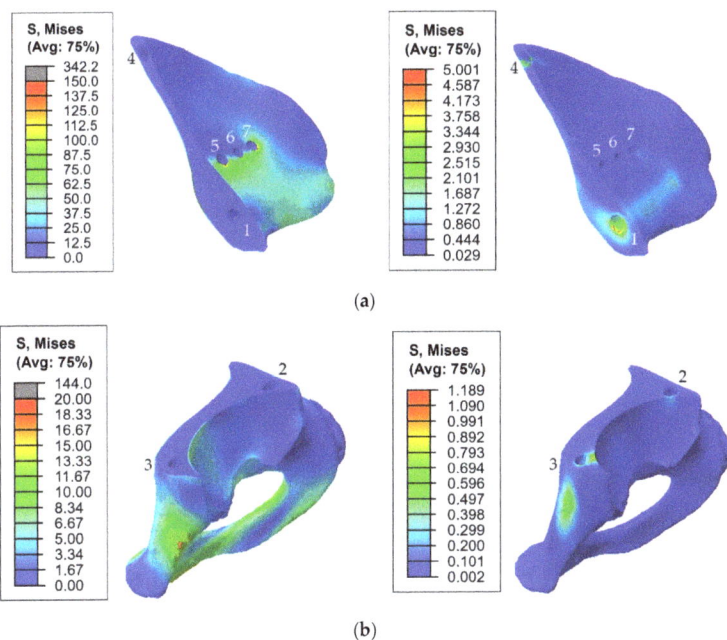

(a)

(b)

Figure 15. Equivalent von Mises stress (MPa) distribution in the bone tissue caused by the pretension load of 1000 N and joint reaction forces at 45% of the gait cycle: (**a**) the upper part of the resected pelvic bone; (**b**) the lower part of the resected pelvic bone. The images on the left show stresses in the cortical tissue, and the images on the right show stresses in the spongy tissue.

4. Discussion

For a more convenient assessment and comparison of the obtained results, a summary of the maximum von Mises stresses (MPa) occurring in the endoprosthesis parts and bone tissue are presented in Table 3 for the two considered values for the screw pretension loads: 500 and 1000 N.

Table 3. Summary of the maximum von Mises stresses occurring in the endoprosthesis parts and bone tissue for screw pretension loads equal to 500 N and 1000 N.

Assembly Components	Maximum Von Mises Stresses (MPa) and Their Location			
	Pretension Force of 500 N		Pretension Force of 1000 N	
	Pretension Stage	Walking Cycle (45% Phase)	Pretension Stage	Walking Cycle (45% Phase)
Screw system	122 MPa, screw 7	192 MPa, screw 7 197 MPa, hole 1	212 MPa, screw 7	263 MPa, screw 7
Implant	151 MPa, hole 1	255 MPa, *hole 1* (65% phase)	253 MPa, hole 4	335 MPa, hole 1
Pelvic cortical tissue, top part of resected bone	29 MPa, hole 7	171 MPa, hole 7 168 MPa, hole 5	48 MPa, hole 7	168 MPa, hole 7 162 MPa, hole 5
Pelvic spongy tissue, top part of resected bone	0.68 MPa, hole 4	4.98 MPa, hole 1 4.3 MPa, hole 4	1.25 MPa, hole 4	5.0 MPa, hole 1 4.1 MPa, hole 4
Pelvic cortical tissue, bottom part of resected bone	2.7 MPa	19 MPa	4.5 MPa	19 MPa
Pelvic spongy tissue, bottom part of resected bone	0.28 MPa, hole 3	1.1 MPa, hole 3	0.47 MPa, hole 3	1.2 MPa, hole 3

The first step in the assessment of the long-term strength and reliability of the biomechanical "bone–endoprosthesis" system is to evaluate the mechanical strength of the implant and its fixation system. According to Figure 7, the stresses occurring in the screws vary mostly linearly depending on the applied pretension force. This finding allows us to obtain the required force in a relatively simple way. The high stresses are distributed quite locally. The highest stress level takes place in the areas of contact between the screws and the screw holes in the implant (Figure 8) and the bone tissue (Figure 10).

However, it should be mentioned that the real stresses in the screws differ significantly from the tensile stress in the metal rod loaded with a similar tensile force, which is calculated as the ratio between the force and the cross-sectional area. For example, for screws 5, 6, and 7 with a diameter of 4.5 mm loaded with a longitudinal force of 1000 N, the sectional stresses are equal to 63 MPa, whereas the total equivalent stresses considering the contact interaction reach values from 117 to 212 MPa. This fact emphasizes the importance of considering the contact interaction of the bodies of the "bone–endoprosthesis" system both for the first stage of screw pretension and for the subsequent walking step.

According to Table 3, the screws and implant at pretension values of 500 N and 1000 N have a safety factor of more than 4.0 based on data in the literature [3] (Table 2). The Mises equivalent contact stresses on the edges of the screw holes in the implant are close to the total stresses in the corresponding screws. However, the stresses in the implant are slightly lower. The difference approximately equals the pretension stresses in the screws.

Analysis of the static strength of the screws allows us to assume that there should be no destruction of the titanium components of the system when the patient is walking. However, this assumption becomes debatable in further analysis of the structural strength under periodic loads occurring in the hip joint during normal human activity.

In the case of walking, the development of fatigue damage is highly possible on the edges of the implant holes for both considered options for screw tightening (Figures 12 and 13). The maximum equivalent stresses exceed the lower bound of the fatigue limit for 3D printed Ti-6Al-4V, which is 200 MPa according to the obtained material data (Table 2). At a pretension force of 500 N, the stresses exceed the limit only in the region of hole 1, whereas at a force of 1000 N, the limit is exceeded for all holes except hole 2. However, most likely, the risk of fracture due to fatigue effects may decrease with successful osseointegration and should not have any significant effect thereafter. Regardless, the use of high-quality titanium powders and the application of advanced manufacturing technologies for producing implants [22] should have a high priority in the planning stage for such surgeries.

In the fixation system, screws 1 and 4–7, which fix the implant to the upper part of the pelvic bone, are mostly affected by the periodic loads that occur during walking. In screws 2 and 3, the stresses do not exceed 80 MPa since these screws do not carry any significant external load. However, these screws cannot be excluded from the system because they connect the implant and the lower part of the pelvic bone, reconstructing the pelvic ring. In all screws, the maximum stresses did not reach the lower bound of the fatigue limit for Ti-6Al-4V according to the obtained material data (Table 2). The highest stress value equals 263 MPa and occurs in screw 7.

The analysis of the stresses in the bone tissue of the analysed biomechanical system requires special attention. As mentioned above, at the stage of screw tightening, the equivalent von Mises stresses on the edges of the screw holes in the cortical and spongy tissues (Figure 9) for tightening forces of 1000 N and above approach the lower bound of the strength limit of the corresponding tissue (Table 2) but do not reach the critical values. The cortical tissue of the upper pelvic bone has a safety margin against the strength limit at the preloading stage for all the considered values of the pretension forces. The spongy tissue also has a significant safety margin in the case of the pretension of 500 N. In the case of 1000 N, the stresses are close to the limit but do not exceed it. The remaining regions of the bone remain mostly unstressed, which is quite reasonable due to the chosen detailed setup of the problem.

In the cortical layer of the lower part of the pelvis, maximum stresses are situated in the pubic joint. These stresses are caused by the rigid connection of the left and right pelvic bones in this area. In other areas, at all stages of the simulation, stresses do not exceed 20 MPa, and there is a safety margin of at least 4.0. A similar situation occurs in the area of the rigid connection between the upper part of the pelvic bone and the sacrum. The increased stresses in these areas might be neglected because they are caused by artificial rigid constraints and occur due to the absence of the soft cartilage layer in the model. Furthermore, these areas are not subjected to surgical intervention, and their physiological condition remains unchanged compared with the healthy biomechanical system of the pelvis, where there are no high mechanical stresses that exceed the strength limits during normal human activities [30].

During the walking phase, areas with extremely high stresses, which exceed the allowed limits, appear both in the cortical and spongy tissues in the upper part of the resected pelvic bone and for both pretension forces considered (Figures 14 and 15). According to Table 3, the most dangerous regions where local destruction might be expected are the edges of screw holes 1 and 4 (spongy layer, equivalent stress of 4–5 MPa) and holes 5 and 7 (cortical layer, equivalent stress of 160–170 MPa). The limits are significantly exceeded in the walking cycle phases of 17% and 45%. The stress near hole 6 equals 95 MPa and exceeds the lower limit of the allowable stress range (80 MPa). This fact confirms the high risk of bone damage.

The stress values obtained in the "bone–endoprosthesis" system during the walking simulation are slightly higher than the stress values reported previously [28], where a similar approach was used. The reported peak stresses in the implant were approximately 105 MPa during the walking phase, and the stresses in the screws and pelvic bones were approximately 50 MPa under the same loading condition. Therefore, the authors [28] expected that the fatigue limit could be reached only in the case of more severe loading scenarios, such as stair climbing. The difference in the results can be explained by the difference in the implant design and the differences in the approaches of finite element modelling of the behaviour of the bone tissue.

Thus, according to the performed analysis, slight destruction of the bone might be expected in local areas near the screw holes in cases of walking. This destruction can affect the stability of the implant fixation to the upper bone. However, bone tissue is capable of regeneration when it is loaded with an external mechanical field, in particular, with periodic loading of a sufficient level. Therefore, in the case of a moderate dynamic compressive load, the process of regeneration could be initiated in the area that might be initially damaged [31].

The current study particularly focuses on the screw tightening process and on the analysis of the values of the forces applied to the screws. This issue is usually neglected in studies, and the number of related publications is very limited [2–4]. Nevertheless, this point is quite important for proper patient surgery planning. The degree of fixation between the implant and the bones directly depends on the value of the screw pretension force. Additionally, fixation affects the reliability of the reconstructed pelvis in terms of cyclic loading and fatigue effects. If the screws are tightened loosely, then the process of bone regeneration might take much more time. If the screws are tightened excessively, local destruction will occur in the bones near the holes. This fact can lead to more serious consequences over time. Considering previous results [2], it should be pointed out that the screw pretension force equals 3000 N. However, according to the current study, increasing the screw tightening force from 1000 to 1500 N may lead to local bone destruction. Comparison of the obtained results with the charts provided [2] shows the compliance of stresses for tightening forces up to 500 N. The results from previous research [2] also confirm the linear growth of stresses in the screws.

The shape of the endoprosthesis is customized, which causes additional problems for the stress-strain state analysis of the structure. It should be mentioned that the methods of designing individual implants differ in complexity from the methods of developing serial endoprostheses [1]. Additionally, the design of the individual endoprosthesis can change during the modelling process. However, the design of serial implants is usually as efficient as possible, as opposed to individual implants, which are designed and produced only once for a specific patient. Using the finite element method, it is possible to predict the areas of stress concentrations and to choose the optimal number and parameters of the screws. In the current study, there were several screws that did not carry much load. This finding indicates that they may be removed from the structure. The authors presume that if the stress concentration does not occur in the screws, such action should not affect the general performance and reliability of the structure. A previous study [32] confirms this statement, saying that such screws accumulate excess stresses and should be removed.

When the current stress-strain state results are compared to results from similar studies [1,2,10], it should be considered that the forces applied to the structure are not universal and vary depending on the patient. For the stress-strain state comparison [2], the stress concentration arises around holes, and the stress values match the results of current research. The same trend takes place in other studies [11], especially for cup-shaped structures.

5. Conclusions

The stress distribution for a pelvis reconstructed by an individual endoprosthesis with four different screw forces was analysed. The obtained optimal screw pretension force for tightening the implant to the bone was from 500 N to 1000 N for this specific model.

Screw tightening with a force less than 500 N seemed to be insufficient for firm fixation of the implant. At the same time, the results show that a tightening force exceeding 1000 N may result in a local bone fracture. Therefore, the optimal and critical screw forces are determined, and the stress states are calculated for the walking condition. The peak stresses occur near the holes in the bones, implant, and screws. Screw tightening with a force of 500–1000 N should be optimal because the stress state of the bones did not exceed the limits globally. This value for screw force provides reliable fixation of the implant to the bones.

When conducting the subsequent surgery, it is strongly recommended to monitor the value of the actual screw pretension force. In this case, the endoprosthesis will be reliable and durable. To prevent the undesirable development of degenerative effects during the patient's recovery process after osteosynthesis surgery, the rehabilitation plan should be adjusted to reduce the loads on the reconstructed bone by providing additional support when the person is walking.

As a result of the arthroplasty described in current research, the patient has fully recovered with no limitations in motion or activities [8]. This fact confirms the relevance of the performed studies and the significance of further development of computer modelling methods and approaches for solving the problems of personalized orthopaedics. The technology of implant development using computer modelling, finite element analysis, and 3D printing makes it possible to create anatomical prostheses with sufficient safety margins, anatomical designs, and reliable fixation methods.

Author Contributions: Conceptualization, L.M.; methodology, L.M. and M.Z.; validation, I.M. and D.S.; formal analysis, I.M. and F.T.; investigation, I.M.; data curation, F.T.; writing—original draft preparation, L.M. and I.M.; writing—review and editing, L.M. and M.Z.; visualization, D.S.; supervision, A.B. and L.M.; project administration, A.B. All authors have read and agreed to the published version of the manuscript.

Funding: This research was partially funded by the Ministry of Science and Higher Education of the Russian Federation as a part of the World-class Research Center program: Advanced Digital Technologies (contract No. 075-15-2020-934 dated 17 November 2020).

Institutional Review Board Statement: Not applicable.

Informed Consent Statement: Not applicable.

Data Availability Statement: The data presented in this study are available on request from the corresponding author.

Acknowledgments: We gratefully thank Evgeniy Sushentsov (N. N. Blokhin NMRCO) for the provided anonymous CAD models and other biomedical data concerning the implant design and useful discussions regarding the paper.

Conflicts of Interest: The authors declare no conflict of interest.

References

1. Iqbal, T.; Shi, L.; Wang, L.; Liu, Y.; Li, D.; Qin, M.; Jin, Z. Development of Finite Element Model for Customized Prostheses Design for Patient with Pelvic Bone Tumor. *Proc. Inst. Mech. Eng. Part H J. Eng. Med.* **2017**, *231*, 525–533. [CrossRef] [PubMed]
2. Dong, E.; Wang, L.; Iqbal, T.; Li, D.; Liu, Y.; He, J.; Zhao, B.; Li, Y. Finite Element Analysis of the Pelvis after Customized Prosthesis Reconstruction. *J. Bionic Eng.* **2018**, *15*, 443–451. [CrossRef]
3. Jia, Y.; Cheng, L.; Yu, G.; Ding, Z. Finite Element Analysis of Pelvic Reconstruction Using Fibular Transplantation Fixed with Rod-Screw System after Type I Resection. In Proceedings of the 2007 1st International Conference on Bioinformatics and Biomedical Engineering, Wuhan, China, 6–8 July 2007; pp. 430–433. [CrossRef]
4. Shim, V.; Höch, A.; Grunert, R.; Peldschus, S.; Böhme, J. Development of a Patient-Specific Finite Element Model for Predicting Implant Failure in Pelvic Ring Fracture Fixation. *Comput. Math. Methods Med.* **2017**, *2017*, 9403821. [CrossRef]
5. Hu, P.; Wu, T.; Wang, H.Z.; Qi, X.Z.; Yao, J.; Cheng, X.D.; Chen, W.; Zhang, Y.Z. Influence of Different Boundary Conditions in Finite Element Analysis on Pelvic Biomechanical Load Transmission. *Orthop. Surg.* **2017**, *9*, 115–122. [CrossRef]
6. Kluess, D.; Wieding, J.; Souffrant, R.; Mittelmeier, W.; Bader, R. Finite Element Analysis in Orthopedic Biomechanics. In *Finite Element Analysis*; Moratal, D., Ed.; Sciyo: Rijeka, Croatia, 2010; pp. 151–170.
7. Hao, Z.; Wan, C.; Gao, X.; Ji, T. The Effect of Boundary Condition on the Biomechanics of a Human Pelvic Joint under an Axial Compressive Load: A Three-Dimensional Finite Element Model. *J. Biomech. Eng.* **2011**, *133*, 101006. [CrossRef]
8. Sushentsov, E.A.; Musaev, E.R.; Maslov, L.B.; Zhmaylo, M.A.; Sofronov, D.I.; Agaev, D.K.; Dzampaev, A.Z.; Romantsova, O.M.; Fedorova, A.V.; Aliev, M.D. Computer simulation, 3d-printing and custom-made prosthetics in treatment of a patient with osteosarcoma of the pelvis. *Bone Soft Tissue Sarcomas Tumors Skin* **2019**, *11*, 53–61.
9. Dalstra, M.; Huiskes, R.; van Erning, L. Development and Validation of a Three-dimensional Finite Element Model of the Pelvic Bone. *J. Biomech. Eng.* **1995**, *117*, 272–278. [CrossRef]
10. Liu, D.; Hua, Z.; Yan, X.; Jin, Z. Design and Biomechanical Study of a Novel Adjustable Hemipelvic Prosthesis. *Med. Eng. Phys.* **2016**, *38*, 1416–1425. [CrossRef] [PubMed]
11. Phillips, A.T.M.; Pankaj, P.; Howie, C.R.; Usmani, A.S.; Simpson, A.H.R.W. Finite Element Modelling of the Pelvis: Inclusion of Muscular and Ligamentous Boundary Conditions. *Med. Eng. Phys.* **2007**, *29*, 739–748. [CrossRef] [PubMed]
12. Viceconti, M.; Olsen, S.; Nolte, L.P.; Burton, K. Extracting Clinically Relevant Data from Finite Element Simulations. *Clin. Biomech.* **2005**, *20*, 451–454. [CrossRef]
13. Zhou, Y.; Min, L.; Liu, Y.; Shi, R.; Zhang, W.; Zhang, H.; Duan, H.; Tu, C. Finite Element Analysis of the Pelvis after Modular Hemipelvic Endoprothesis Reconstruction. *Int. Orthop.* **2013**, *37*, 653–658. [CrossRef]
14. Dalstra, M.; Huiskes, R.; Odgaard, A.; van Erning, L. Mechanical and Textural Properties of Pelvic Trabecular Bone. *J. Biomech.* **1993**, *26*, 523–535. [CrossRef]
15. Kukin, I.A.; Kirpichev, I.V.; Maslov, L.B.; Vikhrev, S.V. Characteristics of the trabecular bone strength properties of people with hip diseases. *Fundam. Res.* **2013**, *7*, 328–333.
16. Morgan, E.F.; Unnikrisnan, G.U.; Hussein, A.I. Bone Mechanical Properties in Healthy and Diseased States. *Annu. Rev. Biomed. Eng.* **2018**, *20*, 119–143. [CrossRef]
17. Tikhilov, R.M.; Shubnyakov, I.I.; Mazurenko, A.V.; Mitryaykin, V.I.; Sachenkov, O.A.; Kuzin, A.K.; Denisov, A.O.; Pliev, D.G.; Boyarov, A.A.; Kovalenko, A.N. Experimental substantiation of acetabular component impaction with undercoverage in arthoplasty of patients with severe hip dysplasia. *Travmatol. Ortop. Ross.* **2013**, *4*, 42–51. [CrossRef]
18. Reilly, D.T.; Burstein, A.H. The elastic and ultimate properties of compact bone tissue. *J. Biomech.* **1975**, *8*, 393–405. [CrossRef]
19. Wirtz, D.C.; Schiffers, N.; Forst, R.; Pandorf, T.; Weichert, D.; Radermacher, K. Critical evaluation of known bone material properties to realize anisotropic FE-simulation of the proximal femur. *J. Biomech.* **2000**, *33*, 1325–1330. [CrossRef]
20. Keaveny, T.M.; Wachtel, E.F.; Kopperdahl, D.L. Mechanical behavior of human trabecular bone after overloading. *J. Orthop. Res.* **1999**, *17*, 346–353. [CrossRef]
21. Long, M.; Rack, H.J. Titanium alloys in total joint replacement—A materials science perspective. *Biomaterials* **1998**, *19*, 1621–1639. [CrossRef]
22. Borovkov, A.; Maslov, L.; Tarasenko, F.; Zhmaylo, M.; Maslova, I.; Solovev, D. Development of elastic–plastic model of additively produced titanium for personalised endoprosthetics. *Int. J. Adv. Manuf. Technol.* **2021**, *117*, 2117–2132. [CrossRef]
23. Greitemeier, D.; Palm, F.; Syassen, F.; Melz, T. Fatigue performance of additive manufactured TiAl6V4 using electron and laser beam melting. *Int. J. Fatigue* **2017**, *94*, 211–217. [CrossRef]

24. MatWeb Material Property Data. Overview of Materials for High Density Polyethylene (HDPE), Injection Molded. Available online: http://www.matweb.com/search/datasheet.aspx?MatGUID=fce23f90005d4fbe8e12a1bce53ebdc8 (accessed on 12 September 2021).
25. Bergmann, G.; Deuretzbacher, G. Hip contact forces and gait patterns from routine activities. *J. Biomech.* **2001**, *34*, 859–871. [CrossRef]
26. Borovkov, A.; Maslov, L.; Zhmaylo, M.; Zelinskiy, I.; Voinov, I.; Keresten, I.; Mamchits, D.; Tikhilov, R.; Kovalenko, A.; Bilyk, S.; et al. Finite element stress analysis of a total hip replacement in a two-legged standing. *Russ. J. Biomech.* **2018**, *4*, 382–400. [CrossRef]
27. Maslov, L.; Surkova, P.; Maslova, I.; Solovev, D.; Zhmaylo, M.; Kovalenko, A.; Bilyk, S. Finite-Element Study of the Customized Implant for Revision Hip Replacement. *Vibroengineering Procedia* **2019**, *26*, 40–45. [CrossRef]
28. Dong, E.; Iqbal, T.; Fu, J.; Li, D.; Liu, B.; Guo, Z.; Cuadrado, A. Preclinical Strength Checking for Artificial Pelvic Prosthesis under Multi-activities—A Case Study. *J. Bionic Eng.* **2019**, *16*, 1092–1102. [CrossRef]
29. Wong, K.C.; Kumta, S.M.; Gee, N.V.L.; Demol, J. One-Step Reconstruction with a 3D-Printed, Biomechanically Evaluated Custom Implant after Complex Pelvic Tumor Resection. *Comput. Aided Surg.* **2015**, *20*, 14–23. [CrossRef] [PubMed]
30. Akulich, Y.V.; Podgayets, R.M.; Scryabin, V.L.; Sotin, A.V. The Investigation of Stresses and Strains in the Hip Joint after Operation of Endoprosthetics. *Russ. J. Biomech.* **2007**, *11*, 9–35.
31. Maslov, L.B. Biomechanical model and numerical analysis of tissue regeneration within a porous scaffold. *Mech. Solids* **2020**, *55*, 1115–1134. [CrossRef]
32. Hao, Z.; Wan, C.; Gao, X.; Ji, T.; Wang, H. The Effect of Screw Fixation Type on a Modular Hemi-Pelvic Prosthesis: A 3-D Finite Element Model. *Disabil. Rehabil. Assist. Technol.* **2013**, *8*, 125–128. [CrossRef]

Article

Structural Design Method for Constructions: Simulation, Manufacturing and Experiment

Pavel Bolshakov [1], Nikita Kharin [2,3], Ramil Kashapov [3] and Oskar Sachenkov [1,2,*]

[1] Department Machines Science and Engineering Graphics, Tupolev Kazan National Research Technical University, 420111 Kazan, Russia; bolshakov-pavel@inbox.ru
[2] Institute of Mathematics and Mechanics, Kazan Federal University, 420008 Kazan, Russia; nik1314@mail.ru
[3] Institute of Engineering, Kazan Federal University, 420008 Kazan, Russia; kashramil.88@mail.ru
* Correspondence: 4works@bk.ru

Abstract: The development of additive manufacturing technology leads to new concepts for design implants and prostheses. The necessity of such approaches is fueled by patient-oriented medicine. Such a concept involves a new way of understanding material and includes complex structural geometry, lattice constructions, and metamaterials. This leads to new design concepts. In the article, the structural design method is presented. The general approach is based on the separation of the micro- and macro-mechanical parameters. For this purpose, the investigated region as a complex of the basic cells was considered. Each basic cell can be described by a parameters vector. An initializing vector was introduced to control the changes in the parameters vector. Changing the parameters vector according to the stress-strain state and the initializing vector leads to changes in the basic cells and consequently to changes in the microarchitecture. A medium with a spheroidal pore was considered as a basic cell. Porosity and ellipticity were used for the parameters vector. The initializing vector was initialized and depended on maximum von Mises stress. A sample was designed according to the proposed method. Then, solid and structurally designed samples were produced by additive manufacturing technology. The samples were scanned by computer tomography and then tested by structural loads. The results and analyses were presented.

Keywords: structural design; porous constructions; additive manufacturing; CT

Citation: Bolshakov, P.; Kharin, N.; Kashapov, R.; Sachenkov, O. Structural Design Method for Constructions: Simulation, Manufacturing and Experiment. *Materials* 2021, *14*, 6064. https://doi.org/10.3390/ma14206064

Academic Editor: Aleksander Muc

Received: 25 August 2021
Accepted: 12 October 2021
Published: 14 October 2021

Publisher's Note: MDPI stays neutral with regard to jurisdictional claims in published maps and institutional affiliations.

Copyright: © 2021 by the authors. Licensee MDPI, Basel, Switzerland. This article is an open access article distributed under the terms and conditions of the Creative Commons Attribution (CC BY) license (https://creativecommons.org/licenses/by/4.0/).

1. Introduction

The modern approach for design implants and prostheses implies patient-oriented solutions. Such an approach involves not only new manufacturing methods but also a new vision of the product. Additive manufacturing allows production constructions with complex geometry. However, the solution for the automation of the design of such products is still open. So, nowadays lattice constructions have become popular for this purpose. Yet, the dependence between the different geometries of the lattice, the mechanical properties, and the biological adaptive is being researched [1–3]. Additionally, a manifestation of the brittle properties and the geometry deviations after manufacturing is still an issue of the day [4–7]. By changing the materials and melting modes [8–10], the mechanical parameters can be improved or vice versa. Despite the aforementioned difficulties, it is obvious that additive manufacturing and patient-oriented design can notably increase the quality of the medical treatments.

This article is focused on an approach for the structural design method. Previously, a method for designing a lattice endoprosthesis for long bones was developed [11]. The endoprosthesis was manufactured and passed clinical experiments. The developed approach was generalized. The main idea is based on the bone adaptation analogy. It is known that adaptation can be formulated by Wolff's law [12]. To describe bone tissue orthotropy, a fabric tensor is used. The fabric tensor is also used to calculate the stiffness tensor [12,13]. The foundation of the adaptation model is an alignment of the stress and stiffness tensors.

In terms of the fabric tensor, it means that the orthotropic directions are equal to the stress principal directions [13,14]. The widespread approach is to use representative volumes to determine the fabric tensor and the effective mechanical properties [15,16].

It has been shown [17–19] that implants interact with bone tissue and that the structure and the microstructure of the implant influence the quality of this interaction. Additive manufacturing allows the generation of solid irregular or lattice geometry [20,21], but on the other hand, local microporosity decreases fatigue resistance. Classical post-processing, such as tempering, allows the counteraction of the negative sides of the technology [22,23]. Despite the aforementioned technological barriers, the opening opportunities are promising. The ability to design material within a product opens up new possibilities in patient-specific prostheses [24,25]. The complexity of such an approach appears in defining the external loads and the formulation criteria of the design [26,27]. A novel approach is the use of additive manufacturing technology for liquid crystal elastomers, the exceptional properties of which show good usage in a range of applications in the fields of biology and medicine [28–30].

In this article, a method of structural design is presented. An example of structurally designed construction is presented. The designed and regular constructions were manufactured and compared in natural experiments.

2. Materials and Methods

2.1. Problem Formulation

The mechanical behavior of the region V in R^3 with the boundary ∂V, within the linear theory of elasticity, can be described by the following system of equations [11]:

$$\nabla \cdot \tilde{\sigma} = 0, \ \forall \vec{x} \in V^0 \tag{1}$$

$$\tilde{\varepsilon} = \frac{1}{2}\left(\nabla \vec{u} + \left(\nabla \vec{u}\right)^T\right), \ \forall \vec{x} \in V^0 \tag{2}$$

$$\tilde{\sigma} = C : \tilde{\varepsilon}, \ \forall \vec{x} \in V^0 \tag{3}$$

$$\vec{u} = 0, \ \forall \vec{x} \in S_{kin} \tag{4}$$

$$\tilde{\sigma} \cdot \vec{n} = \vec{p}, \ \forall \vec{x} \in S_{sta} \tag{5}$$

$$S_{sta} \cup S_{kin} = \partial V \tag{6}$$

where $V° = V \cup \partial V$; u is the displacement vector; σ is the stress tensor; ε is the elastic strain tensor; and C is the stiffness tensor. S_{sta} is the surface on which static boundary conditions are specified, and S_{kin} is the surface on which kinematic boundary conditions are specified (see Figure 1).

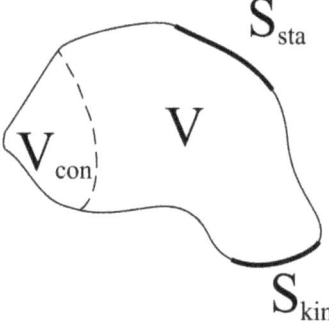

Figure 1. Scheme for problem formulation.

It is necessary to find a distribution of the stiffness tensor C in the volume V such that the stress invariant (in our case, the von Mises stress) reaches a minimum at the constant boundary conditions.

$$\tilde{C} = \tilde{C}(\vec{x}), \max_{\vec{x} \in V'} \|\tilde{\sigma}\| \to \min \tag{7}$$

Applying the design conditions, it is necessary to determine the region V_{con}, in which the components of the tensor of elastic properties remain unchanged:

$$V_{con} \in V^0 \tag{8}$$

Let us call the region V_{con} the constant region. So, V' in (7) can be determined as $V^0 \setminus V_{con}$. Adding Equation (7) to Equations (1)–(6) allows the formulating of the optimization problem for the structure.

2.2. Structural Problem Formulation

The general idea of the method is that the stress-strain state depends on some of the parameters vectors. Assuming that the anisotropy of the material is provided by the microarchitecture, we consider the forming material isotropic [13,16]. The parameters vector λ describes the material microarchitecture and influences the macro-stiffness tensor. On the other side, we should add an additional vector with initializing parameters, which describe the stress-strain state of the microarchitecture. Let us call it the initializing vector γ, which, obviously, depends on the invariants of the stress tensor f. The initializing vector can be interpreted as the control function of the microarchitecture changes. So, we propose that the stiffness tensor can be presented as a function of the parameters vector, the initializing vector, and the spatial coordinate:

$$\tilde{C} = \tilde{C}\left(\vec{\lambda}\left(\gamma, \vec{x}\right), \vec{\gamma}\left(f(\tilde{\sigma}), \vec{x}\right), \vec{x}\right) \tag{9}$$

Let us consider region V as the number of basic cells. For each basic cell we assume:

$$\begin{cases} \vec{\lambda}\left(\gamma, \vec{x}\right) = \vec{\lambda}(\gamma) \\ \vec{\gamma}\left(f(\tilde{\sigma}), \vec{x}\right) = \vec{\gamma}(f(\tilde{\sigma})) \end{cases} \tag{10}$$

This approach considers a basic cell as a micro-construction with constant macro-properties. The parameters vector λ should be changed according to values of the initializing vector γ. So, if we introduce the control function U the problem can be rewritten:

$$\begin{cases} \vec{\lambda}\left(\gamma, \vec{x}\right) = \vec{\lambda}\left(\gamma, U\left(\vec{x}\right)\right) \\ U\left(\vec{x}\right) = f\left(\vec{\gamma}\left(f(\tilde{\sigma}), \vec{x}\right)\right) \\ \tilde{C} = \tilde{C}\left(\vec{\lambda}\left(\gamma, U\left(\vec{x}\right)\right), \vec{x}\right) \end{cases} \tag{11}$$

This means that the state of the initializing vector γ determines the changes of microarchitecture in terms of the parameters vector λ, and the microarchitecture influences the macro-stiffness tensor. Let us consider the investigated region as a composition of basic cells; each one describes the microarchitecture of a material. Each basic cell can be described by the parameters vectors and can be changed according to the initializing vector. To implement such an approach, the basic cell should be determined in order to define the parameters vector and its relationship with the stiffness tensor.

2.3. Basic Cell

In the research unit, a cube with a spheroidal pore was used as a basic cell. In this case, the parameters vector consists of porosity (λ) and the ellipticity coefficient (β). To investigate the dependence between the stiffness tensor and the parameters vector, a representative elements method was used [31–33]. For this purpose, the parameterized finite element model of a cube with a spheroidal pore was implemented. Twenty-node hexahedral finite elements were used. Kinematic loading was used in the numerical simulation. Uniaxial and shear loads in three directions were implemented. To clarify the mechanical properties, additionally combined (uniaxial with shear) loads were implemented [16,26,34,35]. The parameters were investigated in the interval (0; 1). According to the received data, the functions describing the influence of the parameters on the mechanical properties were found:

$$\begin{cases} \vec{\lambda} = \vec{\lambda}(\lambda, \beta) \\ E_{ii} = E_{ii}\left(\vec{\lambda}\right) \equiv C_{iiii}\left(\vec{\lambda}\right) \\ G_{ij} = G_{ij}\left(\vec{\lambda}\right) \equiv C_{ijij}\left(\vec{\lambda}\right) \\ \nu_{ij} = \nu_{ij}\left(\vec{\lambda}\right) \equiv C_{iijj}\left(\vec{\lambda}\right) \end{cases} \quad (12)$$

For approximation, a fourth-degree polynomial function was used with an approximation error threshold of about 0.9 [33–36]. In the calculations, some of the coefficients were equal to zero, so a common form of the final calculated polynomial was as follows:

$$C_{ijkl}(\lambda, \beta) = c_{00} + c_{10}\lambda + c_{01}\beta + c_{11}\lambda\beta + c_{21}\lambda^2\beta + c_{31}\lambda^3\beta + c_{12}\lambda\beta^2 + c_{22}\lambda^2\beta^2 + c_{13}\lambda\beta^3 \quad (13)$$

where λ and β are components of the parameters vector—porosity and ellipticity, respectively, c_{ij} are coefficients of the polynomial, where i shows the power of porosity and j shows the power of ellipticity. The received values of the coefficients for the approximation polynomial are listed in Table 1.

Table 1. The values of coefficients of approximation polynomial for stiffness parameters.

	c_{00}	c_{10}	c_{01}	c_{11}	c_{21}	c_{31}	c_{12}	c_{22}	c_{13}
E_{11}, GPa	109	−3.9	−5.3	−192	287	−115	319	−209	−136
$E_{22,33}$, GPa	102	2.9	10.6	−111	325	−278	−17.8	−18.7	27
$G_{12,13}$, GPa	10.7	−0.1	0.25	−2.7	13	−10	−3.9	−0.1	4.1
G_{23}, GPa	2.5	−0.1	−0.06	−4.4	8	−3.4	6.4	−5	−2.5
$\nu_{12,13}$	0.011	−0.005	−0.009	−0.032	−0.038	0	−0.027	0.464	0
ν_{23}	0.017	−0.049	−0.017	−0.07	0.4	0	0.09	−0.18	0

It should be noted that the polynomial coefficients for Poisson's ratio can be reduced up to c_{00} because the influence of the parameters vector is insignificant. So, $\nu_{12,13} \approx 0.011$ and $\nu_{23} \approx 0.017$.

2.4. Proposed Algorithm

After the principal stress and directions are found, the orthotropic directions can be oriented according to the principal directions. The semi-major axis is directed to the 1st principal stress direction. The porosity is determined by von Mises stress and value $[\sigma]_{\text{inf}}$. The $[\sigma]_{\text{inf}}$ is the infimum of the stress value and determines the value of the underload. So, porosity can be restored by the equation:

$$\lambda\left(\vec{x}\right) = \begin{cases} 1 - \dfrac{[\sigma]_{\text{inf}} - \sigma_{\text{V.M.}}\left(\vec{x}\right)}{[\sigma]_{\text{inf}}}, \sigma_{\text{V.M.}}\left(\vec{x}\right) < [\sigma]_{\text{inf}} \\ 1, \sigma_{\text{V.M.}}\left(\vec{x}\right) \geq [\sigma]_{\text{inf}} \end{cases} \quad (14)$$

To determine the ellipticity coefficient, the 1st and the 3rd principal stresses were used:

$$\beta\left(\vec{x}\right) = \frac{\min\left(\left|\sigma_1\left(\vec{x}\right)\right|, \left|\sigma_3\left(\vec{x}\right)\right|\right)}{\max\left(\left|\sigma_1\left(\vec{x}\right)\right|, \left|\sigma_3\left(\vec{x}\right)\right|\right)} \tag{15}$$

Then, the stiffness constants can be calculated by porosity and the ellipticity coefficient and the stress-state problem can be solved. So, the algorithm can be described:

Algorithm of structural design
1. Load a *mesh* and apply *boundary conditions*.
2. Highlight *elements* from the *constant region* (8)
3. Set initial parameters vector
4. Solve the stress-state problem (1)–(6)
5. **for** each *element* not from *constant region*
6. Calculate the principal stress and directions
7. Calculate the parameters vector (14), (15)
8. Calculate stiffness tensor (13)
9. Orient element coordinate system according to the principal directions.
10. **end for**
11. **if** not *stop* **goto** 4
12. Restore geometry by parameters vector.

2.5. Model Task

A rectangular beam of 140 mm × 28 mm × 14 mm was used for the algorithm implementation. Eight-node hexahedral finite elements were used for the calculations. The kinematic loading of 1 mm was used in the numerical simulation. The length of the kinematic loading region was 20 mm. In Figure 2, the loading scheme is presented; the V_{con} region is marked by a green color. The end faces of the beam were fixed.

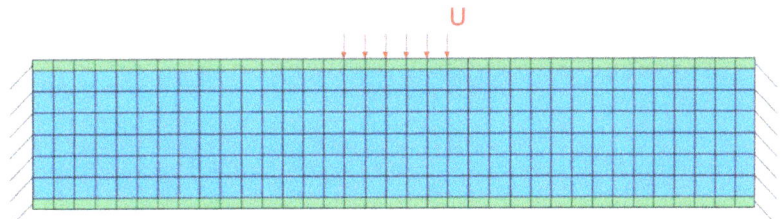

Figure 2. Loading scheme; U is applied displacements; the green region is V_{con} region.

The mechanical properties of acrylonitrile butadiene styrene were used for further production by additive manufacturing. So, Young's modulus was equal to 200 GPa, the shear modulus was equal to 71.5 GPa, and the Poisson ratio was 0.4. For $[\sigma]_{inf}$, 10% of maximum von Mises stress in the construction was used. The stop condition was as follows:

$$\max(|\lambda_i - \lambda_{i-1}|, |\beta_i - \beta_{i-1}|) < \varepsilon \tag{16}$$

where ε was equal to 10^{-3}.

2.6. Experiments

After restoring the geometry, the beam was produced by additive manufacturing technology. Acrylonitrile butadiene styrene was used for the manufacturing. Both the solid and the structural design samples were produced. For every two types of samples, longitudinal and transverse directions of printing were used. Computed tomography (CT) (Vatech PaX-I 3D, Kazan, Russia) was used to estimate the structure. After that, three-point bending was carried out and stress-strain curves were obtained for all samples. For the stress-strain curves, the ultimate force and slope were analyzed.

3. Results and Discussion

The stress-strain state for the initial (solid) and the structurally designed beam were compared. The maximum stress did not change significantly, but the distribution of stress inside the product decreased (see Figure 3). The algorithm showed fast convergence; it was about 38 iterations. The maximum stress was localized in the zones of kinematic constraints for the initial geometry. In addition, for the structurally designed beam the maximum stress was localized in the zones of kinematic boundary conditions (see red regions in Figure 3). On the other hand, zones of stress reduction appeared for the structurally designed beam (see blue regions in Figure 3).

The distribution of the received porosity and ellipticity coefficients is shown in Figure 4. The zones of high porosity are localized where the von Misses stress was minimal (red zones in Figure 4a). In the same zones, the pore's ellipticity coefficient is close to 1 (red zones in Figure 4a), which means that in this region the pore is almost spherical.

The 3D geometry was restored for the following manufacturing (see Figure 5a). The initial and structurally designed samples were manufactured in two ways: longitudinal and transversal printing. After the manufacturing, the samples were scanned by CT (see Figure 5b,c). The deviations of pore geometry in the manufactured samples were noted. They were caused by the cooldown speed of the printing material drop. However, the distribution of the porosity was close enough to the design (deviations about 5%).

In the three-point bending experiments for the initial geometry, which was longitudinally printed, the maximum force was 1675 N, and the maximum displacement was 3.35 mm. For the structurally designed geometry, which was longitudinally printed, the maximum force was 1825 N, and the maximum displacement was 3.56 mm. A crack appeared in the middle, in the longitudinal direction between the kinematics constraints and the applied force (see Figure 6a). In the three-point bending experiments for the initial geometry, which was transversally printed, the maximum force was 7196 N, and the maximum displacement was 10.73 mm. For the structurally designed geometry, which was transversally printed, the maximum force was 6271 N, and the maximum displacement was 4.76 mm. A crack appeared under the applied force. The stress-strain curves for all the cases are shown in Figure 6b,c. The ultimate force deviation for the initial and structurally designed cases was about 10%, and it could be decreased by improving the manufacturing of the samples. A significant difference was noted for the displacements in the case of the transversal printing. The structurally designed sample became more rigid (4.76 mm vs. 10.73 mm).

Comparing the slope (for the longitudinal printing), a 25% increase was noted for the structurally designed sample (1184 N/mm and 1491 N/mm, respectively). The slope in the case of the transversal printing decreased by 20% for the structurally designed sample (710 N/mm and 894 N/mm, respectively).

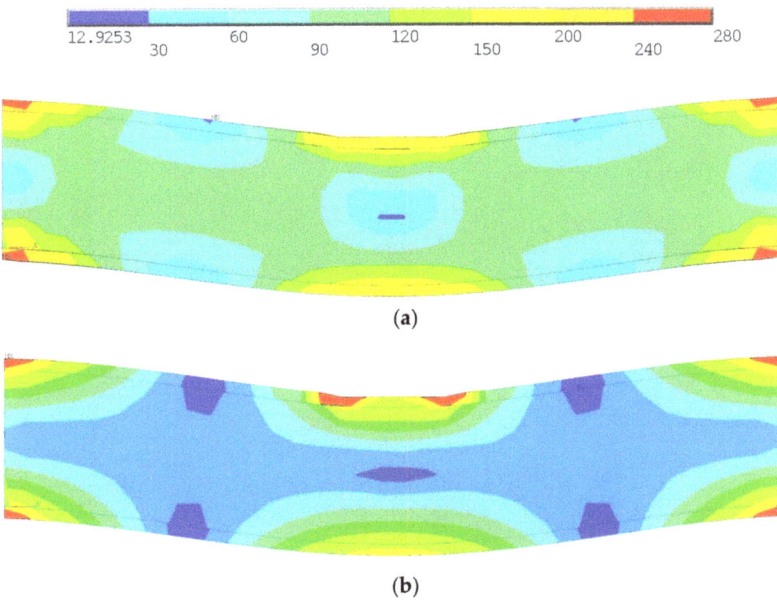

Figure 3. Von Mises stress distribution for initial (**a**) and structurally designed (**b**) beam.

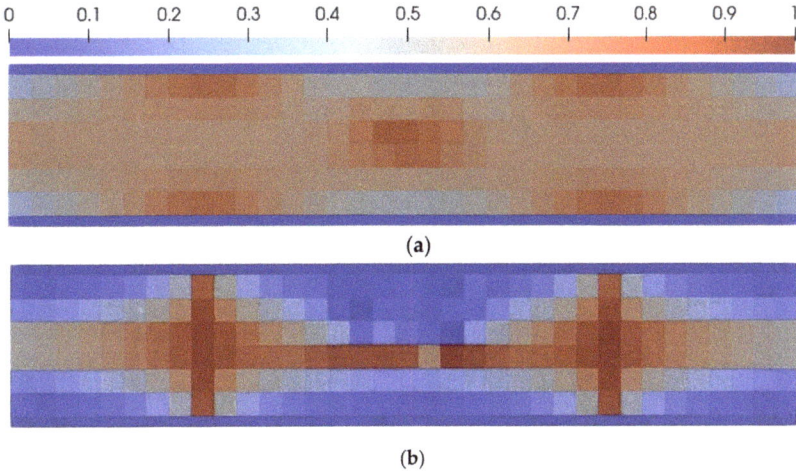

Figure 4. Distribution of porosity(**a**) and ellipticity coefficient (**b**).

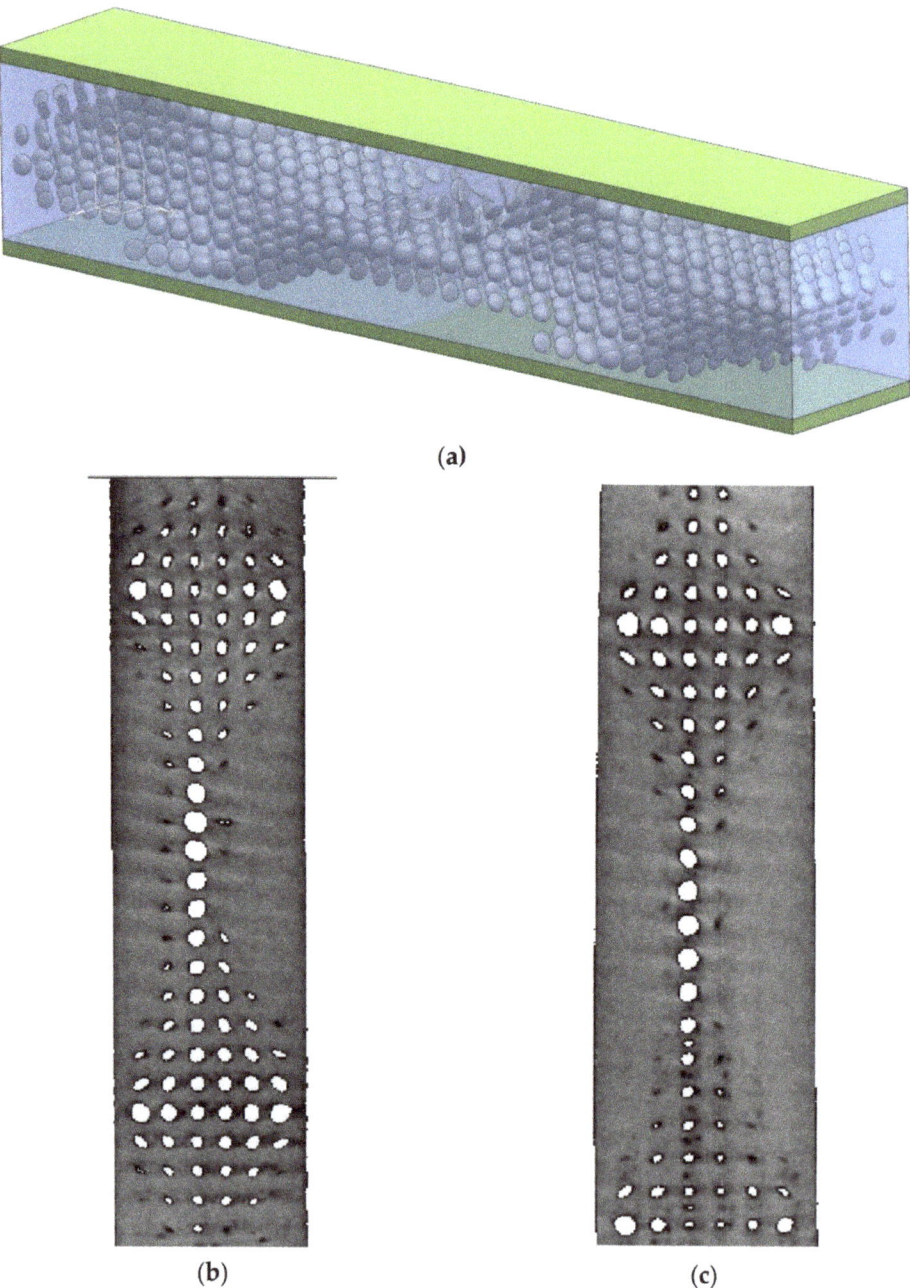

Figure 5. Restored 3D geometry (**a**), CT scans for longitudinal printing (**b**), and transversal printing (**c**).

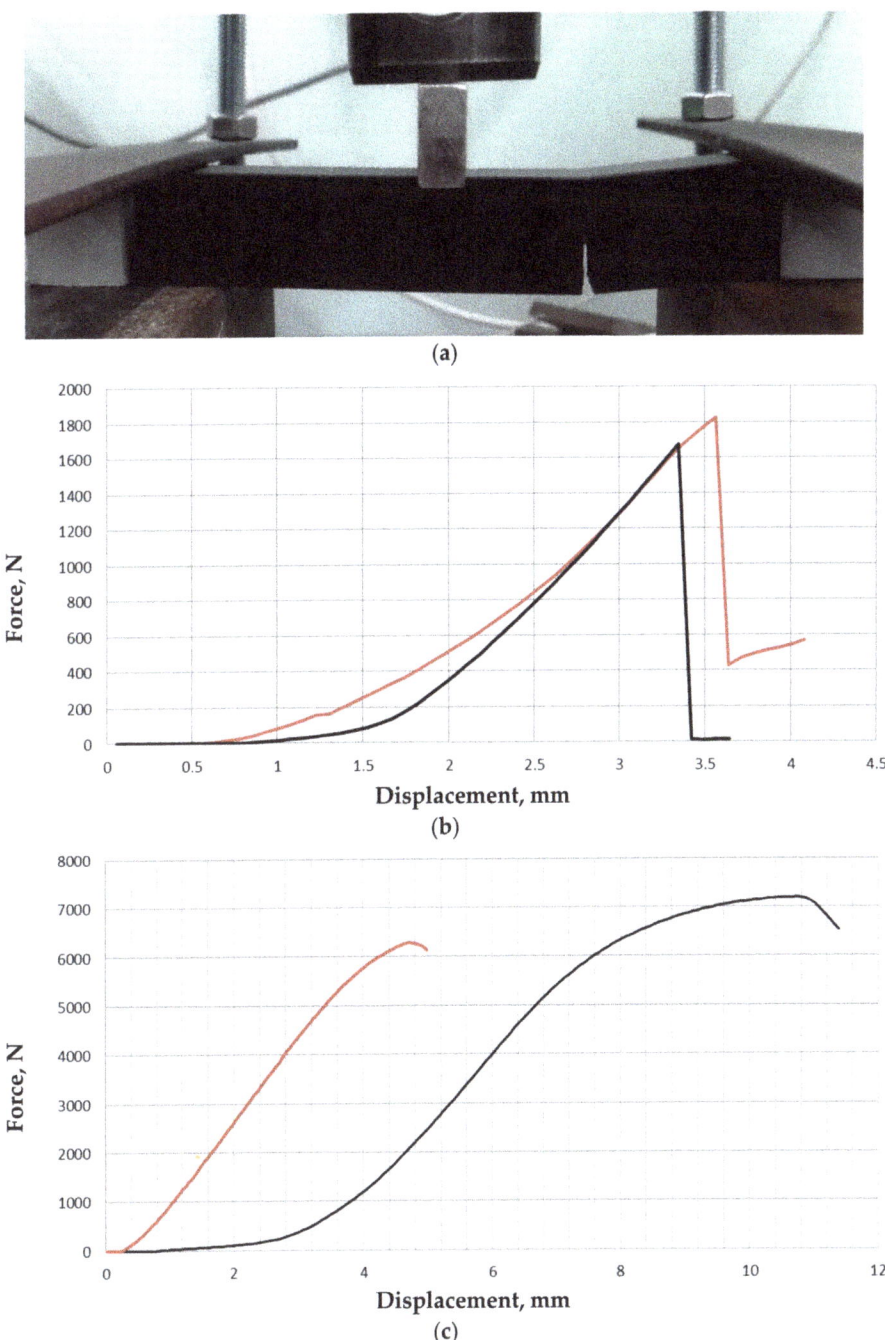

Figure 6. Three-point bending scheme (**a**), a stress-strain curve for longitudinal printing (**b**), and transversal printing (**c**); black lines—initial geometry, red lines—structurally designed geometry.

4. Conclusions

The algorithm for the structural design of the geometry was proposed. In the framework of the study, a porous cube was chosen for the basic cell. The following assumptions were used: calculation was carried out in an elastic zone; the material is isotropic; and anisotropy appears by the porosity of the basic cell. The iterative algorithm for the structural design was presented. The samples were designed, and the verification of the structural simulations was carried out. The comparison of the maximum von Mises stress for all the samples did not show a significant difference. However, for the structurally designed beam, zones of stress reduction appeared.

The manufacturing was provided using additive technologies. The samples were printed using different directions, and three-point bending tests were performed. Stress-strain curves were obtained for all the samples. In the case of the longitudinal direction printing, the ultimate force of the structurally designed sample was about 10% higher. In the case of the transversal direction printing, the rigidness of the structurally designed sample was almost 40% higher. The analysis of the stress-strain curves for all the samples shows the significant influence of the printing directions on the mechanical properties and demonstrated the need for post-processing.

Author Contributions: Methodology and software, P.B., N.K. and O.S.; manufacturing, R.K.; writing—review and editing, O.S.; conceptualization, O.S.; methodology, O.S. and P.B.; software, P.B. and N.K.; validation, O.S. and P.B.; formal analysis, P.B. and N.K.; investigation, P.B.; resources, R.K.; data curation, O.S.; writing—original draft preparation, P.B. and O.S.; writing—review and editing, O.S.; visualization, O.S. and P.B.; supervision, O.S.; project administration, O.S.; funding acquisition, O.S. and R.K. All authors have read and agreed to the published version of the manuscript.

Funding: This research was funded by RFBR, grant number 18-41-160025 p_a.

Institutional Review Board Statement: Not applicable.

Informed Consent Statement: Not applicable.

Data Availability Statement: Data available on request.

Acknowledgments: Special thanks to Independent X-ray diagnostic centers Picasso.

Conflicts of Interest: The authors declare no conflict of interest. The funders had no role in the design of the study; in the collection, analyses, or interpretation of data; in the writing of the manuscript; or in the decision to publish the results.

References

1. Ahmadi, S.; Campoli, G.; Yavari, S.A.; Sajadi, B.; Wauthle, R.; Schrooten, J.; Weinans, H.; Zadpoor, A.A. Mechanical behavior of regular open-cell porous biomaterials made of diamond lattice unit cells. *J. Mech. Behav. Biomed. Mater.* **2014**, *34*, 106–115. [CrossRef]
2. AlKhader, M.; Vural, M. Mechanical response of cellular solids: Role of cellular topology and microstructural irregularity. *Int. J. Eng. Sci.* **2008**, *46*, 1035–1051. [CrossRef]
3. Do, A.-V.; Khorsand, B.; Geary, S.M.; Salem, A.K. 3D Printing of Scaffolds for Tissue Regeneration Applications. *Adv. Health Mater.* **2015**, *4*, 1742–1762. [CrossRef] [PubMed]
4. Goda, I.; Ganghoffer, J.-F. 3D plastic collapse and brittle fracture surface models of trabecular bone from asymptotic homogenization method. *Int. J. Eng. Sci.* **2015**, *87*, 58–82. [CrossRef]
5. Li, P.; Wang, Z.; Petrinic, N.; Siviour, C.R. Deformation behaviour of stainless steel microlattice structures by selective laser melting. *Mater. Sci. Eng. A* **2014**, *614*, 116–121. [CrossRef]
6. Limmahakhun, S.; Oloyede, A.; Sitthiseripratip, K.; Xiao, Y.; Yan, C. 3D-printed cellular structures for bone biomimetic implants. *Addit. Manuf.* **2017**, *15*, 93–101. [CrossRef]
7. Limmahakhun, S.; Oloyede, A.; Sitthiseripratip, K.; Xiao, Y.; Yan, C. Stiffness and strength tailoring of cobalt chromium graded cellular structures for stress-shielding reduction. *Mater. Des.* **2017**, *114*, 633–641. [CrossRef]
8. Bari, K.; Arjunan, A. Extra low interstitial titanium based fully porous morphological bone scaffolds manufactured using selective laser melting. *J. Mech. Behav. Biomed. Mater.* **2019**, *95*, 1–12. [CrossRef]
9. Concli, F.; Gilioli, A. Numerical and experimental assessment of the mechanical properties of 3D printed 18-Ni300 steel trabecular structures produced by Selective Laser Melting—A lean design approach. *Virtual Phys. Prototyp.* **2019**, *14*, 267–276. [CrossRef]

10. Mukhopadhyay, T.; Adhikari, S. Effective in-plane elastic moduli of quasi-random spatially irregular hexagonal lattices. *Int. J. Eng. Sci.* **2017**, *119*, 142–179. [CrossRef]
11. Bolshakov, P.; Raginov, I.; Egorov, V.; Kashapova, R.; Kashapov, R.; Baltina, T.; Sachenkov, O. Design and Optimization Lattice Endoprosthesis for Long Bones: Manufacturing and Clinical Experiment. *Materials* **2020**, *13*, 1185. [CrossRef] [PubMed]
12. Cowin, S.C.; Sadegh, A.M.; Luo, G.M. An evolutionary wolff's law for trabecular architecture. *J. Biomech. Eng.* **1992**, *114*, 129–136. [CrossRef] [PubMed]
13. Maquer, G.; Musy, S.N.; Wandel, J.; Gross, T.; Zysset, P.K. Bone volume fraction and fabric anisotropy are better determinants of trabecular bone stiffness than other morphological variables. *J. Bone Miner. Res.* **2015**, *30*, 1000–1008. [CrossRef]
14. Gross, T.; Pahr, D.H.; Zysset, P.K. Morphology-elasticity relationships using decreasing fabric information of human trabecular bone from three major anatomical locations. *Biomech. Modeling Mechanobiol.* **2013**, *12*, 793–800. [CrossRef] [PubMed]
15. Pahr, D.H.; Zysset, P.K. A comparison of enhanced continuum FE with micro FE models of human vertebral bodies. *J. Biomech.* **2009**, *42*, 455–462. [CrossRef] [PubMed]
16. Kharin, N.V.; Vorobyev, O.V.; Berezhnoi, D.V.; Sachenkov, O.A. Construction of a representative model based on computed tomography. *PNRPU Mech. Bull.* **2018**, *3*, 95–102. [CrossRef]
17. Maslov, L.B. Mathematical Model of Bone Regeneration in a Porous Implant. *Mech. Compos. Mater.* **2017**, *53*, 399–414. [CrossRef]
18. Maslov, L. Mathematical modelling of the mechanical properties of callus restoration. *J. Appl. Math. Mech.* **2015**, *79*, 195–206. [CrossRef]
19. Kirpichev, I.V.; Korovin, D.I.; Maslov, L.B.; Tomin, N.G. Mathematical model of cell transformations at bone tissue regeneration under alterating biochemical medium with possible mechanoregulation. *Russ. J. Biomech.* **2016**, *20*, 187–201.
20. Maslov, L.; Surkova, P.; Maslova, I.; Solovev, D.; Zhmaylo, M.; Kovalenko, A.; Bilyk, S. Finite-element study of the customized implant for revision hip replacement. *Vibroeng. PROCEDIA* **2019**, *26*, 40–45. [CrossRef]
21. Tikhilov, R.M.; Konev, V.A.; Shubnyakov, I.I.; Denisov, A.O.; Mikhailova, P.M.; Bilyk, S.S.; Kovalenko, A.N.; Starchik, D.A. Additive technologies for complete recovery of joint function in revision endoprosthesis surgery (experimental trial). *Khirurgiya. Zhurnal Im. N.I. Pirogova* **2019**, *5*, 52–56. [CrossRef]
22. Hettich, G.; Schierjott, R.A.; Epple, M.; Gbureck, U.; Heinemann, S.; Jovein, M.-; Grupp, T.M.; Mozaffari-Jovein, H. Calcium Phosphate Bone Graft Substitutes with High Mechanical Load Capacity and High Degree of Interconnecting Porosity. *Materials* **2019**, *12*, 3471. [CrossRef] [PubMed]
23. Scheele, C.B.; Pietschmann, M.; Schröder, C.; Lazic, I.; Grupp, T.M.; Müller, P.E. Influence of bone density on morphologic cement penetration in minimally invasive tibial unicompartmental knee arthroplasty: An in vitro cadaver study. *J. Orthop. Surg. Res.* **2019**, *14*, 331–339. [CrossRef] [PubMed]
24. Kaplun, B.W.; Zhou, R.; Jones, K.W.; Dunn, M.L.; Yakacki, C.M. Influence of orientation on mechanical properties for high-performance fused filament fabricated ultem 9085 and electro-statically dissipative polyetherketoneketone. *Addit. Manuf.* **2020**, *36*, 101527.
25. Tilton, M.; Lewis, G.S.; Hast, M.W.; Fox, E.; Manogharan, G. Additively manufactured patient-specific prosthesis for tumor reconstruction: Design, process, and properties. *PLoS ONE* **2021**, *16*, e0253786. [CrossRef]
26. Kuchumov, A.G.; Selyaninov, A. Application of computational fluid dynamics in biofluids simulation to solve actual surgery tasks. *Adv. Intell. Syst. Comput.* **2020**, *1018*, 576–580. [CrossRef]
27. Kamenskih, A.; Astashina, N.B.; Lesnikova, Y.; Sergeeva, E.; Kuchumov, A.G. Numerrical and experimental study of the functional loads distribution in the dental system to evaluate the new design of the sports dental splint. *Ser. Biomech.* **2018**, *32*, 3–15.
28. Traugutt, N.A.; Mistry, D.; Luo, C.; Yu, K.; Ge, Q.; Yakacki, C.M. Liquid-Crystal-Elastomer-Based Dissipative Structures by Digital Light Processing 3D Printing. *Adv. Mater.* **2020**, *32*, 2000797. [CrossRef]
29. Shaha, R.K.; Merkel, D.R.; Anderson, M.P.; Devereaux, E.J.; Patel, R.R.; Torbati, A.H.; Willett, N.; Yakacki, C.M.; Frick, C.P. Biocompatible liquid-crystal elastomers mimic the intervertebral disc. *J. Mech. Behav. Biomed. Mater.* **2020**, *107*, 103757. [CrossRef]
30. Mori, T.; Cukelj, R.; Prévôt, M.E.; Ustunel, S.; Story, A.; Gao, Y.; Diabre, K.; McDonough, J.A.; Freeman, E.J.; Hegmann, E.; et al. 3D Porous Liquid Crystal Elastomer Foams Supporting Long-term Neuronal Cultures. *Macromol. Rapid Commun.* **2020**, *41*, 1900585. [CrossRef] [PubMed]
31. Kharin, N.; Vorob'Yev, O.; Bol'Shakov, P.; Sachenkov, O. Determination of the orthotropic parameters of a representative sample by computed tomography. *J. Phys. Conf. Ser.* **2019**, *1158*, 032012. [CrossRef]
32. Vahterova, Y.A.; Fedotenkov, G.V. The inverse problem of recovering an unsteady linear load for an elastic rod of finite length. *J. Appl. Eng. Sci.* **2020**, *18*, 687–692. [CrossRef]
33. Fedotenkov, G.V.; Makarevskiiy, D.I.; Vahterova, Y.V.; Thang, T.Q. The inverse non-stationary problem of identification of defects in an elastic rod. *INCAS Bull.* **2021**, *13*, 57–66. [CrossRef]
34. Kayumov, R.A.; Muhamedova, I.Z.; Tazyukov, B.F.; Shakirzjanov, F.R. Parameter determination of hereditary models of deformation of composite materials based on identification method. *J. Phys. Conf. Ser.* **2018**, *973*, 012006. [CrossRef]
35. Kayumov, R.A. Structure of nonlinear elastic relationships for the highly anisotropic layer of a nonthin shell. *Mech. Compos. Mater.* **1999**, *35*, 409–418. [CrossRef]
36. Kharin, N.V.; Gerasimov, O.V.; Bolshakov, P.V.; Khabibullin, A.A.; Fedyanin, A.O.; Baltin, M.E.; Baltina, T.V.; Sachenkov, O.A. Technique for determining the orthotropic properties of the bone organ according to computer tomography. *Russ. J. Biomech.* **2019**, *23*, 395–402. [CrossRef]

MDPI
St. Alban-Anlage 66
4052 Basel
Switzerland
Tel. +41 61 683 77 34
Fax +41 61 302 89 18
www.mdpi.com

Materials Editorial Office
E-mail: materials@mdpi.com
www.mdpi.com/journal/materials